Electronic Properties of Solids Using Cluster Methods

FUNDAMENTAL MATERIALS RESEARCH

Series Editor: M. F. Thorpe, *Michigan State University*
East Lansing, Michigan

ELECTRONIC PROPERTIES OF SOLIDS USING CLUSTER METHODS
Edited by T. A. Kaplan and S. D. Mahanti

Electronic Properties of Solids Using Cluster Methods

Edited by

T. A. Kaplan and S. D. Mahanti
Michigan State University
East Lansing, Michigan

Springer Science+Business Media, LLC

Library of Congress Cataloging-in-Publication Data

Electronic properties of solids using cluster methods / T.A. Kaplan and S.D. Mahanti.
p. cm. -- (Fundamental materials research)
Includes bibliographical references and index.

1. Solid state physics. 2. Electronic structure. 3. Cluster analysis. I. Kaplan, T. A. II. Mahanti, S. D. III. Series.
QC176.8.E4S373 1995
530.4'12--dc20 95-35587
CIP

Proceedings of a Summer School at Michigan State University on Electronic Properties of Solids Using Cluster Methods, held July 17–19, 1994, in East Lansing, Michigan

ISBN 978-1-4757-7017-9 ISBN 978-0-306-47063-9 (eBook)
DOI 10.1007/978-0-306-47063-9

Originally published by Plenum Press, New York in 1995.
Softcover reprint of the hardcover 1st edition 1995

10 9 8 7 6 5 4 3 2 1

Professor Joseph Callaway 1931 – 1994

SERIES PREFACE

This series of books, which will appear at the rate of about one per year, address *fundamental problems in materials science*. The topics will cover a broad range of topics from *small clusters of atoms* to *engineering materials* and involve chemistry, physics, and engineering, with length scales ranging from Ångstroms up to millimeters. The emphasis will be on basic science rather than on applications. Each book will focus on a single area of current interest and will bring together leading experts to give an up-to-date discussion of their work and the work of others. Each article contains enough references that the interested reader can access the relevant literature. Thanks are given to the Center for Fundamental Materials Research at Michigan State University for supporting this series.

M.F. Thorpe, *Series Editor*
E-Mail: Thorpe@pa.msu.edu
East Lansing, Michigan

PREFACE

This book records invited lectures given at the workshop on Electronic Properties of Solids Using Cluster Methods, held at Michigan State University July 17-19, 1994. Cluster methods for solid-state studies were introduced a long time ago by Hans Bethe in his important work on the crystal field splitting of the energy levels of impurity ions in insulators [*Ann. Physik* **3**, 133 (1929)]. More recently, it has been realized that the applicability of such an approach is far broader. The intent of the workshop was to bring together a small number of leaders in the field, to give an up-to-date picture of this breadth, as well as, of course, to assess the quality of the results obtained.

The picture that emerged shows a truly remarkable diversity of physical problems, discussed in the various chapters of this book, all in the context of cluster calculations. Namely: hyperfine properties of insulators and semiconductors (T.P. Das); point defects in ionic solids (J.M. Vail); electronic structure of transition metal impurities in bulk Cu (R. Zeller); reaction energetics and adsorbate structure on metal and semiconductor surfaces (J.L. Whitten); electronic and lattice properties of doped La cuprate (R.L. Martin); spin density in transition metal compounds (T.A. Kaplan et al.), characterization of the ionicity and analysis of XPS measurements in transition metal oxides (P.S. Bagus *et al.*); diamond and fullerene-assembled polymers (M. Pederson); compressed nitrogen molecular and atomic solids, and diamond — correlation energies via Monte Carlo calculations (L. Mitáš), geometry of and magnetic correlations in polyacetylene and polyacene (M.A. Garcia-Bach); generation and solution of effective many-body Hamiltonians for rare earth and transition metal compounds (A.K. McMahan); temperature dependence of the uniform magnetic susceptibility, Fermi surface and superconductivity of Sr-doped La cuprate (A. Moreo). We mention two important contributions to the workshop that unfortunately are not fully represented in this book: W.C. Nieuwpoort, University of Groningen: *Clusters in Solids;* and E. Stechel, Sandia National Laboratories: *Normal State Electronic Structure of Doped Copper Oxide as Obtained from Strongly Correlated Models of Periodic Clusters.* Professor J. Callaway, Louisiana State University, who was to have been an invited speaker, unfortunately passed away shortly before the workshop, which was dedicated to his memory.

The works presented divide, in the main, into two types of cluster calculations, involving either (i) rather small clusters with the embedding medium treated approximately or (ii) larger clusters of increasing size extrapolated to approximate the infinite crystal. Type (i) is represented by Chapters 1-8, while Chapters 9, 11, and 12 are along the lines of type (ii).

T.A. Kaplan
S.D. Mahanti

CONTENTS

HARTREE–FOCK CLUSTER PROCEDURE FOR STUDY OF HYPERFINE PROPERTIES OF CONDENSED MATTER SYSTEMS

T.P. Das

Department of Physics
State University of New York at Albany
Albany, New York 12222 USA

I. INTRODUCTION

I would like to start this paper with some brief comments about the relative advantages of the band and cluster approaches to the electronic structures of solid state systems. During the sixties and seventies, our research group worked on the orthogonalized plane-wave procedure, and modifications of it, to study the hyperfine properties of a number of non-magnetic metals including Knight-shifts[1] representing the ratio of the induced magnetic fields at nuclei in metals produced by the electrons and the applied field, and Ruderman–Kittel–Kasuya–Yosida (RKKY) interactions[2] between nuclear spins mediated by the conduction electrons, the interactions between the quadrupole moments of the nuclei and the electric field gradient tensors at the nuclei[3] due to the electrons and nuclear charges in non-magnetic metals, and hyperfine fields in ferromagnetic metals.[4] In the eighties and nineties we have concentrated on the cluster approach for properties of solid state systems. I have been fortunate therefore in having had some experience with both types of methods and would like to discuss some of my views regarding their relative merits for the properties we were interested in.

Advantages and Disadvantages of the Band-Structure Procedure for the Study of Properties of Condensed Matter Systems

First, considering the band-structure approach, its major advantage is that for perfect systems, it takes account of the infinite nature of the systems involved through the use of Bloch functions suitable for their symmetry. Secondly, one is able to determine in a direct and natural way the Fermi-surface[5] and density of states[6] which one always needs for hyperfine properties.

Some of the problems associated with band-structure procedures are the following. It is rather time consuming to carry out first-principle self consistent field Hartree-Fock band calculations for each cycle of iteration[4,7] where one has to sum over all the continuum band states that are occupied in evaluating the Coulomb and exchange potentials given respectively by:

$$V_{Coul,i}(\mathbf{r}_1) = \sum_{j \neq i} \int \left|\Psi_j(\mathbf{r}_2)\right|^2 \frac{e^2}{r_{12}} d\tau_2 \,, \tag{1}$$

$$V_{exch,i}(\mathbf{r}_1) = \sum_{j \neq i}' \int \Psi_j^*(\mathbf{r}_2)\Psi_i(\mathbf{r}_1)\frac{e^2}{r_{12}}\left(\Psi_j(\mathbf{r}_1)/\Psi_i(\mathbf{r}_1)\right) . \tag{2}$$

The potentials in Eqs. (1) and (2) refer to those seen by an electron in a state i and the prime in Eq. (2) means that the summation over the occupied states j with same spin as the state i where Ψ_i and Ψ_j refer to the electronic wave-functions for the states i and j.The summation over the occupied states make the iterations for attaining self-consistency very expensive in terms of time. Secondly, the incorporation of correlation effects in band-structure procedures in a first-principle manner by Many-Body Perturbation Theoretic[8] or Configuration Interaction[9] methods as in atomic and molecular systems is even more time consuming than for Hartree-Fock calculations. However, a few attempts have been made[10] using the Many-Body Perturbation Procedure. Thirdly, it is rather difficult to incorporate lattice relaxation effects in impurity systems in a first-principles or near first-principles manner.

For perfect metallic systems, we have in the past been able to carry out accurate calculations of hyperfine properties of a number of metallic systems using the Hartree-Fock procedure[4] or near-Hartree Fock procedure[1,2,3] using methods based on the Orthogonalized Plane Wave approach. A few typical references are quoted here.

Advantages and Disadvantages of Cluster Procedure

The Cluster procedure[11–17] for simulating solid state systems had its major growth primarily over the past two decades, later than the band–structure approach. We are very fortunate to have at this Workshop some of the scientists who have made pioneering contributions[11–15] to the development of the Cluster Procedure and continue to make important applications, as well as others who have joined the effort more recently and are contributing valuably to the theory and applications of the method making it a viable alternative to the band procedure. There are of course different theoretical approaches used in the cluster procedure for the actual determination of energy-levels and wave functions, such as the Hartree-Fock procedure and local density approximation[17] to it as well as more approximate or semi-empirical methods like the Neglect of Differential Overlap[18] family of procedures and the Self-Consistent Charge Extended Hückel (SCCEH)[19] procedure. All of them use a finite number of atoms, as large a number as can be handled practicably from a computational point of view, to simulate the infinite solid and the major advantages and disadvantages of the cluster procedures are related to this feature.

We shall discuss the main advantages of the cluster procedures first and then its limitations as we have done for the band-structure procedure. The first advantage is connected with the fact that one deals with a finite number of atoms and therefore with finite numbers of electrons and energy-levels. The incorporation of self-consistency effects using the Hartree-Fock procedure, or approximations to it, is much easier than with the band-structure procedure where one has to deal with continuum energy levels for all the energy bands. Secondly, many-body perturbation theory[8] and configuration interaction[9] procedures, using the empty energy levels for excitations from the occupied states, while requiring rather substantial amounts of computational times for large clusters, are more practicable than with band-structure procedures. Lastly, lattice relaxation effects associated with impurities or defects are much less difficult and more practicable to include than with band-structure procedures. This is because the determination of lattice distortion effects essentially requires a study of the variation of total energy of the system, the infinite solid in band-structure procedures, and the finite cluster used to simulate the infinite solid in the cluster procedures, as a function of the positions[20,21] of the atoms or ions in the systems. The calculation of the total energy is significantly less time-consuming for cluster procedures because one has to deal with finite numbers of the occupied one-electron energy levels and wave-functions that are required for such a calculation than in the case of the continuum **k**-states in band structure procedures.[5–7] Additionally, the loss of translation or point-group symmetry in the presence of impurities and defects is much less serious from a computational point of view for the cluster procedures than in the case of band-structure procedures. In the latter case one has to use supercell[22] or Green s function[23] procedures to deal with the impurity or defect system which are very much more time consuming than band-structure procedures[1–7] involving Bloch wave functions associated with perfect systems with the well-known lattice parameters for the latter.

We turn next to some of the problems associated with the cluster procedure. The determination of the Fermi-energies and densities of states, for the cluster procedure using the finite numbers of energy levels for the clusters used to simulate the infinite solid, is much more problematic and less accurate than in the case of band-structure procedures. For the latter, the counting of occupied states using the histogram or related approaches using the dependence of band energies on **k** points in the Brillouin Zone that goes into the Fermi-surface and density of states determination[5,6] is fairly straightforward. Secondly, to study the reliability of the cluster procedure as a good tool to simulate the essentially infinite nature of the solid-state system concerned it is necessary to study the convergence of the calculated results for the properties concerned and the energy levels and eigenfunctions as a function of cluster size and most importantly to use some embedding procedure[24–26] to simulate the effects of the rest of the solid beyond the cluster studied.

There have been elegant and comprehensive discussions in this workshop on embedding procedures in metallic[27] and ionic crystal[28] systems. In our work so far where we have been concerned primarily with localized properties like hyperfine interactions, one would expect these properties to be perhaps less sensitive to the nature of the boundary of the cluster used (provided the cluster is large enough) as compared to other properties such as for instance spectroscopic effects associated with charge transfer electronic transitions or activation energies associated with ion diffusion for transport properties. For the different classes of systems we have studied for hyperfine properties, we have used the embedding procedures described in the following paragraphs.

In ionic crystals, including also our work on high-T_c systems, we have simulated the effect of the influence of the ions outside the cluster by two alternate procedures.[24,29] One is to handle them as point charges[24] and including their Coulomb potentials in the Hartree-Fock Hamiltonian used for the electrons in the cluster. The other procedure we have employed[29] is to use a number of *buffer* ions surrounding the chosen cluster with their influence on the electrons in the cluster represented by pseudopotentials[30] associated with the latter ions and the rest of the lattice represented by point charges and their influence on the electrons in the cluster represented by their Coulomb potential. Both these choices are expected to be suitable for perfect crystals without any impurity or imperfections, with the second one[29,30] being expected to be somewhat more representative of the actual surrounding of the chosen cluster because it includes the influence of the interaction between the electrons of the cluster and its immediate neighbors which would not be included if these neighboring ions were dealt with as point charges.[28] For the case of impurities or imperfections where one would expect to have lattice relaxation effects associated with the displacements of the neighboring ions, only the second procedure would be the appropriate one to apply because with the first procedure, where one considers all the rest of the lattice as point charges only, relaxation effects associated with their displacements would not be stable because without any repulsive effects[31] between those ions between themselves and between them and the cluster, they would collapse into each other and into the cluster. With the second procedure, one allows only the buffer ions to relax, keeping the rest of the ions fixed. The use of the pseudopotentials on the buffer ions allows one to include the repulsive interaction between the electrons of the latter ions and the cluster electrons thus preventing their collapse into the cluster.

For semiconductors, we have terminated[32] the dangling bonds at actual surfaces or at surfaces of the clusters used inside the bulk system by hydrogen atoms as has been done in these systems by others[33] as well. There is some question[34] about the best choice of the X-H bond length between the semiconductor atom and the terminal hydrogen atom, the limits being the bond distance X-H in covalent molecular systems and the X-X bond in the semiconductor. But from our own experience we feel that when one uses sufficiently large clusters as practicable and is dealing with localized properties like hyperfine interactions, there should be[32] no pronounced sensitivity with respect to the terminal X-H bond distance.

For molecular crystals[35] since one deals usually with neutral molecular units interacting relatively weakly with each other, there is no need for embedding as in semiconductors where there is strong covalent bonding between immediate neighbors and in ionic crystals, where there is long range interaction between the ions.

Lastly, regarding metallic systems, we have only recently started working on them by the Hartree-Fock Cluster procedure in both second-group metals[36] and semimetals[37] of the sixth group. In our initial work we are using as large clusters as practicable and no embedding, the reasons for this being the expected neutrality of the surroundings due to the canceling effect of the positive charges on the ions and negative charges on the conduction electrons and the expectation that any departure from neutrality at the surface of a large enough cluster may not be too pronounced in effect, especially since we are dealing with localized hyperfine properties. Should it appear that embedding does have significant influence on our results, we shall make use of embedding choices[26,27] that have been studied before in metallic systems.

Brief Summary of the Systems Studied Using Cluster Procedure

We have been using the cluster procedure in general for study of condensed matter and biological systems for more than two decades. During the late sixties and seventies, we had used the semi-empirical SCCEH procedure,[38] primarily for biological systems (continued into late eighties) including hemoglobin derivatives,[39] the enzyme carbonic anhydrase[40] responsible for hydrolysis of the carbon dioxide in the body and the contents of the reaction center in bacterial photosynthesis.[41] We had also studied by the same procedure a number of condensed matter systems, mainly semiconductors containing normal and anomalous muonium centers[42] and iodine and tellurium impurities[43] in elemental and III-V semiconductors and sixth group systems.[44] Because of the semi-empirical aspect of the SCCEH procedure, one could not put much emphasis on the quantitative nature of the agreement with experiment, but nevertheless, we were able to obtain[39–44] satisfactory semi-quantitative agreement with a variety of hyperfine interaction data from magnetic resonance, Mössbauer and muon spin rotation measurements including trends in related systems. In the case of the semiconductor systems we were able to obtain[42–44] physically meaningful information about the location of impurity atoms as well, besides a satisfactory understanding of hyperfine data.

The Hartree-Fock Cluster procedure has been used by our group since the early eighties. In the field of semiconductors, we have worked on (a) the electronic structures, hyperfine properties and environments of normal and anomalous muonium in the elemental semiconductors,[45] diamond, silicon and germanium; (b) the location, associated ultraviolet and x–ray photoemission properties and nuclear quadrupole interaction properties of adsorbed atoms at silicon surface,[46] the adsorbed atoms including alkali, third group, chalcogen, and halogen atoms; and (c) fluorine atoms[47] in diamond, silicon and germanium with emphasis on the location and nuclear quadrupole interaction of the excited nuclear state of $^{19}F^*$ with spin 5/2 in contrast to spin 1/2 for the ground state of ^{19}F nucleus and no quadrupole moment. The use of the nuclear radiation technique, Time Dependent Perturbed Angular Distribution (TDPAD) [48] associated with the angular distribution of emitted gamma radiation from the excited $^{19}F^*$ nucleus and the direction of the proton beam used to excite the ^{19}F nucleus from its ground-state to the $^{19}F^*$ state has provided[47,49] a valuable tool to use the nuclear quadrupole interaction of $^{19}F^*$ to analyze the nature of the electron distribution over the fluorine atom in different materials, either when it occurs naturally or is implanted.

In the area of ionic crystals, we have investigated the electronic structures of a number of materials, the emphasis being again on the understanding of their hyperfine properties to check the accuracy of their calculated electron distributions. Among the systems we have investigated are Li_3N, a superionic conductor,[50] the oxide systems Al_2O_3 (corundum), [51] Cu_2O and Fe_2O_3 which[52] are important minerals, the latter being important also for its magnetic properties, the zinc compounds[53] ZnF_2 and the zinc chalcogenides[54] where Mössbauer measurements[55] on ^{67}Zn have provided isomer shift and quadrupole interactions (also at $^{19}F^*$ nucleus from TDPAD measurements)[56] that provide valuable tests of the isotropic and anisotropic charge distributions in the vicinity of the nuclei, and spinels[57] such as $ZnAl_2O_4$ and $ZnFe_2O_4$ which are important in geophysical studies and for their magnetic properties.

We have also worked on the high-T_c systems[58,59] La_2CuO_4, $YBa_2Cu_3O_6$ and $YBa_2Cu_3O_7$. Again, the emphasis has been on the analysis of hyperfine properties of the various nuclei in these systems[60] including the magnetic hyperfine properties[61] in the antiferromagnetic state. The motivation of course is to get a good quantitative understanding of the electronic structure which will be useful for quantitative study of the mechanisms proposed for the origin of superconductivity in these systems. In addition to the pure materials, we have also studied[62] the location of muon in La_2CuO_4 and its associated hyperfine properties observed[63] by the muon spin rotation (μSR) technique.

Another broad area of condensed matter physics, that we have studied using the Hartree-Fock Cluster procedure, comprises three related fields, biological physics, medical physics and the physics of drugs. In the area of biological physics, following our past efforts[39,40] on the use of the semi-empirical SCCEH procedure, which were qualitatively and even semi-quantitatively successful in explaining observed magnetic and hyperfine data in large biological systems, we have been working on more quantitative studies of electronic structures[64,65] of hemoglobin, cytochromes and related systems by the Hartree-Fock cluster procedure with satisfactory results for the hyperfine properties of deoxyhemoglobin and hemin. In the area of medical physics, we have worked on the geometry, electron distributions and hyperfine properties of transition metal[66] and rare earth[67] ions in aqueous solutions and with porphyrin substrate.[68] These systems are important in increasing the relaxivity of protons in aqueous solutions, making them valuable as contrast agents in Magnetic Resonance Imaging. In the area of physics of drugs, we have recently started investigations[69] on the electron distributions and nuclear quadrupole interactions in cocaine free base and cocaine hydrochloride. Experimental results have become available[70] in these systems for comparison with theory. It should be remarked here that both in our work on biological systems, involving large molecular groups like the heme unit of hemoglobin[64,65] or the cocaine molecule,[69] we have had to cut off parts of the complete systems to make the electronic structure calculations practicable. We have done this by terminating bonds involving carbon or oxygen at appropriate points by hydrogen atoms, as in the case of semiconductor systems, choosing these points so that they are distant enough from the nucleus whose hyperfine properties are being studied. In addition it is preferable to use terminal sites involving single bonds wherever possible, so as to have less influence of the truncation on the electron distribution within the chosen fragment studied than if multiple bond sites were used.

In the field of molecular crystals, we have worked on $^{19}F^*$ nuclear quadrupole interactions in a number of fluorine systems[35] including solid hydrogen fluoride, a series of compounds involving third, fourth, fifth and sixth group fluorides and fluorobenzenes including some fluoromethanes that involved the formation of complexes between the host molecules and $H^{19}F^*$ molecules[71] formed during the process of implantation of $^{19}F^*$ nuclei.

Lastly, we have recently been carrying out Hartree-Fock investigations[72] in fullerene systems, concentrating on the location of hydrogen and muonium atoms and the associated hyperfine interactions available from electron paramagnetic resonance[73] and muon spin rotation[74] measurements. As in the case of high-temperature superconductors, in addition to understanding the properties of the individual systems we have investigated, our motivation is also the general one to test how well Hartree-Fock procedures can provide accurate information about the electron distributions in these systems. This is important for assessing the applicability of these procedures to understand quantitatively the interaction of metal atoms with fullerenes to explain their resulting interesting magnetic, conducting and superconducting properties. Again, a remark should be made here about the question of possible termination of the fullerene compounds to reduce computation time. In our work[72] on the pure fullerene C_{60} and hydrogen or muonium atoms attached to fullerene, we have found that, perhaps because of the delocalized nature of the π-electrons in graphite to which the fullerene system is related, the influence of the bonding of the hydrogen atoms to one of the carbon atoms is felt over nearly the entire molecule. It is therefore essential to use the complete fullerene system in electronic structure calculations. The situation could of course be different for less symmetrical systems like C_{70} and for systems in which a more strongly bonding or larger group than hydrogen, bonding directly with a number of carbon atoms rather than a single one, is attached to the fullerene.

It will be difficult to discuss here, due to space constraints, our work in all the systems we have just listed. We shall therefore concentrate on one of the areas we have worked on, namely high-T_c systems. We have made this choice both because of the current interest in this class of systems and also because they provide insights into the nature of the success one has in dealing with the electronic structures and properties of complex systems by the Hartree-Fock Cluster Procedure. The interested reader can obtain information about our results and procedure in other systems from the references we have listed. We will conclude this article by discussing, based on our experience, some of the directions in which improvements should be attempted in the future in using cluster procedures in condensed matter systems.

II. HIGH TEMPERATURE SUPERCONDUCTING MATERIALS-THEORY OF ELECTRONIC STRUCTURES AND HYPERFINE PROPERTIES

The major motivation for our work on high-T_c systems has been to understand from first-principles the electronic structures of these systems, since we believe that the successful theory or theories that emerge from among those proposed in the literature for the origin of superconductivity in these systems will have to utilize for quantitative purposes the actual electron energy level structures and electron distributions in these materials. Hyperfine data can provide quantitative information about the isotropic and anisotropic charge and spin distributions in solid state systems including the high–temperature superconductivity systems. We have so far been working on the normal states of these systems and have confined ourselves to the single and double layer systems

La_2CuO_4, $YBa_2Cu_3O_6$ and $YBa_2Cu_3O_7$. I will present our results[75–82] on the hyperfine properties of ^{63}Cu, ^{17}O, ^{135}Ba, ^{139}La nuclei in these systems and the hyperfine properties associated with muon in La_2CuO_4 and make extensive comparison with available experimental data.[83–91] After briefly presenting the Hartree-Fock-Roothaan procedure[92] that we have used in our investigations and the methods for using the calculated electronic structures to obtain the various associated hyperfine properties, we shall present our results and compare them with experiment. The effective charges on the various atoms in these materials will also be presented and discussed because of the important bearing they have on the nature of the covalent bonding in these systems.

Procedure

With the Hartree-Fock procedure,[92] one solves the Hartree-Fock equation:

$$\left(H_\mu - E_\mu\right)\Psi_\mu = 0 \ , \tag{3}$$

where H_μ is the one-electron self-consistent Hamiltonian involving the kinetic energy, nuclear attraction energy from the nuclei in the molecular cluster and the direct and exchange interaction energies between an electron in the state Ψ_μ and electrons in the other occupied states. In the high-T_c systems we have studied,[75–82] as in ionic crystal systems,[93,94] a sizable cluster is chosen around the ion containing the nucleus under study so that its immediate environment up to the third and fourth neighbors is well represented. Additionally the influence of the ions outside the cluster is represented by including in the Hartree-Fock potential, the potential due to the point charges on these external ions. In effect therefore, one includes in the simulation of the solid state system, the whole crystal. To extend the size of the cluster treated in detail in the Hartree-Fock procedure,[93–95] we sometimes use pseudopotentials for some of the ions more distant from the nucleus under study, with the effects of the ions beyond these distant ions being handled in the point charge approximation.

Since, because of the multicenter nature of the cluster system, one cannot solve for the eigen-values and eigen-functions of the Hartree-Fock differential equations (3) by a direct integration procedure, a variational procedure is applied as in the Hartree-Fock-Roothaan formalism,[92, 95–97] expressing the molecular orbital eigen-function, Ψ_μ, in terms of orbitals χ_i based on the atoms in the cluster in the form

$$\Psi_\mu = \sum_i C_{\mu i}\chi_i \tag{4}$$

with the parameters $C_{\mu i}$ being determined variationally. The natures and numbers of basis functions have to be chosen with both accuracy and practicability in mind and in the past, both numerical atomic Hartree-Fock functions as well as exponential Slater-type functions have been used. In our recent investigations by the cluster procedure, as is the common practice currently because of the enhanced speed of computations using them, we have been using Gaussian orbitals[98] and employing the available Gaussian software.[99] The results obtained by the variational procedure have of course to be tested with respect to the sizes of the basis sets in Eq. (4) and this has been done in all our investigations[75–82,93,96,97] so far, including the high-T_c systems. In our high-T_c calculations,[75–82] we have used extensive basis sets for the ions of copper, oxygen, lanthanum and barium involved, such as for instance, 4s (6,2,1,1), 4p (4,1,1,1), and 2d (4,l) Gaussian basis functions for copper and 4s (4,4,3,2) and 2p (4,3) for oxygen. The conventional notations used imply, for instance 4 basis s functions for copper involving respectively contracted (linear combinations) of functions involving 6

Gaussians, 2 Gaussians and two separate Gaussians. These functions had been found to give satisfactory results for nuclear quadrupole interactions in other ionic crystals[100,101] like cuprous and zinc oxides. The convergence of our results with respect to cluster sizes has been studied by using different size clusters and studying the variation of hyperfine properties with cluster size. In handling clusters with finite spin as in the case of magnetic hyperfine properties of antiferromagnetic systems, such as[80] $YBa_2Cu_3O_6$ and[76] La_2CuO_4, we have made use of the Unrestricted Hartree-Fock procedure[102] in which the wave-functions for states of opposite spin are allowed to be different, enabling the inclusion of exchange polarization contributions to the hyperfine fields at the nuclei from paired spin states.

Before describing our results for the hyperfine properties and charge distributions in the systems we have studied, the expressions for these properties in terms of the calculated one-electron wave-functions for the clusters used will be listed. Thus, the effective charge on an atom A in the cluster is given in the Mulliken approximation[103] by:

$$\zeta_{eff,A} = \zeta_A - n_A \tag{5}$$

where ζ_A is the charge on the nucleus A. The term n_A is the total electronic population on atom A given by:

$$n_A = \sum_{\mu} \sum_{i \in A} \left[\sum_{j \notin A} \left(C_{\mu i}^2 + S_{ij} C_{\mu i} C_{\mu j} \right) \right] \tag{6}$$

the summation over μ referring to all the occupied molecular orbitals, $S_{ij} = \langle \chi_i | \chi_j \rangle$ is the overlap integral between the basis functions χ_i and χ_j and the summation over i and j in Eq. (6) referring respectively to basis functions centered on A and other atoms in the cluster.

For nuclear quadrupole interaction, one needs the principal components of the electronic field-gradient tensor at the nucleus under study. To obtain these, one first obtains the components of the field-gradient tensor in a chosen axis system where the tensor will in general be non-diagonal and the component V_{ij} is given by:[104]

$$V_{ij} = \sum_n \frac{\zeta_n e \left(3x_{in} x_{jn} - r_n^2 \delta_{ij} \right)}{r_n^5} - \sum_\mu \int e |\Psi_\mu|^2 \frac{3x_i x_j - r^2 \delta_{ij}}{r^5} d\tau + \sum_{N(ext)} \frac{\zeta_N e \left(3x_{iN} x_{jN} - r_N^2 \delta_{ij} \right)}{r_N^5} \tag{7}$$

the values 1,2,3 for i and j referring to the x,y and z components, the first term in Eq. (7) referring to the contributions from the nuclear charges ζ_n in the cluster, the second to the contributions from the electrons in the cluster and the third from the effective point charges on the ions outside the cluster. Since the influence of the potential due to these point charges on the electron distribution in the cluster has been incorporated in the electronic structure investigations, the influence of antishielding effects[105] is already included in the second term, obviating the need for any antishielding parameters[105] in the third term. Once the components V_{ij} in any chosen coordinate system are calculated, one can obtain the principal axes (X',Y',Z') and principal components $V_{i'i'}$ by diagonalizing Vij. In the accepted convention in the literature,[106] since only two of the principal components are independent in view of Laplace's equation, only two parameters are used to characterize the principal components, namely,

$$q = V_{z'z'} \tag{8}$$

related to the quadrupole coupling constant e^2qQ and the asymmetry parameter

$$\eta = \frac{V_{x'x'} - V_{y'y'}}{V_{z'z'}}, \tag{9}$$

where

$$\left|V_{z'z'}\right| \geq \left|V_{y'y'}\right| \geq \left|V_{x'x'}\right|. \tag{10}$$

The effort for obtaining the principal components can be substantially reduced if one has some knowledge of the principal axes from symmetry considerations.[107]

For the magnetic hyperfine interaction there are two main contributions,[108] isotropic and anisotropic. The isotropic contribution involves a term $A\vec{I}\cdot\vec{S}$ in the spin-Hamiltonian used to analyze the experimental signal in magnetic resonance, Mössbauer effect or other radioactive techniques, the isotropic hyperfine constant A being given by

$$A = \frac{8\pi}{3S}\gamma_e\gamma_I\hbar^2\left[\sum_u\left|\Psi_{u\uparrow}(0)\right|^2 + \sum_p\left\{\left|\Psi_{p\uparrow}(0)\right|^2 - \left|\Psi_{p\downarrow}(0)\right|^2\right\}\right] \tag{11}$$

where Ψ_u refers to the electronic wave–functions for the unpaired spin states and $\Psi_{p\uparrow}$ and $\Psi_{p\downarrow}$ for the paired spin states with spins parallel and antiparallel to the unpaired. The quantities γ_e and γ_I refer to the gyromagnetic ratios for the electronic and nuclear spins and S to the total electronic spin of the system. The anisotropic component, which occurs when there is departure from cubic or tetrahedral symmetry about the nucleus, leads[108] to a term of the form $B\left(3I_ZS_z - \vec{I}\cdot\vec{S}\right)$ for axial symmetry and is given by

$$B = \frac{2}{3S}\gamma_e\gamma_I\hbar^2\left\{\sum_\mu\left\langle\Psi_{u\uparrow}\left|\frac{3\cos^2\theta - 1}{r^3}\right|\Psi_{u\uparrow}\right\rangle + \sum_p\left[\left\langle\Psi_{p\uparrow}\left|\frac{3\cos^2\theta - 1}{r^3}\right|\Psi_{p\uparrow}\right\rangle - \left\langle\Psi_{p\downarrow}\left|\frac{3\cos^2\theta - 1}{r^3}\right|\Psi_{p\downarrow}\right\rangle\right]\right\} \tag{12}$$

In Eq. (12), θ refers to the angle between the radius vector of the electron and the direction of axial symmetry. Usually, the constants A and B are expressed in frequency units, so one has to divide A and B by $2\pi\hbar$. The hyperfine field H_{hyp} is then obtained by applying the factor $2\pi S/\gamma$ to the components (A+2B) and (A-B) parallel and perpendicular to the symmetry axes respectively. We shall be concerned here primarily with axial symmetry for the magnetic hyperfine interaction. For the general case, one needs three parameters B_{xx}, B_{yy} and B_{zz} in a principal axis system which can be obtained by a diagonalization procedure as in the case of the field-gradient tensor discussed earlier in this section.

Results and Discussions

We shall present the electronic structure and hyperfine interaction results and discussion for the compounds $YBa_2Cu_3O_6$ and $YBa_2Cu_3O_7$ first, followed by that for La_2CuO_4 and finally for the La_2CuO_4-muon systems. The interrelationships of the results in the three sets of systems and the insights they provide regarding the electron distributions in them shall be carefully discussed.

$YBa_2Cu_3O_7$ and $YBa_2Cu_3O_6$ Systems. We have studied[75,77,80,81] the ^{63}Cu, ^{17}O and ^{135}Ba nuclear quadrupole interactions in both $YBa_2Cu_3O_7$ and $YBa_2Cu_3O_6$, and the magnetic hyperfine interaction for the ^{63}Cu nucleus in $YBa_2Cu_3O_6$ in the antiferromagnetic state. The experimental results for these properties have been obtained by nuclear magnetic resonance measurements[83–86] in both powdered and single crystal systems.

The nuclear quadrupole interactions will be discussed first. The structure[109] of $YBa_2Cu_3O_7$ is presented in Fig. 1. For the investigation of ^{63}Cu nuclear quadrupole interactions for both Cu(1) and Cu(2), a cluster Cu_3O_{12} was chosen involving two planar Cu(2) ions directly above and below each other and their four oxygen neighbors (O(2), O(3) and their equivalent counterparts) bonded to the same Cu(2) in each case, as well as the chain Cu(1) ion on the c-axis in between the two Cu(2) ions, the four neighboring oxygen ions of Cu (1), two O(4), along the c-axis between Cu(1) and each of the Cu(2) in the cluster and the two O(1) ions on the chain containing the Cu(1) in the cluster. The same cluster is used for the ^{17}O nuclear quadrupole interaction[75] associated with O(4), since it has within the same cluster, the two copper ions Cu(2) and Cu(1) it is directly bonded to as well as their oxygen neighbors. A similar cluster Cu_3O_{10} was considered adequate[80] for $^{63}Cu(1)$, $^{63}Cu(2)$ and $^{17}O(4)$ in $YBa_2Cu_3O_6$ where the chain oxygens O(1) are not present in the lattice and therefore do not occur in the cluster used. For the $^{17}O(2)$ and $^{17}O(3)$ nuclei, clusters Cu_2O_9 were used,[75] which involve the oxygen atom containing the ^{17}O nucleus under study and its two nearest neighbors Cu(2) on the plane and for each of the latter, its three other oxygen neighbors on the plane and the O(4) neighbor on the c-axis.

In Table 1, we have listed the charges we have obtained on the Cu(1), Cu(2) and O(1) through O(4) ions, O(1) of course being absent in $YBa_2Cu_3O_6$. The charges depart significantly from the formal charges +l, +2 and -2 for Cu(1), Cu(2) and O(2) (and O(3))

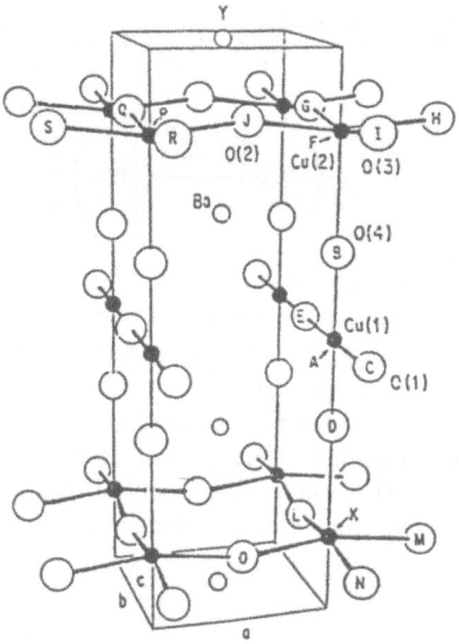

Figure 1. The orthorhombic unit cell of $YBa_2Cu_3O_7$. The various copper and oxygen atoms are designated in the way they are referred to in the text.

respectively in both cases, -l and -2 for O(4) and O(1) in the (1,2,3,7) system and -2 for O(4) in the (1,2,3,6). These departures suggest the presence of significant covalent bonding between copper and oxygen ions in these systems, the covalent bonding between Cu(1) and O(4) in the (1,2,3,7) compound being stronger than all other Cu-O bondings because of the shorter[109] Cu(1)-O(4) distance of 1.843Å as compared to all the other Cu-O distances in this system. For the (1,2,3,6) systems, the O(4) ion carries a charge close to the formal charge -2 than the -1 in the (1,2,3,7) systems due to the different numbers of electrons involved in the two cases because of the different stoichiometry for oxygen.

In Table 2, the ^{63}Cu nuclear quadrupole coupling constants (e^2qQ) and asymmetry parameters η obtained by our theoretical investigations and experiment are listed. For the theoretical values of the e^2qQ in Table 2, we have used the most recent value[100] of Q= -0.18X10^{-24}cm^2 available in the literature. The theoretical results in Table 2 show many features in agreement with experiment[83,85] and some differences. We shall discuss both these successes and the departures from experiment and possible avenues of improvement. Thus, considering the quadrupole coupling constants in $YBa_2Cu_3O_7$, the experimental trend of substantial increase in going from ^{63}Cu(1) to ^{63}Cu(2) is well explained by theory. Also from $YBa_2Cu_3O_7$ to $YBa_2Cu_3O_6$ there is a substantial increase in e^2qQ for ^{63}Cu(1) and an opposite trend of substantial decrease in the case of ^{63}Cu(2), both features being in agreement with experiment. However, considering the e^2qQ in $YBa_2Cu_3O_6$ alone, there is a significant decrease observed experimentally in going from ^{63}Cu(1) to ^{63}Cu(2). On the other hand, the theoretical values are nearly equal. The asymmetry parameters for both Cu(1) and Cu(2) vanish both theoretically and experimentally. This is a consequence of the tetragonal symmetry.

Table 1. Charges on the Copper and Oxygen Ions in $YBa_2Cu_3O_7$ and $YBa_2Cu_3O_6$ Systems

System	$Ba_2Cu_3O_7$	$YBa_2Cu_3O_6$
Ion		
Cu(1)	1.21	0.92
Cu(2)	1.67	1.77
O(1)	-1.84	----
O(2)	-1.95	-1.94
O(3)	-1.96	-1.94
O(4)	-1.11	-1.97

Table2. ^{63}Cu Nuclear Quadrupole Interaction Parameters in $YBa_2Cu_3O_7$ and $YBa_2Cu_3O_6$

System	$YBa_2Cu_3O_7$				$YBa_2Cu_3O_6$			
Nucleus	Theory [a]		Experiment [a,b]		Theory [a]		Experiment [a,c]	
	e^2qQ	η	$\lvert e^2qQ \rvert$	η	e^2qQ	η	$\lvert e^2qQ \rvert$	η
^{63}Cu(1)	56.6	0.11	38.4	1.00	86.5	0.0	59.1	0.0
^{63}Cu(2)	105.1	0.07	62.9	0.02	89.5	0.0	47.6	0.0

a. e^2qQ in MHz.
b. Ref. 83, C.H. Pennington et. al. and T. Shimizu *et al.*
c. Ref. 85.

For the $YBa_2Cu_3O_7$ system, the asymmetry parameter for $^{63}Cu(2)$ is found to be small but finite and the same is found from theory. For $^{63}Cu(1)$, on the other hand, η is found to be very large experimentally, close to the maximum value of unity. The theoretical value of η increases in going from $^{63}Cu(2)$ to $^{63}Cu(1)$ but only by a small amount. So there is a very significant difference between theory and experiment in the magnitude of η for $^{63}Cu(1)$ in the (1,2,3,7) system. Additionally, the theoretical values of e^2qQ for both the (1,2,3,7) and (1,2,3,6) systems are significantly larger than experiment, the ratio being about 1.5 for $^{63}Cu(1)$ in both systems, and significantly larger for $^{63}Cu(2)$, about 1.7 and 1.9 in the two systems.

We have examined[80] the case of e^2qQ for $^{63}Cu(2)$ in $YBa_2Cu_3O_6$ in some detail and found that a small but significant admixture of d_{z^2}-like hole character to the dominant $d_{x^2-y^2}$-like hole character could lead to agreement with experiment. Similar comments could be made for the other Cu systems. Thus, in particular in the case of Cu(1) for $YBa_2Cu_3O_7$, differences in the admixtures of d_{xz} and d_{yz} characters to the d_{z^2} and $d_{x^2-y^2}$-like orbitals or $4p_x$ and $4p_y$ characters to the $4p_z$-like orbitals (since the 4p-like orbitals make major contributions to the electric field gradient tensor in this ion with close to Cu^+ like character and a nearly filled 3d shell), could explain the large observed η. The question is what types of effects can lead to these types of orbital admixtures. One type of effect that can be suggested is the influence of spin-orbit interaction which has been found to be very effective[110] in influencing the nuclear quadrupole interactions for the Mössbauer isotope ^{57m}Fe in heme compounds among biological systems. The other possibility is the influence of many-body effects, for instance, correlation between 3d-like orbitals and between 3d-like and 4s-like and 4p-like orbitals, effects of the latter type having been found to be rather important in copper atom for the magnetic hyperfine constant.[111] These effects[112] need to be explored quantitatively in the future. However the fact that the difference with experiment is stronger for e^2qQ for planar $^{63}Cu(2)$ which is associated with the superconducting process, as compared to chain $^{63}Cu(1)$ in $YBa_2Cu_3O_7$, suggests that part of the correlation effects may be associated with the mechanisms connected with the origin of superconductivity.

Table 3 lists the nuclear quadrupole interaction parameters e^2qQ and η for the ^{17}O nuclei in $YBa_2Cu_3O_7$ from experimental nuclear magnetic resonance measurements[84] and from our cluster calculations.[75] No calculations were carried out on the ^{17}O nuclei in $YBa_2Cu_3O_6$, since at the time of our work,[80] no experimental data on the nuclear quadrupole interactions in these nuclei were available. For $YBa_2Cu_3O_7$, since $^{17}O(2)$ and $^{17}O(3)$ corresponding to the oxygen neighbors of planar Cu(2) are nearly equivalent, both from their geometrical parameters[109] and the experimental results in Table 3, we have carried out[75] calculations for only $^{17}O(2)$. The cluster Cu_2O_9 chosen for this work has already been described earlier in this section. Of the other two ^{17}O nuclei, only $^{17}O(4)$ has been investigated in our work.[75] For this nucleus, the immediate environment that one needs to include in a cluster investigation is covered in the cluster chosen[75] for investigations on the nuclear quadrupole interaction in the ^{63}Cu nuclei. No theoretical investigations were carried out on the chain $^{17}O(1)$ which are the neighbors of the chain Cu(1) ions. As in the case of $^{17}O(2)$, one would require a different cluster than that for the associated ^{63}Cu nucleus of Cu(1). It would be helpful to have theoretical results for $^{17}O(1)$ to compare with experiment[84] in view of the very significant differences from experiment found for the $^{63}Cu(1)$ nucleus, especially for the asymmetry parameter η.

It should be noted that while the signs of e^2qQ cannot be obtained from NMR measurements and therefore only the magnitudes are given in Table 3, the theoretical values for $^{17}O(4)$ as $^{17}O(2)$ (and one expects the same sign in $^{17}O(3)$ as in $^{17}O(2)$) are both negative. The value of Q (^{17}O) is taken from the literature[113] as-$0.026\times10^{-24}cm^2$ to obtain e^2qQ from our calculated values of q.

Table 3. ^{17}O Nuclear Quadrupole Interaction Parameters in $YBa_2Cu_3O_7$

	Theory[b]		Experiment[c]	
Nucleus[a]	e^2qQ	η	$\|e^2qQ\|$	η
$^{17}O(1)$	--	--	10.0	0.1
$^{18}O(2)$	-8.0	0.27	6.6	0.22–0.23
$^{17}O(3)$	--	--	6.4	0.22–0.23
$^{17}O(4)$	-14.2	0.20	7.3	0.32

a. All e^2qQ are expressed in MHz.
b. Ref. 75.
c. Ref. 84.

Our results for $^{17}O(2)$ in Table 3 show quite satisfactory agreement with experiment.[84] The value of η agrees rather well with experiment. The value of e^2qQ is somewhat larger than experiment;[84] the overestimation, by a factor of about 1.2, being smaller than that (about 1.7) for the neighboring Cu(2) ion on the plane (Table 2). The calculations on the cluster containing O(2) were carried out with a smaller basis set than for the cluster containing the ^{63}Cu nuclei, for economy in computing effort, and it would be helpful in the future to test the result using the larger basis set. For $^{17}O(4)$, the theoretical result[75] in Table 3 is in less satisfactory agreement with experiment than for $^{17}O(2)$ and is more similar in this respect to the results for $^{63}Cu(1)$, although the asymmetry parameter is closer to experiment than in the case of $^{63}Cu(1)$ in Table 2. However, the magnitude of e^2qQ is about 1.95 times that from experiment.

In looking for causes to bridge the differences between theory and experiment for the ^{17}O nuclei, one could consider the same causes as discussed for ^{63}Cu nuclei. The influence of spin-orbit interactions[110] at the Cu(1) and Cu(2) sites leading to mixing of molecular orbitals with different d-like characters can of course influence the adjacent oxygen atoms O(1), O(4), O(2) and O(3). The ion O(4) is adjacent to both Cu(1) and Cu(2), but closer to Cu(1) and more covalently bonded as indicated by the effective charges in Table 1. It is therefore interesting that there is greater departure for O(4) from experimental quadrupole coupling parameters than for O(2) which is bonded to Cu(2), the same trend that is observed in this respect for Cu(1) and Cu(2). This situation is partly understandable since any causes that influence the nuclear quadrupole interactions for the copper ions are expected to also influence the oxygen ions to which they are bonded. In addition to the influence of spin-orbit effects[110] associated with the copper ions, one also expects contributions to the ^{17}O nuclear quadrupole interactions from the influence of conventional many-body effects[111,112] associated with both the copper and oxygen ions, as well as special many-body effects associated with the origin of the superconductivity of the high-T_c systems based on these materials.

Table 4 lists our calculated values[77,81] of e^2qQ and η in ^{135}Ba in $YBa_2Cu_3O_6$ and $YBa_2Cu_3O_7$ which are compared with experimental values[86,90] from nuclear magnetic resonance measurements. The cluster used for both the systems in these calculations[77,81] was a BaO_4 cluster involving a Ba^{2+} ion above the center of a rectangle containing four O(4) ions. The rectangle in the case[81] of $YBa_2Cu_3O_7$ was close to a square because of the slight orthorhombic character of the lattice and in the case[77] of $YBa_2Cu_3O_6$ which had a tetragonal structure, it was exactly a square. For $YBa_2Cu_3O_7$, we also used[81] a BaO_6 cluster in which, in addition to the four O(4) ions mentioned above, we also included the next nearest O(1) neighbors which were along the a axis below the plane containing the O(4) ions. The results for the latter cluster differed very little from the BaO_4 cluster indicating that it did not make too much difference whether one used the two O(1) ions as point charges or ions with extended electron distributions.

From Table 4, it can be seen that the result[77] for e^2qQ in $YBa_2Cu_3O_6$ is in good agreement with experiment,[90] the asymmetry parameter of course being zero because of tetragonal symmetry at the Ba^{2+} site. For $YBa_2Cu_3O_7$, the theoretical e^2qQ had[81] a magnitude of about a factor of 1.5 times the experimental value, similar to the situation[75] for Cu(1). However, the asymmetry parameter η which is rather large,[81] namely 0.82, is in very good agreement with experiment.[90] What is also interesting is that the Z-direction in the principal axis system, corresponding to the largest principal component of the field-gradient tensor, is in the direction of the a-axis in agreement with experimental data[90] which indicate that the Z-axis is perpendicular to the c-axis in the crystal. The large value of η indicates that the electric field-gradients along the principal Z and X axes are nearly equal in magnitude and that for the Y-direction is small. This situation is expected to be a result of the interplay of the presence[109] of the two next nearest O(1) ions which removes the equivalence of the b and a directions, and the hole in the p-shell in O(4) which has a configuration close to 0^-. The situation is totally different for the $YBa_2Cu_3O_6$ lattice[114] where there is $\eta=0$ with an axially symmetric environment about the ^{135}Ba nucleus, there being no O(1) ions and the four O(4) ions having spherical O^{2-} configurations and arrayed symmetrically on a square in the plane below Ba^{2+}. The charges[77,81] on Ba in the (1,2,3,7) and (1,2,3,6) systems are 1.88 and 1.94 respectively, indicating stronger ionic characters than for Cu(1) and Cu(2) as suggested by the charges in Table 1.

We turn next to the magnetic hyperfine interaction in the $YBa_2Cu_3O_6$ system in the antiferromagnetic state. The antiferromagnetism appears to arise from the planar Cu(2) ions and so there are associated unpaired spin distributions in their vicinity. This would lead to magnetic hyperfine fields at the sites of the $^{63}Cu(2)$ nuclei and the neighboring ^{17}O nuclei. The fields at the $^{63}Cu(2)$ nucleus have been measured in $YBa_2Cu_3O_{6+x}$ for zero[115] and small values[116] of x. These measurements suggest that the experimental value for $x = 0$ is about 80 kiloGauss and it is oriented in the ab plane. This orientation would be expected because neutron diffraction measurements[117] indicate that the magnetic moments in the antiferromagnetic state are aligned in the ab plane with no definitive information available about the exact orientation in the ab plane. In our work,[76] we have used the orientation of the above spin as along the bisector of the a and b axes, not an unlikely orientation. From our UHF calculations,[80] one can obtain, using Eqs. (11) and (12), and the discussion following them, the hyperfine field at $^{63}Cu(2)$ in the direction perpendicular to the symmetry axis. This is the appropriate procedure to use because the axial symmetry direction in the tetragonal $YBa_2Cu_3O_6$ lattice is along the c-axis and the hyperfine field would be oriented along the direction of the magnetic moment on the Cu(2) ion.

For a proper calculation of the electron distribution in the antiferromagnetic state we have to deal with a cluster including a number of Cu(2) ions with aligned antiparallel spins. Apart from the fact that such a cluster would involve a large number of atoms, it is difficult to include the charge and spin transfer mechanisms that lead to the antiferromagentic alignment. We have therefore used[80] the approximation involving a CuO_5 cluster with a

Table 4. ^{135}Ba Nuclear Quadrupole Interaction Parameters in $YBa_2Cu_3O_7$ and $YBa_2Cu_3O_6$

System	Theory		Experiment	
	e^2qQ	η	$\lvert e^2qQ \rvert$	η
$YBa_2Cu_3O_7$	-24.2[b]	0.82[b]	16.0[c]	0.80[c]
$YBa_2Cu_3O_6$	-51.6[d]	0[d]	54.0[e]	0[e]

a. e^2qQ in MHz.
b. Ref. 81.
c. Ref. 90.
d. Ref. 77.
e. Ref. 86.

Cu(2) ion, its two O(2) neighbors, two O(3) neighbors and the O(1) neighbor in Fig. 1, with the influence of the rest of the lattice being approximated as point charges for their Coulomb potentials. The unpaired spin on the Cu(2) ion in the cluster leads[80] to the major contribution to the hyperfine field at the ^{63}Cu nucleus and one has to make an approximate analysis for the contribution from the spin distributions outside the cluster.

By this procedure, the contributions from the isotropic contact and anisotropic dipolar mechanisms for the ^{63}Cu nucleus from within the cluster come out as -20.9 kiloGauss and 137.8 kiloGauss respectively, leading to a net hyperfine field of 116.9 kiloGauss. The contribution to the hyperfine field from the magnetic dipole moments at the other copper ions outside the cluster, considering them as point dipoles and summing over the entire lattice, comes out as rather small, less than 2 kiloGauss and can be neglected. The experimental result[115,116] is about 25 percent lower than the calculated net hyperfine field. The mixing of d_{z^2} like character with the $d_{x^2-y^2}$ -like character of the hole on the Cu(2) ion does not influence the hyperfine field as sensitively as the nuclear quadrupole interaction discussed earlier. Thus, spin-orbit interaction[110] and many-body effects[111,112] responsible for this mixing are not as effective in influencing the hyperfine field at the ^{63}Cu nucleus.

One therefore has to examine other sources that could bridge the 25 percent gap between the theoretical and experimental hyperfine fields. Thus, the magnetic moment on the Cu(2) ion we have obtained[80] from a Mulliken population analysis[103] of the unpaired spin population on this ion is 0.85 Bohr magnetons as compared to the value of 0.6 Bohr magnetons suggested by neutron diffraction measurements.[117] Since the hyperfine field at the Cu(2) ion arises from the unpaired spin population, a simple scaling by this factor leads to a reduced theoretical value of about 83 kiloGauss in good agreement with the experimental value. It has recently been shown[118] that the Mulliken approximation[103] in antiferromagnetic systems leads to an overestimate and that careful analysis including the magnetic moment distribution including the influence of the antiparallel spin alignment on adjacent ions, and the effect of spin fluctuations,[119] leads to good agreement with experiment.[115,116] A similar analysis of the contributions to the hyperfine field from the antiparallel spin distributions on the adjacent clusters centered on the corresponding Cu(2) ions would be expected to lead to a quantitative reduction in the hyperfine field. An analysis of this type would represent the quantitative evaluation of the influence of the mechanism suggested by Mila and Rice,[120] and is distinct from the negligibly small point dipole contribution from these and other copper ions in the lattice pointed out in the preceding paragraph.

La_2CuO_4 System. For this system, we have studied[76,77,82] the ^{63}Cu, ^{17}O and ^{139}La nuclear quadrupole interactions as well as the magnetic hyperfine interaction[76] at ^{63}Cu in the antiferromagnetic state. Experimental data[87,88,89,121] are available for all these properties to allow comparison between theory and experiment as in the Yttrium Barium systems. Considering the nuclear quadrupole interaction results first, for the ^{63}Cu nucleus, we have made use of the bipyramidal cluster CuO_6 represented by the copper ion P, the planar oxygens K,M,L and O and apical oxygens B and N in Fig. 2 which shows the lattice structure in La_2CuO_4. The influence of the rest of the lattice is included, as described earlier, by treating the ions as point charges in terms of their potential as seen by the electrons in the cluster. For studying the convergence with respect to cluster size, a cluster Cu_5O_6 was employed, including the nearest neighbor copper ions of the planar oxygens K, M, L and O. The charges on the ions in the cluster changed by no more than 5 percent, indicating satisfactory convergence. The averages of the charges on the copper, the planar and apical oxygens using the two clusters were found to be 1.55, -1.89 and -1.95. The departures of these results from the formal charges of 2, -2 and -2 indicate significant covalent binding as in the case of the YBa Copper Oxide systems (Table 1). The charges on the planar and apical oxygens indicate that there is stronger covalent bonding of the former with copper than the latter. This is expected since the distance[122] between the copper and planar oxygens is 1.904Å, significantly smaller than the corresponding distance of 2.397Å for the apical oxygen. The antiferromagnetic coupling between the neighboring copper ions is expected to take place through super-exchange involving the planar oxygen ions.

Nuclear magnetic resonance measurements[K7] with both a little excess or deficiency of oxygen compared to La_2CuO_4 indicate that the value of the quadrupole coupling constant for ^{63}Cu in the latter system is about 65 MHz, the asymmetry parameter being very small, of the order of 0.03, but not zero, because of the slight departure of the lattice from tetragonal symmetry. Our calculation for the CuO_6 and Cu_5O_6 clusters lead respectively to $e^2qQ = 86.5$ MHz and $\eta = 0.020$ and $e^2qQ = 84.1$ MHz, $\eta = 0.027$. Like the effective charges, these results indicate that there is reasonable convergence with respect to cluster size. The ratio of the theoretical and experimental[K7] quadrupole coupling constants is about 1.3, the agreement being significantly better than in the cases of the YBa Copper Oxide systems (Table 2). In looking for mechanisms to improve agreement with experiment we can consider the same ones as for those discussed for Cu(2) ion in both $YBa_2Cu_3O_6$ and $YBa_2Cu_3O_7$, since for the copper ion in La_2CuO_4, the electronic configuration is close to Cu^{2+} as for the planar copper ions in the other two systems. The mechanisms such as spin-orbit interaction[110] and many-body effects leading to admixture of d_{z^2} and $d_{x^2-y^2}$ like hole configurations are thus expected to be significant in effect for the La_2CuO_4 system.

For the ^{17}O nuclei, as in the case of the planar oxygens in the YBa Copper Oxide systems, one has to consider clusters centered around the oxygen atoms of interest. For the ^{17}O nucleus in the apical oxygen ion B, we have taken[K2] the cluster containing all the labeled atoms A through P, in Fig. 2. It thus involves a La_5CuO_{10} cluster involving the bipyramid to which B belongs and the four oxygen ions C through F and the lanthanum ions A, G, H, I, and J. This allows us to include up to the sixth sets of neighbors of the oxygen ion B. For the planar oxygen nuclei, typically that in the oxygen ion L in Fig. 2, we have used[K2] a

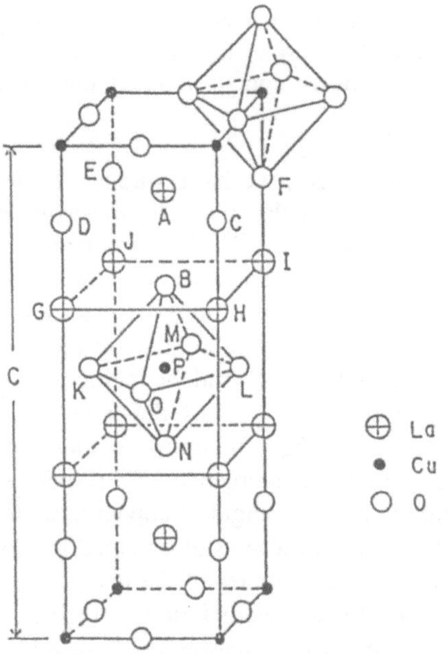

Figure 2. The crystal structure of La_2CuO_4 lattice in the tetragonal phase which is similar to the orthorhombic structure used in our work.

cluster $La_4Cu_2O_{11}$ involving the bipyramid with the copper ion P at the center and the corresponding one on the right to which L also belongs and the four La^{3+} ions H and I and their counterparts in the bc plane below L. In Table 5, we have presented the results obtained for the quadrupole coupling constants e^2qQ and asymmetry parameters η for the apical and planar ^{17}O nuclei and the directions of the Z' axis corresponding to the largest component and compared them with experimental results.

The theoretical results for the ^{17}O quadrupole interaction properties are seen from Table 5 to be in generally good agreement with experiment. For the planar ^{17}O, the theoretical vales of e^2qQ and are very close to the experimental values, with the principal Z' direction being along the a-axis for both theory and experiment. For the apical ^{17}O nucleus the Z' principal axis is found to be along the c-axis from both theory and experiment.[88] The asymmetry parameter comes out as zero because of the near tetragonal symmetry of its environment and the unit cell. The magnitude of e^2qQ however is significantly larger than experiment, the ratio being about 1.75. This difference could possibly be bridged by the influence of the mechanisms discussed earlier for $^{17}O(2)$ in the case of $YBa_2Cu_3O_7$, namely those involving spin-orbit effects[110] associated with the neighboring copper ion and the oxygen ion itself as well as many-body effects.[111,112] However, the fact that there is excellent agreement with experiment[88] for the planar ^{17}O nucleus while there is significant difference for apical ^{17}O suggests another possible contribution. Thus in our work, in common with other investigations on electronic structures of ionic crystals, only the influence of the potential from charges on the ions outside the cluster has been included. However, for the highly deformable ions like the doubly negative oxygen ions, significant dipole moments are expected to be present on them, especially when there is no reflection symmetry at the oxygen site as is the case for the apical oxygens. Since the influence of these dipole moments has not been included on the potential seen by the electrons in the cluster, the contribution of the dipole moments[123] to the electric field gradient has also not been included. In the La_2CuO_4 system, as may be seen from Fig. 2, there are four apical oxygen ions C, D, E and F quite close to the apical oxygen ion B and therefore a significant influence on the electric field gradient from the dipole moments is expected. In general, such effects are expected to be effective for all the nuclei in the three different systems we have studied and in particular could be significant contributors for the $^{17}O(2)$ and $^{17}O(4)$ nuclei in $YBa_2Cu_3O_7$ and perhaps lead to a narrowing of part of the difference between theory and experiment in Table 3.

For the study of ^{139}La nuclear quadrupole interaction, we have used the LaO_5 cluster which involves the La^{3+} ion A and its five apical oxygen ion type neighbors, C, D, E and F on a plane parallel to the ab plane and B in Fig. 2. Using the value[124] of 0.21×10^{-24} cm^2 for the nuclear quadrupole moment of ^{139}La and our calculated result for the largest of the principal components of the field gradient tensor, which is oriented along the c-axis, the value of the quadrupole coupling constant e^2qQ comes out as -90.81MHz. The asymmetry

Table 5. Nuclear Quadrupole Interaction Parameters for Planar and Apical ^{17}O Nuclei in La_2CuO_4

Nucleus	Theoretical Results[a]			Experimental Results [a,b]		
	e^2qQ	η	Z'-axis	e^2qQ	η	Z'-axis
$^{17}O_{(p)}$	-4.5	0.40	a	4.67	0.36	a
$^{17}O_{(a)}$	-2.3	0.00	c	1.30	0.00	c

a. Results for e^2qQ are in MHz.
b. Ref. 88.

parameter is small, 0.02, but non-vanishing because of the slight orthorhombicity[122] of the La_2CuO_4 lattice. These results are in good agreement with the experimental results of $|e^2qQ|$ = 89.4 MHz and η = 0.027. This observed better agreement with experiment for ^{139}La, as compared to the ^{63}Cu and ^{17}O nuclei, indicates that if many–body effects[112,113] represent the major contributors to the difference from experiment, then they are less pronounced near the La^{3+} ion, away from the active region involving the copper and oxygen ions which are considered to be responsible for the superconducting character of the La_2CuO_4 based systems. The same observation can also be made for the YBa Copper Oxide systems. The charge obtained for the lanthanum ion from our cluster investigations is 2.98 indicating that, like barium in the YBa Copper Oxide systems, the lanthanum ion is strongly ionic and has a configuration close to La^{3+}.

For the magnetic hyperfine field at the ^{63}Cu nucleus, we have followed the same procedure for antiferromagnetic La_2CuO_4, as was described for $YBa_2Cu_3O_6$ earlier in this article. The same bipyramidal cluster with a copper ion at the center was used as for the investigation of the ^{63}Cu nuclear quadrupole interaction. The contact hyperfine field was found[76] to be -31 kiloGauss and was dominated by the dipolar field of 136 kiloGauss oriented along the direction of the unpaired spin on the copper ion at the center of the cluster, which has been found from neutron diffraction measurements[125] to be in the direction of the b-axis in the orthorhombic axes system,[126] that is at 45° to the line joining the lanthanum ion and a planar oxygen. The contribution from the magnetic dipole moments on the copper ions outside the cluster, considering them as point dipole moments and summing over the entire lattice, again comes out negligible, similar to the case of antiferromagnetic $YBa_2Cu_3O_6$ discussed earlier. The total hyperfine field 105 kiloGauss is 25 kiloGauss higher than the experimental value of about 80 kiloGauss (close to that for the $YBa_2Cu_3O_6$ system) suggested from nuclear magnetic resonance measurements[127] for slightly higher and lower contents of oxygen as compared to La_2CuO_4. The same discussion as for $YBa_2Cu_3O_6$ considered earlier would apply about the sources that could improve agreement with experiment. Thus, while many-body[111,112] and spin-orbit effects[110] could have some influence on the hyperfine field, although perhaps not a major one, the main reduction in the theoretical hyperfine field is expected to arise from the influence of the opposite spin distributions in adjacent clusters. The magnetic moment on the copper ion from our electronic structure calculations using the Mulliken approximation was found to be 0.88 Bohr magneton, substantially higher than the value 0.5 found from neutron diffraction measurements.[125] The explanation discussed for bridging a similar gap between our result[80] and experiment[117] in $YBa_2Cu_3O_6$ should apply here. This would involve the incorporation of spin fluctuations[119] and a more accurate treatment as for the magnetization[118] for including the influence of the opposite spin distributions on neighboring clusters as discussed for the ^{63}Cu hyperfine field in $YBa_2Cu_3O_6$.

La_2CuO_4-Muon System. I would like to finish this discussion of our electronic structure investigations in high-T_c materials by briefly describing our recent work[78,79] on the interaction between the positive muon and La_2CuO_4. This work was stimulated by the extensive muon spin rotation measurements[128,129,130] on this system by a number of groups including the group at University of Tokyo.

There are a number of reasons for studying the muon position and associated hyperfine properties. The first is to test how well we can explain the interaction of the muon with La_2CuO_4 and thus obtain a check on the understanding of the electron distribution in the latter which is complimentary to that one gets by working on the pure material as just discussed. The second reason is to see how well theory can explain the experimentally observed hyperfine field at the muon from muon spin rotation measurements[129] in the antiferromagnetic state. The third reason is to make comparison with the position derived from the relaxation effect[130] in zero-field muon spin rotation measurements,[131] this effect arising from the distribution in the magnetic field[130] at the muon site due to the distribution in orientations of the copper nuclear spins in the presence of the nuclear quadrupole interactions at their sites.

Earlier attempts to determine the position of the muon based on approximate theoretical analyses[132,133] of the energy of interaction of the muon with La_2CuO_4 and the estimation of the hyperfine field[134] assuming that it arose from localized moments on the copper ions had suggested that it is located on the ac plane, the a and b axes in this work being taken as the tetragonal system of axes pointing along the Cu-O lines in Fig. 2 for the planar oxygens instead of the orthorhombic axes system used earlier for our discussion of the ^{63}Cu hyperfine field in La_2CuO_4 in the antiferromagnetic phase. A similar conclusion regarding the location of the muon on the ac plane was made from the analysis of relaxation effects in zero-field muon spin rotation measurements[131] although the data also seem to fit a position on a plane involving the c-axis and the bisector of a and b axes.

In view of the above results and since a more extensive search for muon position based on energy minimization would be rather time-consuming, we have primarily concentrated our analysis[78,79] with the muon on the ac plane. Since the muon, like hydrogen or a proton, would tend to attach itself to an oxygen negative ion, we have explored the vicinities of both the apical and planar oxygens. For muon next to the apical oxygen B in Fig. 2, we have used a seventeen atom cluster, involving the muon together with the sixteen atom cluster used[82] for studying the ^{17}O nuclear quadrupole interaction as described earlier. For the vicinity of the planar oxygen ion L in Fig. 2, we have used an eighteen atom cluster involving the muon in addition to the seventeen atom cluster used[82] for the study of planar ^{17}O nuclear quadrupole interaction. The energies of each of these two clusters was studied as a function of the angle made by the line Bμ with the c-axis for the vicinity of the apical oxygen B and between the line Lμ and the line LP joining L and the copper ion P for the vicinity of the planar oxygen L. The results show a number of energy minima, the deepest one being found near the apical oxygen with the line Bμ making an angle of 25° with the c-axis. This position is denoted by U1 in Fig. 3, the minimum position found near the planar oxygen being indicated by U2, which corresponds to a higher energy than that for U1. The total energy of the cluster was also minimized as a function of the Bμ bond distance and positions perpendicular to the ac plane, the latter procedure showing that the position U1 on the ac plane is the lowest in energy and therefore the most likely position for the muon from our investigations.

The positions T1 and T2 in Fig. 3 represent the positions found from zero-field μSR investigations[131] in the paramagnetic state, H from fitting[134] the hyperfine field at the muon from μSR measurements[128] in the antiferromagnetic state using a point dipole approximation for the magnetic moments on the copper ions and M and S from earlier theoretical investigations[132,133] of energy minima. We have calculated the hyperfine field at the muon at the position U1 following the steps described in the procedure at the beginning of Sec. II and find the net hyperfine field to have components -559.2, -657.3 and 515.3 Gauss along the a, b and c directions in the antiferromagnetic state. Use has been made of the fact that the magnetic moments on the copper ions are aligned parallel and antiparallel to the 110 direction, from neutron diffraction measurements.[125] A weighting factor has been applied as in the case of ^{63}Cu hyperfine field to reduce the calculated results to take account of the fact the observed magnetic moments[125] on the copper ions are 0.5 Bohr magnetons, rather than the 0.85 Bohr magnetons found from our cluster calculation. A more sophisticated estimate including the contributions from neighboring clusters with antiparallel spins on the copper ions could be done later, as pointed out earlier for the hyperfine interaction in ^{63}Cu in this system, to incorporate the Mila-Rice type of effect,[120] which would obviate the use of the approximate weighting procedure.

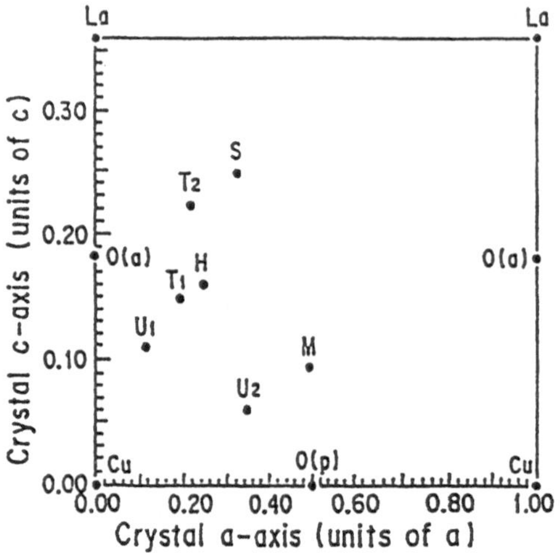

Figure 3. Position of muon on the ac plane of the tetragonal unit cell determined in earlier work[131–134] and present work.

In making comparisons with experiment, for the muon position, the experimental reference points most appropriate are expected to be the ones, T1 and T2, determined from zero-field μSR measurements[131] because the point H in Fig. 3 is based on an estimate[134] of the hyperfine field from point dipoles on the copper ion and does not take account of the local contributions from the electron spin distribution in the vicinity of the muon. This local contribution is found to be about 30 percent of the distant contribution from our work.[78,79] Our muon position U1 in Fig. 3 is found to be closer to T1 than T2 and our calculated hyperfine field is found[78,79] to have a magnitude of 1005 Gauss somewhat more than twice as large as the experimental value[128,129] of 400 to 430 Gauss. The direction for the hyperfine field described by the polar and azimuthal angles $\theta = 59^\circ$ and $\varphi = 49.6^\circ$ are in reasonable agreement with single crystal measurements[135] leading to $\varphi=65^\circ\pm5^\circ$ and $\varphi=28^\circ\pm8^\circ$. Possible sources that could lead to improved agreement with experiment are many-body effects and the lattice distortion effects associated with the presence of the muon. We are in the process of exploring the latter effect which should also influence the position T1 determined from zero-field μSR data in the paramagnetic state based on the magnetic field distribution[131] at the muon site from the copper nuclear moments. These latter contributions are dependent on the nuclear quadrupole interactions at the ^{63}Cu nuclei. The earlier analysis[131] for T1 used the ^{63}Cu nuclear quadrupole interaction for the pure La_2CuO_4. Revised analysis is under way in our group in collaboration with the Tokyo group including the changed ^{63}Cu nuclear quadrupole interaction in the presence of the muon as determined from our cluster investigations.[78,79]

In summary, the results of all our Hartree-Fock cluster investigations[75–82] described here in Sec. II have provided an overall satisfactory explanation[83–91,128,135] of the hyperfine properties of the family of high-temperature superconducting systems La_2CuO_4 (including that for muon in La_2CuO_4), $YBa_2Cu_3O_6$ and $YBa_2Cu_3O_7$. The agreement with experiment is more satisfactory at the ^{139}La and ^{135}Ba nuclei which are away from the active region for superconductivity, as compared to the ^{63}Cu and ^{17}O nuclei in the vicinity of the latter region. It appears that both standard many-body effects[111,112] as among the electrons in conventional atomic, molecular and solid state systems, as well as many-body effects associated with the origin of superconductivity in this family of systems may be responsible. In this respect, the agreement with experiment seems to be overall somewhat better for the La_2CuO_4 system[76,77,82] than the $YBa_2Cu_3O_6$ and $YBa_2Cu_3O_7$ systems,[75,77,80,81] which suggests that the role of many-body effects may be different in the two cases. This should be useful for the understanding of the origin of the superconductivity in these systems. Also expected to be useful in this respect are the general conclusions that there is significant covalent bonding among the copper and oxygen ions in these systems, especially between the chain copper and bridge oxygen in $YBa_2Cu_3O_7$ systems, the charges on the various ions being significantly different from the expected formal charges.

III. GENERAL CONCLUSIONS AND SUGGESTED IMPROVEMENTS

Our investigations on the broad spectrum of systems ranging from ionic crystals, high-T_c systems, fullerenes, molecular crystals, semiconductors (bulk and surfaces) and large biological molecules, dealing with magnetic hyperfine fields, nuclear quadrupole interactions and isomer shifts, indicate that in general one obtains satisfactory agreement with experiment using the Hartree-Fock Cluster procedure. Notable exceptions are the transition metal ionic compound Fe_2O_3 and the high-T_c systems La_2CuO_4, $YBa_2Cu_3O_6$ and $YBa_2Cu_3O_7$. In the former system we have found[136] reasonably good agreement with experiment for the magnetic hyperfine field[80] and isomer shift[54,55] but an underestimation of the nuclear quadrupole interaction. This suggests that satisfactory results for the isotropic components of the charge and spin distributions have been obtained but that the anisotropic component of the charge distribution requires improvement. This may be symptomatic of the nuclei in transition metal oxides because they are surrounded by diffuse anions which do not help in their localization and it may be essential to use larger clusters. The influence of the electrostatic interactions due to the effective positive charges on the cations in the rest of the lattice outside the cluster do provide an attractive potential which helps to contract the diffuse oxygen ion charge distributions within the cluster but that may not be enough and one would have to consider the repulsive influence of the electrons on the cations outside the cluster by including the pseudopotentials[137] associated with them in the embedding procedure.[57] The confining effect on the diffuse charge distributions on the oxygen anions of these influences of the cations outside the cluster can eventually influence the cations in the cluster including the central one and hence affect the hyperfine properties. It will be interesting to see if these effects influence the nuclear quadrupole interaction of the ^{57m}Fe nuclei appropriately to improve the agreement with experiment[138] and do not change the isotropic properties like the isomer shift and hyperfine field which are already in good agreement[139] with experiment. Of course there is the possibility that the ^{57m}Fe nuclear quadrupole moment[140] used to calculate the nuclear quadrupole coupling constants may have been underestimated in the literature and efforts are under way in our group to test this possibility. For the high-T_c systems, which involve the ^{63}Cu nuclei, the same observations regarding the influence of the diffuse oxygen on electron distribution might apply. But as mentioned in Sec. II, there is the added possibility that special effects associated with the origin of the superconducting mechanism in these materials may also be influencing the electron distributions on the copper ions and hence their hyperfine properties.

Reasonable agreements have been found[141] between the results of Hartree-Fock investigations and those involving calculations using the local density approximation (LDA) for the exchange interaction in a few molecules. However, for magnetic hyperfine properties, significant differences are expected because these latter properties are influenced quite strongly by the exchange interactions. Large differences have been found[4,144] between the results of the two procedures for alkali atoms[142,143] and the ferromagnetic metals iron, cobalt and nickel. While we therefore suggest similar reservations for LDA cluster calculations in semiconductor and ionic crystal systems it would be helpful to have some careful comparisons between the results of Hartree-Fock and LDA cluster calculations in these systems in the future.

Another factor that needs to be considered in future investigations by the Hartree-Fock Cluster Procedure is the influence of multipole moments, especially dipole moments,[31,145,146] that may be present on the ions outside the cluster depending on the symmetry of their locations. The Coulomb potentials due to the charges on the ions outside the cluster have been included in the embedding procedure we have used[24] for ionic crystals but not the effect of their multipole moments. This is a particularly important consideration for systems with highly deformable ions, such as the O^{2-} ions in oxides of course but also other negative ions like the halogen ions especially heavier ones like the Br^- and I^- ions. The influence of the multipole moments on ions inside the cluster are already included through the electronic molecular orbital wave-functions that are based on the atomic orbitals on these ions. In addition to s, p and d orbitals, it is desirable in this respect to include f and g orbitals in the basis set to be able to properly include polarization effects associated with the p and d orbitals on the atoms.

Use of larger clusters as far as practicable is of course recommended for taking account of the diffuseness of the negative ions in the lattice and also for better inclusion of the influence of their multipole moments by having more of them within the cluster chosen. It is also important to use larger clusters because they allow better simulation of the infinite solid and make the results less sensitively dependent on the embedding procedure.

For accurate results, it is also important to include correlation or many-body effects as well as practicable. It is currently the practice in molecular calculations to use for the excited states needed in the many-body perturbation-theoretic and configuration interaction[9] procedures, the empty states outside the occupied states that become available in the course of the variational Hartree-Fock-Roothaan calculation. These states correspond to excited Hartree-Fock states, or what are referred to in the many-body perturbation theory literature,[8] excited states in the V^N approximation with position energies. In the V^{N-1} approximation,[8] on the other hand, where one omits in the one-electron potential used to generate the excited states the self-coulomb and self-exchange contributions for one of the occupied states, usually the highest occupied one, one obtains in the spectrum of unoccupied excited states, a number with negative energies, or what are called bound excited states in many–body literature. The V^{N-1} potential is a more physical one[8,147] than the V^N potential because the latter makes an electron in a neutral molecule (or atom) experience a potential at large distances from the nuclei (or nucleus in the case of an atom) of the molecule due to effectively a charge close to zero. The V^{N-1} potential makes the neutral molecule (or atom) look like a single positive charge to the electron at large distances. The latter corresponds to a more physical picture because an electron of a neutral system should indeed see an effective single positive charge at large distances since the nuclear charges are shielded by one less electron than the total number of electrons in the system. The physical nature of the V^{N-1} potential makes the perturbation approach to many-body effects more convergent[8,147] than when one uses the V^N potential. Additionally, the availability of excited states close in energy to the highest occupied states in the V^{N-1} approximation, which can make important contributions to the perturbation terms or diagrams as discussed in many-body perturbation theory in atomic physics,[147] is expected to make the departure from completeness of the excited states used in many-body perturbation theory less important a factor than in the V^N approximation where the excited states start with positive energies. This would not be an important consideration if one had a really complete set of excited states. However, in practice this is not the case, since one has to use as excited states the empty one-electron states obtained from the variational calculation for the occupied states, the number of available excited states being limited in number and determined by the size of the basis set of atomic like states used in the variational Hartree-Fock-Roothaan calculation for the occupied one–electron states.

Lastly, we would like to point out the need to use relativistic methods in dealing with the electronic structures of systems composed of even moderately heavy atoms. From our calculations on atomic systems,[148] by the relativistic linked cluster perturbation–theoretic methods,[149] we have found enhancements by factors of as much as 1.2 and 1.5 for the magnetic constants in rubidium and cesium respectively. Both for this influence of atomic-like relativistic effects and indirect but important effects characteristic of molecular systems, it is necessary to include relativistic effects for hyperfine interactions studied by the Hartree-Fock Cluster procedure. The indirect effects have to do with the feature that relativistic effects which occur near the nuclei where the electrons move with greatest speed, also influence the electron densities in regions away from the nucleus, such as the space between atoms, because of normalization effects. This will in turn affect the bonding between the atoms, which influence such features as covalency and charge transfer effects which would affect hyperfine properties. Similar features have been seen[150] in the study of the influence of relativistic effects on exchange interactions between valence, 4f and core electrons in rare-earth ions and atoms which affect the exchange core polarization contributions to the hyperfine constants. Inclusion of relativistic effects in a systematic manner have been pioneered in cluster calculations by Professor Niewupoort's group[151] which it is hoped will be increasingly applied to study of hyperfine interactions in solid state systems.

These suggested improvements are of course necessary and desirable to provide better agreement with experiment for hyperfine properties and in general for other properties. It should, however, be emphasized that the Hartree-Fock Cluster Procedure for condensed matter systems has already provided satisfactory quantitative agreement with experiment for a large variety of properties including hyperfine interactions, as has been described by a number of experts and pioneers in the field who are attending the Conference. It has become in many respects a viable alternative to band-structure procedures for study of perfect systems and has particular potential for dealing with imperfect systems where the symmetry of the host system is destroyed by the presence of an impurity, vacancy or other defects, especially for localized properties. It is expected that it will continue to be used to add to our understanding of the properties of an increasing number and variety of condensed matter systems including biologically important systems.

ACKNOWLEDGMENT

I would like now to acknowledge all my collaborators in the field of Cluster studies in all the condensed matter systems that we have worked on and which I have surveyed in Sec. 1 and in particular my collaborators in the field of high-T_c materials, namely, Dr. N. Sahoo, Dr. Shuleri, B. Sulaiman, Ms. Sudha Srinivas, Ms. Sigrid Markert, Dr. F. Hagelberg, Prof. K. Nagamine, Prof. E. Torikai, and Prof. O. Donzelli. The remarks about the merits and disadvantages of the Cluster Procedure for condensed matter systems and the improvements needed in the future have resulted from the collaborative investigations, interactions and discussions I have been fortunate to have had with many scientists, both in this country and abroad.

I am grateful for the help and advice of Dr. Xing Yuan in the preparation of this article.

REFERENCES

1. G.D. Gaspari and T.P. Das, Phys. Rev. **167**, 660 (1968); P. Jena and T. P. Das, Phys. Rev. B4, 3931 (1971), and references therein.
2. L. Tterlikkis, S. D. Mahanti and T. P. Das, Phys. Rev. Lett. 21, 1796 (1968); S. D. Mahanti and T. P. Das, Phys. Rev. 170, 426 (1968).

3. M. Pomerantz and T. P. Das, Phys. Rev. 119, 70 (1960); 123, 1070 (1961); P.C. Schmidt, P.C. Pattnaik and T. P. Das, Phys. Rev. B29, 3066 (1984), and references therein.
4. T. P. Das, Hyperfine Interaction, 6, 53 (1979); K. J. Duff and T. P. Das, Phys. Rev. B3, 2294 (1971); B12, 3870 (1975); B. Krawchuk, C. M. Singal and T. P. Das, Phys. Rev. B16, 5108 (1977).
5. See for instance: K. J. Duff and T. P. Das, Phys. Rev. B3,192 (1971).
6. C. M. Singal and T. P. Das, Phys. Rev. B16, 5068 (1977). See Ref. 5 also.
7. C. Pisani, R. Dovesi and C. Roetti, *Hartree-Fock Ab Initio Treatment of Crystalline Systems* (Springer-Verlag, Berlin, 1988). See also C. M. Singal and T. P. Das, Ref. 6.
8. E.S. Chang, R.T. Pu and T.P. Das, Phys. Rev. **174**, 1 (1968); T. P. Das, Hyperfine Interaction, 34, 149 (1987); R. W. Dougherty, S. N. Panigrahy, J. Andriessen and T. P. Das, Phys. Rev. A47, 2710 (1993), and references therein.
9. T. L. Barr and E. R. Davidson, Phys. Rev. A1, 24 (1970); R. K. Nesbet, Phys. Rev. A2, 661, 1208 (1970); C. F. Bunge and E. M. A. Peixoto, Phys. Rev. A1, 1277 (1970).
10. C. M. Singal and T. P. Das, Phys. Rev. B8, 3675 (1973); Phys. Rev. B12, 795 (1975).
11. W. C. Nieuwpoort (Talk at this Workshop).
12. P. S. Bagus, F. Illas, C. Sousa, and G. Pacchioni (Talk at this Workshop); M. Seel and P. S. Bagus, Phys. Rev. B28, 2023 (1983); I. P. Batra, P. S. Bagus and K. Hermann, Phys. Rev. Lett. 52, 384 (1984).
13. J. L. Whitten (Talk at this Workshop); J. L. Whitten, Chem. Phys. 177, 387 (1993) and references therein.
14. J. M. Vail (Talk at this Workshop); J. Meng, J. M. Vail, A. M. Stoneham and P. Jena, Phys. Rev. B42, 1156 (1990) and references therein.
15. P. S. Bagus, F. Illas et. al., (Talk by Dr. Bagus at this Workshop); P. S. Bagus, F. Illas, C. Sousa, J. Chem. Phys. 100, 2943 (1994) and references therein; F. Illas, J. Casanovas, M. A. Garcia-Bach, R. Caballol, and O. Castell, Phys. Rev. Lett. 71, 3549 (1993).
16. J. Sauer, Chem. Rev. 89, 199 (1989); S. M. Mohapatra, B.N. Dev, T. M. Thundat, L. Luo, W. M. Gibson, K. C. Mishra, N. Sahoo and T. P. Das, *Reviews of Solid State Science* (World Scientific Publishing Company, Singapore), 4, 873 (1990) and references therein; N. Sahoo, S. B. Sulaiman, K. C. Mishra and T. P. Das, Phys. Rev. B39, 13389 (1989); P. C. Kelires, K. C. Mishra and T. P. Das, Hyperfine Interactions, 34, 289 (1987); D. W. Mitchell, W. Potzel, G. M. Kalvius, H. Karzel, W. Schiessl, M. Steiner, M. Kofferlein and T. P. Das, Phys. Rev. B48, 16449 (1993); O. Donzelli, Tina Briere and T.P. Das, Solid State Communication 90, 663 (1994).
17. D. E. Ellis and G. S. Painter, Phys. Rev. B2, 2887 (1970); B. Lindgren and D. E. Ellis, Phys. Rev. B26, 636 (1982); B. Lindgren, Phys. Rev. B34, 648 (1986).
18. For MNDO procedure see M. J. S. Dewar and W. Thiel, J. Am.Chem. Soc. 99, 4899 (1977); W. Percival and S. Wlodek, Chem. Phys. Lett. 196, 317 (1992); For PRDDO procedure see for instance S. K. Estreicher, C. D. Latham, M. I. Hegie, R. Jones and S. Oberg, Chem. Phys. Lett. 196, 311 (1992) and references therein.
19. M. Van Rossum, G. Langouche, K. C. Mishra and T. P. Das, Phys. Rev.B28, 6086 (1983).
20. S. B. Sulaiman, S. Swingle Nunes, N. Sahoo and T. P. Das in *Physics and Chemistry of Finite Systems-Clusters to Crystals* ed. P. Jena, B. K. Rao, S. N. Khanna, Kluwer Press, Dordrecht, Netherlands, page 699 (1992).
21. See for instance K. S. Song and R. C. Baetzold, Phys. Rev. B46, 1960 (1992); J. M. Vail, J. Phys. Chem. Solids 51, 589 (1990).
22. R. A. Evarestor, Phys. Stat Solidi B72, 569 (1975); K. C. Mishra, P. C. Schmidt, K. H. Johnson, B. G. DeBoer, J. K. Berkowitz and E. A. Dale, Phys. Rev. B42, 1423 (1990); K. C. Mishra, K. H. Johnson, P. C. Schmidt, B. G. DeBoer, J. Olson and E. A. Dale, Phys. Rev. B43, 14188 (1991).
23. A. Zunger, and U. Lindfelt, Phys. Rev. B26, 846, 5989 (1982).
24. K. A. Colbourn and J. Kendrick, in *Computer Simulation in Solids*, ed. C. R. A. Catlow and W. C. Mackrodt, Springer-Verlag; P. C. Kelires and T. P. Das, Hyperfine Interactions, 34, 285 (1987); P. C. Kelires et. al., Ref. 16; D. W. Mitchell et. al., Ref. 16, (1991) and (1993).
25. M. Seel and P. Bagus, Ref. 12; I. P. Batra et. al., Ref. 12; F. Illas and J. Rubio, Phys. Rev. B31, 8068 (1985); A. Redondo and W. A. Goddard III, and T. C. McGill, J. Vac. Sci. Tech. 21, 649 (1982) and references therein;S.M. Mohapatra *et al.*, Ref. 16; N. Sahoo et. al., Ref. 16; L. Luo, N. Sahoo, G. A. Smith, W. M. Gibson and T. P. Das, J. Phys. Soc. (Japan) 60, 3139 (1991).
26. B. Lindgren and D. Ellis, Ref. 17; B. Lindgren, Ref. 17; J. C. Malvido and J. L. Whitten, Phys. Rev. B26, 4458 (1982); J. L. Whitten, Ref. 13.
27. J. L. Whitten, Ref. 13.
28. J. M. Vail (Talk at this Workshop), Ref. 13.
29. D. W. Mitchell, Ph. D. Thesis, State University of New York at Albany (1993) (Unpublished); D. W. Mitchell, S. B. Sulaiman, N. Sahoo, T. P. Das, W. Potzel, G. M. Kalvius, W. Schiessl, H. Karzel and M. Steiner, Hyperfine Interactions, 78, 403 (1993).

30. D. W. Mitchell (private communication).
31. B. G. Dick and T. P. Das, Phys. Rev. 127, 1053, 1063 (1962); T. P. Das, Phys. Rev. 140, A1957 (1965); R. J. Quigley and T.P. Das, Phys. Rev. 164, 1185 (1967); 177, 1340 (1969).
32. N. Sahoo et. al., Ref. 16; L. Luo et. al., Ref. 25 (1991).
33. P. Bagus, Ref. 25; W. A. Goddard, Ref. 25.
34. N. Sahoo et. al., Ref. 16; W. A. Goddard, Ref. 25.
35. N. Sabirin Mohamed, N. Sahoo, K. C. Mishra, P. C. Kelires, M. Frank, Ch. Ott, B. Roseler, G. Weeske, W. Kreische, M. Van Rossum and T. P. Das, Hyperfine Interactions, 60, 857 (1990).
36. G. Gowri, H. S. Cho, T. Briere, S. Srinivas, F. Hagelberg and T. P. Das, *Cluster Investigations on ^{9}Be Hyperfine Properties in Beryllium Metal* (Unpublished).
37. H. S. Cho, G. Gowri, T. Briere, S. Srinivas, F. Hagelberg and T. P. Das, *Cluster Investigations on ^{77}Se Hyperfine Properties in Selenium* (Unpublished).
38. R. Hoffmann and W. N. Lipscomb, J. Chem. Phys. 36, 2179 (1962); M. Zerner, M. Gouterman and H. Kobayashi, Theo. Chim. Acta 6, 636 (1966) and references therein; P. S. Han, T. P. Das and M. F. Rettig, Theo. Chim. Acta 16, 1 (1970).
39. J. N. Roy, S. K. Mishra, K. C. Mishra and T. P. Das, J. Phys. Chem. 93, 194 (1989); J. Chem. Phys. 90, 7273 (1989), and references therein.
40. S. K. Mun, Ph. D. Thesis, State University of New York atAlbany (1979) (Unpublished).
41. J. C. Chang and T. P. Das, J. Chem. Phys. 68, 1462 (1978); A. Coker, K. C. Mishra and T. P. Das, Federation of American Society of Experimental Biology Procedings 39, No. 6, p. 1802 (1980); Bull. Am. Phys. Soc. 25, 417 (1980).
42. A. Coker, T. Lee, A. Glodeanu and T. P. Das, Hyperfine Interactions, 4, 821 (1978).
43. M. Van Rossum et. al., Ref. 19 (1982 and 1983).
44. A. Coker, T. Lee and T.P. Das, Phys. Rev. B21, 416 (1980).
45. N. Sahoo et. al., Ref. 16.
46. S. M. Mohapatra et. al., Ref. 16 (1990); L. Luo et. al., Ref. 25; B. N. Dev, S. M. Mohapatra, K. C. Mishra, W. M. Gibson and T. P. Das, Phys. Rev. B36, 2666 (1987); T. Thundat, S. M. Mohapatra, B. N. Dev, W. M. Gibson and T. P. Das, J. Vac. Sci. Tech. A6, 681 (1988).
47. S. B. Sulaiman, Ph. D. Thesis, State University of New York at Albany (1992) (Unpublished); S. B. Sulaiman et. al., Ref. 20.
48. H. Fraunfelder and R. M. Steffen in *Alpha-, Beta- and Gamma-Ray Spectroscopy*, Ed. K. Siegbahn (Amsterdam, North Holland, 1966).
49. K. Bonde Nielsen, H. K. Schou, T. Lauritsen, G. Weyer, I. Stensgaard and S. Damgaard, J. Phys. C17, 3519 (1984); E. Bertholdt, M. Frank, F. Gubitz, W. Kreische, B. Lösche, C. Ott, B. Röseler, M. Schneider, F. Schwab, K. Stammler, and G. Weeske, J. Mol. Struct. 192, 199, 383 (1989).
50. P. C. Kelires et. al., Ref. 16.
51. P. C. Kelires, Ph. D. Thesis, State University of New York at Albany (1987) (Unpublished).
52. P. C. Kelires et. al., Ref. 24 (1987); J. Stein, S. B. Sulaiman, N. Sahoo and T. P. Das, Hyperfine Interactions, 60, 849 (1990).
53. H. H. Klauss, N. Sahoo, P. C. Kelires, T. P. Das, W. Potzel, M. Kalvius, M. Frank, W. Kreische, Hyperfine Interactions, 60, 853 (1990); M. Steiner, W. Potzel, M. Kofferlein, H. Karzel, W. Schiessl, G. M. Kalvius, D. W. Mitchell, N. Sahoo, H. H. Klauss, T. P. Das, R. Feigelson and G. Schmidt, Phys. Rev. B50, 13555 (1994).
54. D. W. Mitchell et. al., Ref. 16 (1993).
55. W. Potzel, in *Mössbauer Spectroscopy Applied to Magnetism and Materials Science*, Vol. 1, Eds. E. J. Long and F. Grandjean (Plenum Press, New York, 1993); M. Steiner, W. Potzel, C. Schafer, W. Adlassnig, M. Peter, H. Karzel and G. M. Kalvius, Phys. Rev. B41, 1750 (1990).
56. H. Barfuss, G. Böhnlein, H. Hohenstein, W. Kreische, M. Meinhold, H. Niedrig and K. Reuter, J. Mol. Struct. 58, 503 (1980).
57. D. W. Mitchell, Ref. 29; D. W. Mitchell, et. al., Ref. 29.
58. S. B. Sulaiman, N. Sahoo, T. P. Das, O. Donzelli, E. Torikai and K. Nagamine, Phys. Rev. B44, 7028 (1991); S. B. Sulaiman, N. Sahoo, T. P. Das and O. Donzelli, Phys. Rev. B45, 7383 (1992); S. Srinivas, S. B. Sulaiman, N. Sahoo, T. P. Das, E. Torikai, and K. Nagamine, Bull. Am. Phys. Soc. 39, 839 (1994).
59. N. Sahoo, Sigrid Markert, T. P. Das and K. Nagamine, Phys. Rev. B41, 220 (1990); S. B. Sulaiman, et. al., Ref. 58 (1992); S. B. Sulaiman, S. Srinivas, N. Sahoo, T. P. Das, O. Donzelli, E. Torikai, and K. Nagamine, Bull. Am. Phys. Soc. 38, 174 (1993).
60. C. H. Pennington, D. J. Durand, C. B. Zax, C. P. Slichter, J. P. Rice, and D. M. Ginsberg, Phys. Rev. B37, 7944 (1988); T. Shimizu, H. Yasuoka, T. Imai, T. Tsuda, T. Takabatake, Y. Nakazawa, and M. Ishikawa, J. Phys. Soc. (Japan), 57, 2494 (1988); M. Mali, I. Mangelschots, H. Zimmermann, and D.

Brinkmann, Physica C175, 581 (1991); C. Coretsopoulos, H. C. Lee, E. Ramli, L. Reven, T. B. Rauchfuss, and E. Oldfield, Phys. Rev. B39, 781 (1989); J. Shore, S. Yang, J. Haase, D. Schwartz, and E. Oldfield, Phys. Rev. B46, 595 (1992); A. Yakabowskii, A. Egorov, and H. Lütgemeier, Appl. Magn. Reson. 3, 665 (1992); H. Lütgemeier, V. Florentiev, and A. Yakubowskii, in *Electronic Properties of High-Tc Supercondtors and Related Compounds*, Springer-Verlag, Berlin, Proceedings of the conference at Kirchberg, (1990); K. Ueda, T. Sugata, Y. Kohara, Y. Oda, M. Yamada, S. Kashiwai, and M. Motoyama, Solid State Comm., 64, 267 (1987); K. Ishida, Y. Kitaoka, and K. Asayama, J. Phys. Soc. Japan, 58, 36 (1989); K. Ishida, Y. Kitaoka, G. Zheng, and K. Asayama, J. Phys. Soc. Japan, 60, 3516 (1991); H. Lütgemeier, and M. W. Pieper, Solid State Comm. 64, 267 (1987).

61. H. Yasuoka, T. Shimizu, Y. Ueda and K. Kosuge, J. Phys. Soc. Japan 57, 2659 (1988); Y. Yamada, K. Ishida, Y. Kitaoka, K. Asayama, H. Takagi, H. Iwabuchi, and S. Uchida, J. Phys. Soc. Japan 57, 2663 (1988); P. Mendels and H. Alloul, Physica C156, 355 (1988); T. Tsuda, T. Shimiza, H. Yasuoka, K. Kishio and K. Kitazawa, J. Phys. Soc. Japan 57, 2908 (1988).
62. S. B. Sulaiman, N. Sahoo, S. Srinivas, F. Hagelberg, T. P. Das, E. Torikai, and K. Nagamine, Phys. Rev. B.49, 9879 (1994); Hyperfine Interactions, 79, 901 (1993); Hyperfine Interactions, 84, 87 (1994).
63. J. I. Budnick, A. Golnik, Ch. Niedermayer, E. Recknagel , A. Weidinger, B. Chamberland, M. Filipowski, and D. P. Yang, Phys. Lett. A124, 103 (1987); B. Hitti, P. Birrer, K. Fischer, F. N. Gygax, E. Lippelt, H. Malletta, A. Schenck, and M. Weber, Hyperfine Interactions, 67, 287 (1990); L. P. Le, G. M. Lake, B. J. Sternlieb, Y. J. Uemura, J. H. Brewer, T. M. Riseman, D. C. Johnston, L. L. Miller, Y. Hidaka, and H. Murakami, Hyperfine Interactions, 63, 279 (1990); E. Torikai, H. Ishihara, K. Nagamine, H. Kitazawa, I. Tanaka, H. Kojima, S. B. Sulaiman, S. Srinivas, and T. P. Das, Hyperfine Interactions, 79, 915 (1993).
64. N. Sahoo, K. Ramani Lata and T. P. Das, Theo. Chim. Acta 82, 285 (1992).
65. K. Ramani Lata, Ph. D. Thesis, State University of New York at Albany, (1993) (Unpublished); K. Ramani Lata, N. Sahoo and T. P. Das. Under preparation for J. Am. Chem. Soc.
66. N. Sahoo and T. P. Das, J. Chem. Phys. 91, 7740 (1989); 93, 1200 (1990); K. Ramani Lata, N. Sahoo, and T. P. Das, J. Chem. Phys. 94, 3715 (1991).
67. N. Sahoo, K. Ramani Lata and T. P. Das. Under preparation for J. Chem. Phys.
68. N. Sahoo and T. P. Das, Hyperfine Interactions, 61, 1197 (1990).
69. R. Pati, S. N. Ray and T. P. Das, (To be published).
70. J. P. Yesinowski, M. L. Buss, A. N. Garroway, (To be published).
71. M. Frank, F. Gubitz, W. Kreische, A. Labahn, C. Ott, B. Rseler, F. Schwab, G. Weeske, Hyperfine Interactions, 34, 193 (1987).
72. T. Briere, O. Donzelli and T.P. Das, Indian J. Phys. 67 (spl.), 35 (1993); O. Donzelli, T. Briere and T. P. Das, Solid State Comm. 90, 663 (1994).
73. J. A. Howard, Chem. Phys. Lett. 203, 540 (1993); J. R. Morton, K. F. Preston, P. J. Krusie, and L. B. Knight, Jr., Chem. Phys. Lett. 204, 481 (1993).
74. E. J. Ansaldo, C. Niedermayer and C. E. Stronach, Nature, 353, 129 (1991); E. J. Ansaldo, I. J. Boyle, C. Niedermayer, G. D. Morris, J. H. Brewer, C. E. Stronach, and R. S. Carey, Z. Phys. B86, 317 (1992); R. F. Kiefl, J. W. Schneider, K. Chow, T. L. Duty, T. L. Estle, B. Hitti, R. L. Lichti, E. J. Ansaldo, C. Schwab, P. W. Perciral, G. Wei, S. Wlodek, K. Kojima, W. J. Romanow, J. P. Maclanley, Jr., N. Coustel, J. E. Fischer, A. B. Smith III, Phys. Rev. Lett. 68, 1347 (1992); R. F. Kiefl, T. L. Duty, J. W. Schneider, A. MacFarlane, K. Chow, J. W. Elzey, P. Mendels, G. D. Morris, J. H. Brewer, E. J. Ansaldo, C. Niedermayer, D. R. Noakes, C. E. Stronach, B. Hitti, and J. E. Fisher, Phys. Rev. Lett. 69, 2005 (1992).
75. N. Sahoo, Sigrid Markert, T. P. Das and K. Nagamine, Phys. Rev. B41, 220 (1990); S.B. Sulaiman, N. Sahoo, S. Markert, J. Stein, T.P. Das and K. Nagamine, Bull. Mat. Sci. (India) 14, 149 (1991).
76. S. B. Sulaiman, N. Sahoo, T. P. Das, O. Donzelli, E. Torikai, and K. Nagamine, Phys. Rev., B44, 7028 (1991).
77. S. B. Sulaiman, N. Sahoo, T. P. Das, and O. Donzelli, Phys. Rev., B45, 7383 (1992).
78. S. B. Sulaiman, N. Sahoo, Sudha Srinivas, F. Hagelberg, T. P. Das, E. Torikai, and K. Nagamine, Phys. Rev., B 49, 9879 (1994); Hyperfine Interactions, 84, 87 (1994).
79. S. B. Sulaiman, N. Sahoo, Sudha Srinivas, F. Hagelberg, T.P. Das, E. Torikai, and K. Nagamine, Hyperfine Interactions, 79, 901 (1993).
80. S. B. Sulaiman, Sudha Srinivas, N. Sahoo, T. P. Das, O. Donzelli, E. Torikai, and K. Nagamine, Bull. Am. Phys. Soc., 38, 174 (1993).
81. Sudha Srinivas, S. B. Sulaiman, N. Sahoo, T. P. Das, E.Torikai, and K. Nagamine, Bull. Am. Phys. Soc., 38, 176 (1993).

82. Sudha Srinivas, S. B. Sulaiman, N. Sahoo, T. P. Das, E. Torikai, and K. Nagamine, Bull. Am. Phys. Soc., 39, 839 (1994).
83. M. Mali, D. Brinkmann, L. Pauli, J. Ross, and H. Zimmermann, Phys. Lett., A124, 112 (1987); R. E. Walstedt, W. W. Warren, Jr., R. F. Bell, G. F. Brennert, G. P. Espinosa, J. P. Remeika, R. J. Cava, and E. A. Rietman, Phys. Rev., B36, 5727 (1987); Y. Kitaoka, S. Hiramatsu, T. Kondo, and K. Asayama, J. Phys. Soc. Japan, 57, 30 (1988); C. H. Pennington, D. J. Durand, C. B. Zax, C. P. Slichter, J. P. Rice, and D. M. Ginsberg, Phys. Rev., B37, 7944 (1988); T. Shimizu, H. Yasuoka, T. Imai, T. Tsuda, T. Takabatake, Y. Nakazawa, and M. Ishikawa, J. Phys. Soc. Jpn., 57, 2494 (1988).
84. C. Coretsopoulous, H. C. Lee, E. Ramli, L. Reven, T. B. Rauchfuss, and E. Oldfield, Phys. Rev., B39, 781 (1989).
85. M. Mali, I. Mangelschots, H. Zimmermann, and D. Brinkmann, Physica, C175, 581 (1991).
86. H. Lütgemeier, V. Florentiev, and A. Yakubowskii, in Electronic Properties of High-Tc Superconductors and Related Compounds, Springer-Verlag, Berlin, Proceedings of the Conference at Kirchberg, (1990).
87. K. Ueda, T. Sugata, Y. Kohara, Y. Oda, M. Yamada, S. Kashiwai, and M. Motoyama, Solid State Comm., 73, 49 (1990); K. Ishida, Y. Kitaoka, and K. Asayama, J. Phys. Soc. Japan, 58, 36 (1989).
88. K. Ishida, Y. Kitaoka, G. Zheng, and K. Asayama, J. Phys. Soc. Japan, 60, 3516 (1991).
89. H. Lütgemeier, and M. W. Pieper, Solid State Comm., 64, 267 (1987).
90. J. Shore, S. Yang, J. Haase, D. Schwartz, and E. Oldfield, Phys. Rev., B46, 595 (1992); A. Yakubowskii, A. Egorov, and H. Lütgemeier, Appl. Magn. Reson., 3, 665 (1992).
91. J. I. Budnick, A. Golnik, Ch. Niedermayer, E. Recknagel,, M. Rossmanith, A. Weidinger, B. Chamberland, M. Filipowski, and D. P. Yang, Phys. Lett., A124, 103 (1987); B. Hitti, P. Birrer, K. Fischer, F. N. Gygax, E. Lippelt, H. Malletta, A. Schenck and M. Weber, Hyperfine Interactions, 63, 287 (1990); E. Torikai, H. Ishihara, K. Nagamine, H. Kitazawa, I. Tanaka, H. Kojima, S. B. Sulaiman, Sudha Srinivas, and T. P. Das, Hyperfine Interactions, 79, 915 (1993).
92. D. R. Hartree, The Calculation of Atomic Structures, New York: John Wiley and Sons Inc., (1957); C. C. J. Roothaan, Revs. Mod. Phys., 23, 69 (1951).
93. P. C. Kelires, K. C. Mishra, and T. P. Das, Hyperfine Interactions, 34, 289 (1987); D. W. Mitchell, S. B. Sulaiman, N. Sahoo, T. P.Das, W. Potzel, and G. M. Kalvius, Phys. Rev., B44, 6728 (1991).
94. Ravindra Pandey and A. Barry Kunz, Phys. Rev., B38, 10150 (1988); John M. Vail, R. Pandey and A. B. Kunz, Revs. Solid State Science, 5, 241 (1991); D. W. Mitchell, S. B. Sulaiman, N. Sahoo, T. P. Das, W. Potzel, G. M. Kalvius, W. Schiessl, H. Karzel, and M. Steiner, Hyperfine Interactions, 78, 403 (1993).
95. K. A. Colbourn and J. Kendrick, in Computer Simulation of Solids, C. R. A. Catlow and W. C. Mackrodt, (Ed.), New York: Springer-Verlag, (1982); S. M. Mohapatra, B. N. Dev, T. M. Thundat, L. Luo, W. M. Gibson, K. C. Mishra, N. Sahoo and T. P. Das, Reviews of Solid State Science, World Scientific Publishing Company, Singapore, 4, 873 (1990); N. Sahoo, S. B. Sulaiman, K. C. Mishra, and T. P. Das, Phys. Rev., B39, 13389 (1989); D. W. Mitchell, T. P. Das, W. Potzel, G. M. Kalvius, H. Karzel, W. Schiessl, M. Steiner, and M. Kofferlein, Phys. Rev., B48, 16449 (1993).
96. N. Sahoo et. al., Ref. 95.
97. S. M. Mohapatra, et al., Ref. 95.
98. Gaussian Basis Sets for Molecular Calculations, S. Huzinaga, (Ed.), Amsterdam: Elsevier (1984).
99. Gaussian 80, 86, 88, 90, and 92 Series of Programs, Gaussian Inc., Pittsburgh, Pennsylvania.
100. J. Stein, S. B. Sulaiman, N. Sahoo and T. P. Das, Hyperfine Interaction 60, 849 (1990).
101. D. W. Mitchell et. al., Ref. 93.
102. R. E. Watson, and A. J. Freeman, in Hyperfine Interaction, A. J. Freeman and R. B. Fraenkel, (Ed.), New York: Academic Press (1967).
103. R. S. Mulliken, J. Chem. Phys., 23, 1833 (1955).
104. S. B. Sulaiman et. al., Ref. 76.
105. H. M. Foley, R. M. Sternheimer, and D. Tycko, Phys. Rev., 93, 734 (1954); S. N. Ray, T. Lee, and T. P. Das, Phys. Rev., 49, 1108 (1974); R. M. Sternheimer, Zeitschrift fr Naturforschung, 41a, 24 (1985); T. P. Das and P. C. Schmidt, Zeitschrift fr Naturforschung, 41a, 47 (1986).
106. T. P. Das, and E. L. Hahn, Nuclear Quadrupole Resonance Spectroscopy, New York: Academic Press, (1957), Chap. 1.
107. M. H. Cohen and F. Reif, in Solid State Physics, F. Seitz and D. Turnbull, (Ed.), New York: Academic Press, (1957), Vol. 5.
108. T. P. Das, Relativistic Theory of Electrons, Chap. 7, New York: Harper and Row, (1973), Chap. 8.
109. M. A. Beno, L. Sederholm, D. W. Capone II, D. G. Hinks, J. D. Jorgensen, J. D. Grace, I. K. Schuller, C. U. Segre, and K. Zhang, Appl. Phys. Lett., 51, 57 (1987).
110. J. N. Roy, S. K. Mishra, K. C. Mishra, and T. P. Das, J. Chem. Phys., 90, 7273 (1989).

111. R. W. Dougherty, S. N. Panigrahy, T. P. Das, and J. Andriessen, Phys. Rev., A47, 2710 (1993).
112. S. N. Panigrahy, R. W. Dougherty, T. P. Das, and J. Andriessen. Phys. Rev., A44, 121 (1991); S. N. Ray, and T. P. Das, Phys. Rev., B16, 4790 (1977).
113. H. P. Schaefer III, R. A. Klemm, and F. E. Harris, Phys.Rev., 181, 138 (1969).
114. J. S. Swinnea, and H. Steink, Journal of Materials Research, 2, 442 (1987).
115. H. Yasuoka, T. Shimizu, Y. Ueda, and K. Kosuge, J. Phys. Soc. Japan, 57, 2659 (1988).
116. Y. Yamada, K. Ishida, Y. Kitaoka, K. Asayama, H. Takagi, H. Iwabuchi, and S. Uchida, J. Phys. Soc. Japan, 57, 2663 (1988); P. Mendels and H. Alloul, Physica, C156, 355 (1988).
117. J. M. Tranquada, A. H. Moudden, M. S. Alvarez, A. J. Jacobson, J. T. Lewandowski, and J. M. Newsam, Phys. Rev., B38, 2477 (1991).
118. T. A. Kaplan, S. D. Mahanti and Hyunju Chang, Phys. Rev., B45, 2565 (1992); S. D. Mahanti, T. A. Kaplan, Hyunju Chang, and J. F. Harrison, J. Appl. Phys. 73, 6105 (1993).
119. T. A. Kaplan and S. D. Mahanti, J. Appl. Phys., 69, 5382 (1991).
120. F. Mila and T. M. Rice, Physica, C157, 561 (1989).
121. T. Tsuda, T. Shimizu, H. Yasuoka, K. Kishio, and K. Kitazawa, J. Phys. Soc. Japan, 57, 2908 (1988).
122. J. M. Longo and P. M. Raccah, Journal of Solid State Chemistry, 6, 526 (1973); C. P. Poole, Jr., T. Datta, and H. A. Frach, Copper Oxide Superconductors, New York: John wiley and Sons, (1988).
123. R. R. Sharma and T. P. Das, J. Chem. Phys., 41, 3582 (1964).
124. P. Raghavan, Atomic and Nuclear Data Tables, 42, 248 (1989).
125. D. Vaknin, S. K. Sinha, D. E. Moncton, D. C. Johnston, J. M. Newsam, C. R. Safinya, and H. E. King, Jr., Phys. Rev. Lett., 58, 2802 (1987); T. Frelthoft, J. E. Fischer, G. Shirane, D. E. Moncton, S. K. Moncton, S. K. Sinha, D. Vaknin, J. P. Remeika, A. S. Cooper, and D. Harshmann, Phys. Rev., B36, 826 (1987); Bingxin X. Yang, S. Mitsuda, G. Shirane, Y. Yamaguchi, H. Yamauchi, and Y. Syono, J. Phys. Soc. Japan,56, 2283 (1987); Y. Yamaguchi, H. Yamauchi, M. Ohashi, H. Yamamoto, N. Shimoda, M. Kikuchi, and Y. Syono, Japan J. Appl. Phys., 26, L447 (1987).
126. J. M. Longo and P. M. Raccah, Ref. 122; C. P. Poole, Jr. et. al., Ref. 122. See also S. B. Sulaiman, Ref. 96, p.111.
127. T. Tsuda et. al., Ref. 121.
128. J. I. Budnick et. al., Ref. 91; B. Hitti et. al., Ref. 91.
129. L. P. Le, G. M. Luke, B. J. Sternlieb, Y. J. Uemura, J. H. Brewer, T. M. Riseman, D. C. Johnston, L. L. Miller, Y. Hidaka, and H. Murakami, Hyperfine Interactions, 63, 279 (1990); E. Torikai et. al., Ref. 91.
130. A. Schenck, Muon Spin Rotation Spectroscopy, Boston: Adam Hilger Publishers, (1985).
131. E. Torikai, K. Nagamine, H. Kitazawa, I. Tanaka, H. Kojima, S. B. Sulaiman, Sudha Srinivas, and T. P. Das, Hyperfine Interactions, 79, 921 (1993).
132. R. Saito, H. Kamimura, and K. Nagamine, Physica, C185–189, 1217 (1991).
133. T. McMullen, P. Jena, and S. N. Khanna, International Journal of Modern Physics, B5, 1579 (1991).
134. B. Hitti et. al., Ref. 91.
135. E. Torikai et. al., Ref. 91.
136. P. C. Kelires and T. P. Das, Ref. 24.
137. P. J. Hay and W. R. Wadt, J. Chem. Phys. 82, 270 (1985).
138. J. O. Artman, A. H. Muir and H. Wiedersich, Phys. Rev. 173, 337 (1968).
139. O. C. Kistner and A. W. Sunyar, Phys. Rev. Lett. 4, 412 (1960); F. Van der Woude, Phys. Stat. Solidi 17, 417 (1966).
140. K. J. Duff, K. C. Mishra and T. P. Das, Phys. Rev. Lett. 46, 1611 (1981).
141. B. Lindgren, N. Sahoo, S. B. Sulaiman and T. P. Das,Hyperfine Interactions, 61, 1189 (1990).
142. M. J. Amoruso and W. R. Johnson, Phys. Rev. A3, 6 (1971).
143. S. D. Mahanti, T. Lee, D. Ikenberry and T. P. Das, Phys. Rev. A9, 2238 (1974); A conclusion similar to that in this paper was reached by M. J. Amoruso, Ph. D. Thesis, University of Notre Dame (Unpublished).
144. S. Wakoh and J. Yamashita, J. Phys. Soc. Japan 21, 1712 (1966); 25, 1272 (1968). See also remarks by K. J. Duff et. al., Ref. 4, p. 2302 and 2303.
145. R. R. Sharma and T. P. Das, J. Chem. Phys. 41, 3581 (1964).
146. B. G. Dick and T. P. Das, Ref. 31; T. P. Das, Ref. 31; R. J. Quigley and T. P. Das, Ref. 31 (1967, 1969).
147. E. S. Chang, R. T. Pu and T. P. Das, Phys. Rev. 174, 16 (1968).
148. M. Vajed-Samii et. al., Ref. 8 (1979); M. Vajed-Samii, J. Andriessen, B. P. Das, S. N. Ray and T. Lee, J. Phys. B15, L379 (1982).
149. T. P. Das, Ref. 8; M. Vajed-Samii et. al., Ref. 8 (1979).
150. J. Andriessen, K. Raghunathan, S. N. Ray and T. P. Das, J. Phys. B10, 1979 (1979).
151. W. J. Nieuwpoort, Ref. 11.

EMBEDDING THEORY AND QUANTUM CLUSTER SIMULATION OF POINT DEFECTS IN IONIC CRYSTALS

John M. Vail

Department of Physics
University of Manitoba, Winnipeg, MB
R3T 2N2, Canada

I. INTRODUCTION

In this work, we address the problem of simulating the properties of solids that depend on electronic structure localized at the atomic scale. Since the individual atoms of terrestrial solids are experimentally discernible, perfect solids are comprised within our discussion. However, we shall emphasize point defects. Examples are impurities, vacancies and color centers (involving electrons or holes localized at vacancies), and aggregates of small numbers of these. Important processes include optical transitions, diffusion, aggregation, dissociation, etc. Simulation is usually easiest for point defects isolated in the bulk of a crystal, but processes and properties associated with surfaces are of great interest, as are those associated with interfaces and dislocations.

The basic feature of a point defect is that it disrupts the crystal in a highly localized way, with the notable exception of charged defects in ionic crystals, to be discussed later. This leads to the expectation that computational analysis of the defect can be limited to a small number of atoms: a molecular cluster. This expectation is predicated on the assumption that the remaining majority of atoms of the crystal, constituting an embedding region around the molecular cluster, can be taken into account in a relatively simple way. In the next two sections we discuss this matter, first in a fairly general formulation, and then in the specific context of ionic materials, and giving, in Sec. IV, examples of calculated results for a variety of experimentally studied processes. Before that, we shall outline the mathematical framework that we use for electronic structure simulation.

We shall apply the methods of quantum chemistry, as realized in terms of Hartree-Fock theory. This means that the molecular cluster of atoms in and around the point defect will be strictly localized in terms of electronic structure. The embedding problem consists of trying to represent the electronic structure of the rest of the crystal in a way that is consistent with this.

To establish notation and terminology, we sketch the Hartree-Fock method. A good reference for this topic is Szabo and Ostlund.[1] A many-electron wave function may be rigorously expressed as a linear combination of Slater determinants, each of the form

Electronic Properties of Solids Using Cluster Methods
Edited by T.A. Kaplan and S.D. Mahanti, Plenum Press, New York, 1995

$$\Psi(\underline{r}) = (N!)^{-1/2} \det \begin{bmatrix} \phi_1(\underline{r}_1) & \phi_1(\underline{r}_2) & \cdots & \phi_1(\underline{r}_N) \\ \phi_2(\underline{r}_1) & \phi_2(\underline{r}_2) & \cdots & \phi_2(\underline{r}_N) \\ \vdots & & & \\ \phi_N(\underline{r}_1) & \phi_N(\underline{r}_2) & \cdots & \phi_N(\underline{r}_N) \end{bmatrix} \tag{1}$$

In Eq. (1), $\underline{r}_j \equiv (\vec{r}_j, s_j)$, where $\vec{r}_j$ and s_j are position and spin of electron j; the molecular cluster consists of N electrons; and $\underline{r} = (\underline{r}_1, \underline{r}_2, \ldots \underline{r}_N)$. The single-particle functions $\{\phi_j\}$ are linearly independent. Under the constraint of orthonormality of the set $\{\phi_j\}$, the variational principle applied to the total energy expectation with respect to Ψ yields the Fock equation

$$F(\vec{r})\phi_j(\vec{r}) = \varepsilon_j \phi_j(\vec{r}) \tag{2}$$

In Eq. (2), we assume that $\phi_j(\underline{r})$ is a spin eigenstate: this is the unrestricted Hartree-Fock approximation. In Dirac notation, where ϕ_j is represented $|j>$, the Fock operator F has the form

$$F = \left\{ -\frac{1}{2}\nabla^2 - \sum_\mu Z_\mu |\vec{r} - \vec{R}_\mu|^{-1} + \sum_j <j||\vec{r} - \vec{r}'|^{-1}(1 - P)|j> \right\}. \tag{3}$$

In Eq. (3) we assume Bohr-Hartree atomic units; Z_μ and $\vec{R}_\mu$ are nuclear charges and positions, and $P(\vec{r}, \vec{r}')$ is the pairwise interchange operator for $\vec{r}$ and $\vec{r}'$.

We consider a variational solution of the Fock Eq. (2), based on a molecular orbital expansion of $|j\rangle$ as a linear combination of atomic orbitals $\{|k>\}$

$$|j> = \sum_k a_j(k)|k>. \tag{4}$$

The linear coefficients $a_j(k)$ are to be determined variationally, for a given choice of the atomic orbital basis functions $\{|k>\}$, which are taken to be of the form

$$|k> = n_k\left[\exp\left(-\alpha_k |\vec{r} - \vec{R}_{\mu(k)}|^2\right)\right] Y_{l(k)}^{m(k)}(\theta, \phi). \tag{5}$$

In Eq. (5), n_k is a normalizing factor, α_k is a constant, and Y_l^m is a spherical harmonic. The Fock Eq. (2) can be expressed in matrix notation in the basis of the atomic orbital set $\{|k>\}$ as follows

$$(\underline{\underline{F}} - \varepsilon_j \underline{\underline{S}}) \cdot \underline{a}_j = 0. \tag{6}$$

In Eq. (6) the elements of the matrix $\underline{\underline{F}}$ are $<k|F|k'>$, and of the eigenvectors $\underline{a}_j$ are $a_j(k)$ [see Eq. (4)]; the overlap matrix $\underline{\underline{S}}$ has elements $<k|k'>$. If the number N' of atomic orbitals is greater than the number of electrons N, then the eigenvectors define an occupied manifold of N dimensions and a virtual manifold of $(N'-N)$ dimensions.

We note that the Fock operator, Eq. (3), contains a self-consistent field term

$$V_{scf}(\vec{r}) = \sum_{j(occ)} < j \left| |\vec{r} - \vec{r}\,'|^{-1} (1-P) \right| j >. \tag{7}$$

This requires knowledge of the solution $\{|j>_{(occ)}\}$ of the occupied molecular orbitals. Thus the solution must be obtained by an iterative procedure, with V_{scf} updated and held fixed at each successive iterative step. In the matrix notation of Eq. (6), V_{scf} has matrix elements as follows

$$(\underline{\underline{V}}_{scf})_{kk'} = \left\{ \sum_{j(occ)} \sum_{K,K'} a_j^*(K) a_j(K') < kK|g|K'k'> \right\}, \tag{8}$$

where in Eq. (8), g is defined by

$$g(\vec{r},\vec{r}\,') = |\vec{r} - \vec{r}\,'|^{-1} [1 - P(\vec{r},\vec{r}\,')] \tag{9}$$

The self-consistent nature of $\underline{\underline{V}}_{scf}$ is clearly seen in Eq. (8) in the linear coefficients [occupied eigenvectors from Eq. (6)], $a_j(K)$.

The total Hartree-Fock energy E is given by

$$E = < \Psi | H | \Psi > \tag{10}$$

where Ψ is the Slater determinant of Eq. (1), and H is the N-electron Hamiltonian. It can be evaluated from the solution of the Fock equation as follows

$$E = \frac{1}{2} \sum_{j(occ)} \left\{ \varepsilon_j + < j | H^{(1)} | j > \right\} \tag{11}$$

In Eq. (11), $H^{(1)}$ is the one-electron part of the Fock operator; i.e., the first two terms of Eq. (3).

Once a given Hartree-Fock manifold, defined by the functions $\{|k>\}$, has been chosen, one can consider special linear combinations of them, in order to achieve particularly convenient visualization or computation. This will only have physical validity if the linear combinations maintain the physical features of the Hartree-Fock method. The method of so-called localizing potentials achieves this. This topic has been reviewed by Gilbert.[2] It is based on the properties of the Fock-Dirac density operator ρ_F

$$\rho_F = \sum_{j(occ)} | j >< j |. \tag{12}$$

This operator projects onto the occupied manifold. Consequently, the trace of the Fock matrix $\underline{\underline{F}}$ and the total energy E, Eqs. (6) and (11), are invariant if the following term is added to $\underline{\underline{F}}$

$$\underline{\underline{V}}_L = \left(\underline{\underline{\rho}}_F \cdot \underline{\underline{A}} \cdot \underline{\underline{\rho}}_F \right), \tag{13}$$

where $A(\vec{r})$ is an arbitrary single-particle operator. A common use of A is to enhance *scf* convergence by reducing atom-atom overlap within the manifold $\{|k>\}$, whence the terminology *localizing potential* for V_L, Eq. (13).

The total Hartree-Fock energy E can be corrected to take account of more than one Slater determinant. This is called a correlation correction. The approach, in terms of Rayleigh-Schrödinger perturbation theory, is to treat the unperturbed Hamiltonian H_0 as a sum of Fock operators

$$H_0 = \sum_{j=1}^{N} F(\vec{r}_j). \tag{14}$$

Its eigenstates include all possible Slater determinants formed from N Fock eigenvectors from the N'-dimensional occupied-plus-virtual manifolds. The perturbation is then $(-V_{scf}/2)$. The resulting second-order energy correction is then found to be

$$\sum_{\substack{k<k' \\ (occ)}} \sum_{\substack{K<K' \\ (virt)}} \frac{|< kk'|g|KK'>|^2}{(\varepsilon_k + \varepsilon_{k'} - \varepsilon_K - \varepsilon_{K'})}. \tag{15}$$

II. GENERAL EMBEDDING THEORY

For a point defect in a crystal or an atomic cluster on a surface, we want to use the molecular cluster method described in Sec. I. The cluster then consists of a manageably small number of electrons, localized about the nuclei of the cluster. This localization is expressed by Gaussian decay of the electronic density with increasing distance from the cluster [see Eq. (5)]. For a real host crystal or substrate, there will be electronic overlap between the cluster and the atoms of the host. The embedding problem, described in this section, addresses this effect.

In metals and semiconductors overlap effects are substantial, with directional character in the latter case. In ionic crystals, overlap effects may be less, allowing somewhat rough approximation. In all three cases, long-range effects are possible, associated with charged defects: Friedel oscillation in metals, and dielectric polarization in insulators. We shall not deal with these effects in this section. In Secs. III and IV, long-range polarization in ionic crystals will be dealt with fully.

We begin with the Hartree-Fock method applied to the whole crystal, containing $(N+N_1)$ electrons, where N is the number that will be associated with an embedded finite atomic cluster. This implies a limitation on the choice of basis set for treatment of a finite cluster embedded in a larger crystal which is common to a large body of published work: it defines a particular variational approach. The occupied Fock eigenstates $\{|n>\}$, $n = 1, 2, \ldots (N+N_1)$ are assumed to consist of two subsets, one of which is localized in the cluster region $\{|j>\}$, $j = 1, 2, \ldots N$. We denote the other set $\{|J>\}$, $J = 1, 2, \ldots N_1$. **Note**: Upper case letters will be used to indicate Fock eigenstates associated with the embedding region (E); lower case will indicate the cluster region (C). The occupied manifold of the crystal now consists of $\{|n>\} \equiv \{|j>, |J>\}$, and the Fock equation can be written in two parts

$$F|j> = \varepsilon_j|j>,\ j = 1, 2, \ldots N, \tag{16}$$

and

$$F|J> = \varepsilon_J|J>, J = 1, 2, \ldots N_1. \tag{17}$$

This approach has been introduced by Vail and Rao.[3]

We now introduce atomic orbitals (see Eq. (5)), $\{|k>\}$ in (C) and $\{|K>\}$ in (E). We expand the cluster Fock eigenstates:

$$|j>=\sum_k a_j(k)|k>. \quad (18)$$

We similarly expand the embedding eigenstates:

$$|J>=\sum_k a_J(k)|k>+\sum_K B_J(K)|K>. \quad (19)$$

The terms $a_J(k)$ in Eq. (19) are required to ensure orthogonality between $\{|j>\}$ and $\{|J>\}$, as assumed in the derivation of the Fock equation in standard form, Eqs. (16) and (17). This orthogonality condition is:

$$<j|J>=0. \quad (20)$$

In matrix representation in the atomic orbital basis $\{|k>, |K>\}$, we shall identify submatrices associated with terms of the types (CC), (CE), (EE). For example, an operator θ has elements of the form $<k|\theta|k'>$, $<k|\theta|K>$, and $<K|\theta|K'>$, for the three submatrices. The orthogonality condition, Eq. (20), is then satisfied by:

$$\underline{a}_J=-\left[\underline{\underline{S}}^{(CC)}\right]^{-1}\cdot\underline{\underline{S}}^{(CE)}\cdot\underline{B}_J \quad (21)$$

This is an important result. It shows that the cluster contribution $\underline{a}_J$ to an embedding function $|J>$ is determinable from the overlap matrix and from the embedding contribution $\underline{B}_J$. We shall show that the latter may be determined once and for all for a given crystal, and it then follows that the cluster part of $\{|J>\}$ does not contribute to the self-consistent field iterative procedure for the cluster.

We now concentrate on the cluster Fock equation (16). Referring to Eq. (3), the Fock operator in Eq. (16) applies to the whole crystal. It has the form:

$$F=\{-\tfrac{1}{2}\nabla^2-\sum_{\mu(C)}Z_\mu|\vec{r}-\vec{R}_\mu|^{-1}+\sum_{j(C)}<j|g|j>$$

$$-\sum_{\mu(E)}Z_\mu|\vec{r}-\vec{R}_\mu|^{-1}+\sum_{J(E)}<J|g|J>\}. \quad (22)$$

The cluster sub-matrix for the Fock equation (see Eq. (6) and (16)) is

$$\left\{\underline{\underline{F}}^{(CC)}-\varepsilon_j\underline{\underline{S}}^{(CC)}\right\}\cdot\underline{a}_j=0, j=1,2,\ldots N. \quad (23)$$

From Eq. (22), Eq. (23) can be written

$$\left\{\left[\underline{\underline{F}}_C^{(CC)}+\underline{\underline{U}}_E^{(CC)}\right]-\varepsilon_j\underline{\underline{S}}^{(CC)}\right\}\cdot\underline{a}_j=0. \quad (24)$$

Here, F_C is given by the first line of Eq. (22), and U_E is given by the second line. F_C is what we would have for the free cluster and U_E represents the embedding potential seen by the cluster (C) due to the surrounding crystal. In spite of the cluster part $\{a_J(k)\}$ of the embedding Fock states $\{|J>\}$, Eq. (19), U_E depends only on the embedding vectors $\underline{B}_J$, because of the orthogonality condition, Eq. (21). It also involves the overlap sub-matrices

$\underline{S}^{(CE)}$ and $\underline{S}^{(CC)}$. Thus, although the matrix dimensionality of $\underline{V}_{scf}$ for the cluster is unchanged by the addition of embedding, the number of overlap (and other) matrix elements is increased to include both of the basis sets {|*k*>} and {|*K*>}. Specifically, the most general form of the total-energy algorithm involves (*CC*), (*CE*), and (*EE*) submatrices of all terms in the Fock operator *F*. From the first line of Eq. (22), however, the cluster self-consistent field term, in matrix notation, denoted $\underline{\underline{g}}$, has the (*k,k'*) element

$$\sum_{j(C)} \sum_{k'',k'''} a_j(k'')a_j(k''') < kk''|g|k'''k'>. \tag{25}$$

We have said that the cluster shall contain N electrons. The boundary within which these *N* electrons are found is not defined, though it must lie about mid-way between cluster and embedding region nuclei. Because of the orthogonality requirement, Eqs. (19) and (21), there will be a contribution from the embedding states {|*J*>} to the electrons in the cluster. Thus cluster normalization is *not* simply

$$N = \sum_j < j|j> = \sum_j \underline{a}_j^T \cdot \underline{\underline{S}}^{(CC)} \cdot \underline{a}_j. \tag{26}$$

In fact, we begin by requiring the integral of total electronic density within the cluster to be equal to *N*. We then fix the boundary by equipartition of cluster-embedding overlap between cluster and embedding regions. An approximation is then introduced in which overlap between pairs of cluster orbitals is neglected in the embedding region, and vice versa. This approximation rigorously maintains the number of electrons in (*C*), and leads to an explicit, physically reasonable, and mathematically convenient normalization for both $\{\underline{a}_j\}$ and $\{\underline{B}_J\}$.

The coefficients $\{\underline{B}_J\}$ may be determined for a given crystal by a method that we have devised, using the foregoing embedding method, applied to a cluster that consists of a unit cell of the crystal, embedded in a small number of such cells. An iteration procedure is then carried out to self-consistency between cluster and embedding unit cells. The result is an approximation to the electronic structure of the crystal, based on point-localized properties, without strict periodic repetition.

III. THE EMBEDDING PROBLEM FOR IONIC CRYSTALS

In ionic crystals, the atoms are usually more localized than in covalent or metallic materials. Furthermore, the ionic charges play a dominant role in determining the nature of the material. Thus an extreme, but to some extent plausible, approximation is to treat the ions of the embedding region as point charges. This gets the Coulomb part of the crystal's energy to reasonable accuracy. Short-range repulsion between ions, which is intrinsically quantum-mechanical in nature, may be expressed by a classical potential. A commonly used functional form is the Buckingham potential, used for all of the applications to be described below in Sec. IV

$$V(R) = B\exp(-R/\rho) - C/R^6 . \tag{27}$$

In Eq. (27), R is the ion-ion separation, and (B, ρ, C) are constants. This classical embedding still omits an important quantum-mechanical effect, namely, in the terms of Sec. II above, part of the role of occupied embedding orbitals { | J>} in the cluster orbitals {|j>}; namely, the deviation of the embedding potential U_E, second line of Eq. (22), from the point-charge plus Buckingham form, Eq. (26). In the rest of this section, we discuss embedding in ionic crystals from the above viewpoint. First, we introduce the shell model, a necessary generalization of the classical point-ion model. We then discuss how to achieve consistency between the charge distribution of the cluster and the configuration of the embedding shell-model crystal. A specific form of localizing potential, Eq. (13), is then introduced, which includes most of the quantum effects of embedding: the Kunz-Klein localizing potential (KKLP). Thereafter, a computer program is described, based on the above methodology, referred to by the acronym ICECAP. Finally, a systematic method of limiting the cluster basis set to reflect embedding is briefly described.

Dielectric polarization occurs in ionic materials as a result of two mechanisms, which are mutually dependent. One is ionic displacement, the other is ionic polarization. The latter is allowed for by modelling the ion by two point charges, a core of charge (Q-Y), and a *shell* of charge Y, where Q is the total charge of the ion. In the shell model[4] the core and shell are harmonically coupled, with force constant K that is assumed to include shell-core Coulomb interaction within the ion. In defect applications of the shell model, shells of nearby ion pairs are subject to short-range repulsion [e.g., Eq. (27)], and interionic Coulomb forces are included fully.

The shell model is inadequate for the electronic properties of ionic materials. For localized electronic properties, for example of a point defect, we may use a quantum molecular cluster, as described in Sec. II. However, as we mentioned in Sec. II, the long-range polarization that can occur in ionic crystals is not included in the embedding scheme given there. Instead, we shall embed the quantum cluster in an infinite shell-model crystal. This leaves the question of residual cluster-embedding quantum effects to be discussed later.

We shall evaluate the total energy of a crystal consisting of a quantum molecular cluster embedded in an infinite shell model crystal. For a stationary state we apply the variational principle to both the cluster wave function, as described in Sec. I, where the exponential coefficients α_k of Eq. (5), and the linear coefficients a_j (k) of Eq. (4) may be subject to variation, and to the positions of the point charges that constitute the rest of the model system. These point charges include the nuclei of the molecular cluster, and the shells and cores of all shell model ions. The methodology for this variational procedure has been discussed in some detail elsewhere, e.g. Ref. [5]. Briefly, one simulates the cluster up to a specified order of electrostatic multipole by a set of point charges; this induces distortion and polarization of the embedding shell-model region which is used in a first determination of the cluster electronic configuration for fixed nuclear positions; corrected multipole point-charge simulators are calculated from the quantum cluster solution, and the process is iterated to self consistency. Separate variation of the cluster nuclear positions, with cluster-embedding consistency achieved at each step, leads to overall minimization of the total energy.

We now address the question of residual quantum effects from the embedding region upon the cluster. In order to apply the full embedding scheme described in Sec. III, it would be necessary to incorporate ionic polarization into the solution $\underline{B}_J$; see Eq. (19). In that case, the perfect-crystal solution could not be used. We have not yet worked out how to do this. In cases where the defect is not charged, and where the polarization that it produces in short-ranged, it would be possible to apply the method of Sec. III directly. Alternatively, if we introduce the approximation for an ionic crystal that recognizes that ion-ion overlap is weak, then we might neglect the terms in a_J (k) in Eq. (19); see also Eq. (21), where this amounts to neglect of cluster-embedding overlap $S^{(CE)}$. If that procedure is combined with a particular form of localizing potential, Eq. (13), we have a very useful form of quantum embedding, which we refer to as Kunz-Klein localizing potential (KKLP), Ref. [6]. With KKLP, the operator, $A(\vec{r})$ in Eq. (13) is taken to be the negative of the short-range part of the cluster-embedding interaction; i.e. the part which omits classical Coulomb effects. This KKLP then produces eigenstates for which the modified Fock operator $F' = (F + V_L)$ has short-range effects essentially cancelled; i.e. a solution which largely satisfies the original assumption of weak cluster-embedding overlap. This method is applied not only to the embedding of the defect cluster, but in the derivation of the perfect-crystal solution where both anions and cations are treated self consistently. The method is preferable to those that use pseudopotentials which are derived only for atomic cores, i.e. for cations. A rough approximation that extends this method to polarized ions is to superimpose point charges on embedding ions described by KKLP, to simulate the dipole moment of the ion as calculated for a corresponding shell-model ion. In this way we get Coulomb effects correct, but leave a residual error in the quantum-overlap effects.

The physical model described above for a quantum cluster embedded in a shell-model crystal, including KKLP, and the analytical procedure described with it, are referred to as the ICECAP methodology. It is embodied in a user-friendly computer program of the same acronym.[7] ICECAP has a wide variety of options, many of which have been briefly described in Ref. [5]. Amongst them is the ability to describe unrelaxed states; i.e. states where the ionic positions and shell-model ion polarizations correspond to one electronic state (e.g. the ground state), whilst the electrons of the cluster are in a different state (e.g. an optically excited state). The latter example corresponds to a Franck-Condon transition of an electronic defect. Because of its substantial flexibility, ICECAP can be used to calculate properties of a very wide range of point defects. Because exactly the same methodology is used in all cases, it becomes possible to assess it critically. Evidence of this sort is reviewed in Sec. IV, below.

While the use of KKLP to describe residual quantum embedding effects does not increase the matrix dimensionality of the Fock eigenvalue problem [see discussion following Eq. (21)], it does increase the time required for a given calculation. If this increase is unacceptable, one must use less accurate means of including these effects. One such method is judicious limitation of the cluster basis set. We illustrate this by referring to the problem, a substitutional Na^+ ion in LiF. Suppose we define the quantum cluster to consist of the impurity Na^+ ion and its six nearest-neighbor F^- ions. A reasonably flexible basis set for the F^- ions will lead to a solution in which the electronic density is unrealistically spread out amongst ions of the embedding shell-model crystal, because no Pauli repulsion is encountered from that region. One can determine the actual localization of F^- ions in LiF by considering a quantum cluster centered on an F^- ion, surrounded by six Li^+ ions. For the latter, a free-ion Li^+ basis set will be appropriate, because the ion consists exclusively of a tightly bound core. Once an optimized F^- basis set has been obtained from calculations with this embedded cluster, the relative amplitudes of some of the primitive atomic orbitals $|k\rangle$, Eq. (5), can be fixed (the process is called *contraction*). This limits the F^- basis set flexibility, restricting the solution of the impurity problem to one which indirectly reflects the quantum-mechanical nature of Li^+ nearest-neighbor ions in the shell-model embedding region. The contractions may be chosen in such a way as to compromise between the need for ionic localization in the crystalline environment, and the ion's corresponding polarizability. Clearly such an approach is intuitive, but it is better than completely ignoring the problem.

IV. POINT DEFECTS IN IONIC CRYSTALS

We shall now present results of computer simulation of some point defects in ionic crystals, based on the ICECAP methodology described in the preceding section. Only works that the author has been personally involved in will be cited. This nevertheless represents a wide range of cases, providing extensive comparison with experiment. The presentation here closely parallels another recent publication.[8] A substantial further body of work associated with R. Pandey is partly reviewed in Ref. [5]. The following physical processes will now be illustrated: the charge-state stability of impurity nickel ions with respect to hole creation and electron loss in a host crystal MgO, local modification of the valence band by an impurity as shown by hole-trapping of Li^+ in NiO; effective short-range forces on Cu^+ and Ag^+ in alkali halides, with application to classical diffusion; the electronic structure of the F center and the F_2^+ center in NaF and of the F^+ center in MgO, in terms of isotropic hyperfine constants, and for the latter, optical excitation.

Our study of substitutional nickel in MgO[9] indicates that in charge state +1 it actually is more like Ni^{2+}, with an electron donated to the Mg^{2+} - 3s band. Correspondingly in charge state +3, it again resembles Ni^{2+}, with a hole created in the O^{2-} - 2p band. These interpretations are based on Mulliken population analysis, in which the nickel contains close to 26 electrons in all cases, with 14 of one spin and 12 of the other. For Ni^{2+}, crystal field splitting and spin polarization effects are clearly illustrated. While the Mulliken population can be a poor indicator of electron number for an atom in a crystal,[10] comparison of total energies for the given charge states supports the view that +2 is the stable charge state of nickel in MgO. Specifically, for ions that are spatially well separated, we find the following:

$$Ni^{+} + Ni^{3+} \rightarrow 2Ni^{2+} + 10eV. \qquad (28)$$

We note that, experimentally, neither Ni^{3+} nor Ni^+ have been found in MgO without associated stabilizing defects.

The effect of lithium impurity in NiO illustrates a subtle and technologically important effect. In pure NiO, the top of the valence band is dominated by nickel 3d levels. [11] Substitutional lithium has net charge +2, consisting of Li^+ and a trapped hole which resides in oxygen 2p levels. [12] Thus, the effect of the lithium impurity is to locally depress the nickel 3d levels relative to the oxygen 2p levels, exposing the latter at the top of the valence band. This rather complicated local deformation of the band structure by the impurity is qualitatively illustrated in our simulation of the system. [13] Specifically, we show that heavy mixing of nickel T_{2g} states with oxygen 2p states at the top of the valence *band* for a $Ni^{2+}.(O^{2-})_6$ cluster, embedded in perfect NiO crystal, is converted in a corresponding $Ni^{2+}.(O^{2-})_6.Li^{+}$.(hole) cluster so that pure oxygen 2p states occur at the top of valence set, which consists of 18 of one spin and 17 of the other. The calculations suggest that the hole localizes on a single oxygen site, reducing the defect to C_{4v} symmetry.

Embedded quantum clusters can be used to derive effective short-range potentials [see Eq. (27)] for impurity-host interaction. Briefly, a classical shell model of the defect region can have its energy fitted to that of the quantum cluster, as a function of cluster geometry. In this way, we have derived the values of parameters (B, ρ, C), Eq. (27), for interaction of Cu^+ and Ag^+ impurities with the halide ions in a range of alkali halide crystals. Once a classical shell-model description has been obtained for the impurity, it is relatively easy to calculate its activation energy for both vacancy and interstitial diffusion mechanisms. In KCl, experimental values have been obtained for Cu^+ vacancy and interstitial diffusion, and for Ag^+ interstitial diffusion. The values are 1.1, 0.83, and 0.95 eV given in Refs. 14, 15, and 15, respectively. Our calculated values for the corresponding processes are 1.19, 0.78, and 0.83 eV. [16] We note both good quantitative agreement with experiment, and also correct ordering. Amongst our other results, we predict that vacancy diffusion of Ag^+ in KCl and in RbCl proceeds by a non-collinear mechanism, with the diffusion paths deviating by 0.33 and 0.35 (units of host nearest-neighbor spacings) from the straight-line jump, respectively.

F-type centers play a central role in radiation damage of ionic crystals, and have important optical properties. They consist of one or more electrons bound in anion vacancies. We now discuss three such centers.

In NaF and other alkali halides, the F center is a single electron bound by an anion vacancy. Our calculations[17] show that in NaF it is in fact well-localized in the vacancy, and not shared in the 3s orbitals of neighboring Na^+ ions. The quantum cluster analysis enables us to determine the isotropic hyperfine constants of neighboring ions included in the cluster, from calculated spin densities at their nuclei. For nearest-neighbor Na^+ and second-neighbor F^- ions the calculated (and experimental) values are 80(107) [18] and 32(97) [18] respectively (units: h.Mc, where h is Planck's constant and 1 Mc = $10^{-6}s^{-1}$). The agreement for nearest-neighbors is good, considering the state of the art, but the disagreement at second neighbors indicates the limitations of the calculation.

A pair of adjacent anion vacancies can bind a single electron in what is referred to as an F_2^+ center. In Ref. [17] we have also analysed this defect, calculating the isotropic hyperfine constants of its three inequivalent sets of nearest-neighbor Na^+ ions. The results are 189, 12, and 31 h.Mc, the first of these referring to the two ions nearest to the defect's center. While experimental values are not available, electron spin resonance results[19] indicate that the F_2^+-center electron is strongly concentrated about the defect's center. Our calculations predict a 5% outward displacement of the vacancies' nearest neighbors, due to the repulsion by the defect's net positive charge.

An O^{2-}-vacancy in MgO has net charge +2 relative to the rest of the crystal. Thus, when it binds a single electron, it is called an F^+ center. Our calculations[20] show that the electron is well-localized in the vacancy, and that the defect does not attract a second electron from the oxygen band, leaving a hole there, although this result is only obtained at the most accurate level of treatment. The calculated excited state is also well-localized in the vacancy. The comparison between calculated and experimentally measured electron spin density at nearest-neighbor Mg^{2+} nuclei is 0.171 A^{-3} as against 0.274 A^{-3}.[21] Calculated and experimental optical excitation energies (Franck-Condon approximation) are 7.59 and 4.95 eV respectively.[22] Since these calculations were fairly refined, a significant role for other factors than those considered is indicated.

Finally, we mention three works in progress, as examples of more complicated systems to which these methods apply. One is the quantum diffusion of muonium in alkali halides. The underlying theoretical work has been published.[23,24] The ICECAP methodology is used to evaluate the muon-phonon coupling and the potential seen by the muon throughout the crystal. Our calculated results[24a] for the muon time-of-stay in quantum diffusion, vs. temperature, in NaF, compared with experimental results in KCl[25] indicate that renormalization effects ignored in the basic theory are crucially important. A second case is the comparison of four diffusion activation energies related to vacancies and F-type centers in CaF_2, pure and with Na^+ substitutional impurities. The calculations are fully quantum-mechanical, rather than based on classical potentials as in the case of Ca^+ and Ag^+ in alkali halides, described earlier. Experimental values are available for the comparison.[26] The third case is a defect complex in NaF, consisting of an F_2^+ center and Mg^{2+} substitutional impurity with its charge-compensating Na^+ vacancy. This defect complex is denoted $(F_2^+)^*$. By representing the F-center electron's interactions with host and impurity ions by classical Buckingham potentials, we investigate the question of its most stable configurations. The defect has been discussed by Hofmann et al.;[19] our preliminary results have been published.[17]

In summary, we have described a wide range of phenomena for point defects in ionic crystals for which simulations have been undertaken. Our results and those of others to date show much satisfactory agreement with experiment, and predict a number of as yet unobserved details of the defect structures and processes. This range of applications and of satisfactory and novel results has been obtained by a single methodology; i.e. with the unified theoretical and computational approach embodied in the ICECAP program.[7]

ACKNOWLEDGMENTS

The author is grateful to his collaborators, whose names appear in the citations, and to Michigan Technological University and Virginia Commonwealth University for hospitality. Financial support is acknowledged from the Natural Sciences and Engineering Research Council of Canada and from the University of Manitoba.

REFERENCES

1. A. Szabo and N. S. Ostlund, *Modern Quantum Chemistry* (McGraw-Hill, New York, 1982).
2. T. L. Gilbert, in *Molecular Orbitals in Chemistry, Physics and Biology*, edited by P.-O. Löwdin and B. Pullman, p. 405 (Academic, New York, 1964).
3. J. M. Vail and B. K. Rao, Int. J. Quantum Chem. **53**, 67 (1995).
4. B. G. Dick and A. W. Overhauser, Phys. Rev. **112**, 90 (1958).
5. J. M. Vail, R. Pandey and A. B. Kunz, Revs. Solid State Science **5**, 241 (1991).
6. A. B. Kunz and D. L. Klein, Phys. Rev. **B17**, 4614 (1978).
7. J. H. Harding, A. H. Harker, P. B. Keegstra, R. Pandey, J. M. Vail and C. Woodward, Physica **131B**, 151 (1985).
8. J. M. Vail, in *International Symposium of Local Order of Solids*, Jekyll Is., GA, 14-17 June 1993, edited by S. D. Mahanti and P. Jena (Nova, New York, 1994), in press.
9. J. Meng, J. M. Vail, A.M. Stoneham and P. Jena, Phys. Rev. **B42**, 1156 (1990).
10. See the article by P. S. Bagus, in these proceedings.
11. D. E. Eastman and J. L. Freeouf, Phys. Rev. Lett. **34**, 395 (1975).
12. P. Kuiper, G. Kruizinga, J. Ghijsen, G. H. Sawatzky and H. Verweij, Phys. Rev. Lett. **62**, 221 (1989).
13. J. Meng, P. Jena and J. M. Vail, J. Phys.: Condens. Matter **2**, 10371 (1990).
14. K. P. Henke, M. Richtering and W. Ruhrberg, Solid State Ionics **21**, 171 (1986).
15. Ya. N. Pershits and T. A. Kallenikova, Sov. Phys.-Solid State **23**, 1497 (1981).
16. J. Meng, R. Pandey, J. M. Vail and A. B. Kunz, J. Phys.: Condens. Matter **1**, 6049 (1989).
17. J. M. Vail and Z. Yang, J. Phys.: Condens. Matter **5**, 7649 (1993).
18. H. Seidel and H. C. Wolf, in *Physics of Color Centers*, edited by W. B. Fowler, Ch. 8 (New York, Academic, 1968).
19. D. M. Hofmann, F. Lohse, H. J. Paus, D. Y. Smith and J.-M. Spaeth, J. Phys. C: Solid State Phys. **18**, 443 (1985).
20. R. Pandey and J. M. Vail, J. Phys.: Condens. Matter **1**, 2801 (1989).
21. W. P. Unruh and J. M. Culvahouse, Phys. Rev. **154**, 861 (1967).
22. B. Henderson and R. D. King, Phil. Mag. **13**, 1149 (1966).
23. T. McMullen and B. Bergerson, Solid State Commun. **28**, 31 (1978).
24. J. M. Vail, T. McMullen and J. Meng, Phys. Rev. **B49**, 193 (1994).
24a. T. McMullen, J. Meng, J. M. Vail, and P. Jena, Phys. Rev. **B51** (1995), in press.
25. R. F. Kiefl, R. Kadono, J. H. Brewer, G. M. Luke, H. K. Yen, M. Celio and E. J. Ansaldo, Phys. Rev. Lett. **62**, 792 (1989).
26. J. M. G. Tijero and F. Jaque, Phys. Rev. **B41**, 3832 (1990).

DENSITY-FUNCTIONAL FULL-POTENTIAL MULTIPLE-SCATTERING CALCULATIONS FOR FREE AND EMBEDDED CLUSTERS

R. Zeller

Institut für Festkörperforschung
Forschungszentrum Jülich GmbH
Postfach 1913, D-52425 Jülich, Germany

I. INTRODUCTION

The fundamental objective of electronic-structure calculations is to study the behavior of the electronic system in the field of the nuclei and to determine the properties of atoms, molecules and solids from first principles. This is essentially a many-body problem and obviously a hopeless task without simplifying approximations. By the Born-Oppenheimer approximation the motions of electrons and nuclei can be decoupled and the problem is reduced to solve a many-electron Schrödinger equation if, as is done here, relativistic effects, in particular the spin-orbit interaction, are neglected.

Except for the simplest cases the solution of the many-electron Schrödinger equation requires additional approximations and various electronic-structure techniques have been developed in the past in quantum chemistry and physics. Compared to the configuration-interaction method, to the Hartree-Fock method supplemented by perturbative treatment of electron correlations, and to quantum-Monte-Carlo calculations, density-functional theory seems to be best suited for complex systems and has been applied in the work reported below to calculate the electronic structure of free clusters and clusters embedded in solids.

The embedding is treated within the Green-function multiple-scattering formalism.[1] By using full potentials[2] instead of the traditional muffin-tin potentials this formalism allows for an exact solution of the one-particle Schrödinger equation and the numerical convergence is easily controlled by a single parameter ℓ_{max}, i.e. the highest angular momentum used in the calculations.

One advantage of the Green-function multiple-scattering formalism is that embedded clusters in solids and free clusters can be treated on equal footing and with essentially the same numerical details in the calculations. This feature allows for a direct comparison between free and embedded clusters and thus reveals how far clusters can be used to investigate the electronic properties of solids.

The present investigation was particularly addressed at the electronic properties of third-row transition-metal impurities in bulk Cu. These systems are prominent examples for a controversy whether they can be described by free clusters or not.[3–6] Results for local density of states, magnetic moments and hyperfine fields will be discussed below.

Edited by T.A. Kaplan and S.D. Mahanti, Plenum Press, New York, 1995

II. THEORETICAL ASPECTS

Density Functional Theory

In 1964 Hohenberg and Kohn[7] showed that for non-degenerate ground states there exists a unique functional F[n] of the electron density n(**r**), such that for a given external potential $V_{ext}(\mathbf{r})$ the minimum of the functional

$$E[n] = F[n] + \int V_{ext}(\mathbf{r})\, n(\mathbf{r})\, d\mathbf{r} \tag{1}$$

occurs when n(**r**) is the ground-state density of the interacting electron system and that this minimum is the ground-state energy. For practical applications the (unknown) functional F[n] is usually written as

$$F[n] = T_s[n] + E_H[n] + E_{xc}[n] \tag{2}$$

where T_s is the kinetic energy of a system of non-interacting electrons with density n(**r**) and E_H is the Hartree energy

$$E_H = \frac{e^2}{2} \int \frac{n(\mathbf{r})\, n(\mathbf{r}')}{|\mathbf{r} - \mathbf{r}'|}\, d\mathbf{r}\, d\mathbf{r}' \tag{3}$$

which includes self-interaction terms. The remaining part $E_{xc}[n]$ is called the exchange-correlation functional and provides the many-body corrections to make (1) an exact equation. Of course, the exchange-correlation functional is not known exactly and is usually approximated in the local density approximation

$$E_{xc}[n] = \int n(\mathbf{r})\, \varepsilon_{xc}(n(\mathbf{r}))\, d\mathbf{r} \tag{4}$$

or its spin-polarized generalization. Here ε_{xc} is the exchange-correlation energy for the homogeneous electron gas of density n(**r**) and can be obtained from many-body calculations.[8] For the calculations presented here the form of Vosko, Wilk and Nusair[9] is taken for ε_{xc}.

Density-functional theory has been used with great success in order to determine structure and geometries of solids and molecules, their vibrational properties and elastic constants, and in particular to describe the itinerant magnetism in solids composed of transition-metal elements. Lattice constants, bond lengths and magnetic moments can be calculated within typical errors of some percents in comparison to experiments. However, there are also shortcomings of density-functional theory. Highly correlated systems like solids with rare-earth elements, like transition-metal oxides and high-temperature superconductors are described poorly. The problems occur because of the self-interactions contained in the Hartree energy (3) and because of the choice of the homogeneous electron gas as a starting point to calculate the exchange-correlation energy (4). The lack of systematic improvements to the local-density approximation (4), however, does not much concern the transition-metal system considered here since they are well described within the local-spin-density approximation.

Multiple-Scattering Theory

The minimization of (1) is usually achieved by writing the density n(**r**) in terms of one-particle wave-functions of non-interacting electrons as

$$n(\mathbf{r}) = 2\sum_i |\psi(\mathbf{r},E_i)|^2 \tag{5}$$

where the factor 2 comes from the electronic spin and the sum is over all states with energy E_i less than the chemical potential μ which is determined by the condition of charge neutrality. The choice (5) leads to the Kohn-Sham (Schrödinger) equation

$$(-\nabla_{\mathbf{r}}^2 + V_{eff}(\mathbf{r}) - E_i)\,\psi(\mathbf{r},E_i) = 0 \tag{6}$$

in which the effective potential $V_{eff}(\mathbf{r})$ is the sum of the external potential $V_{ext}(\mathbf{r})$, the Hartree potential $V_H(\mathbf{r})$, and the exchange-correlation potential $V_{xc}(\mathbf{r})$, both given as functional derivatives of $E_H[n]$ and $E_{xc}[n]$ with respect to the density. (Rydberg atomic units $\hbar^2 = 1$, $2m = 1$ will be used throughout this article.)

Many methods for solving (6) expand the wavefunctions in some basis set and use the Rayleigh-Ritz variational principle. Until the basis functions are well chosen for the system under consideration, very many of them may be needed for accurate results. Here this disadvantage is avoided by using an alternative: multiple-scattering theory as originally proposed by Lord Rayleigh[10] and introduced into quantum mechanics by Korringa[11] and Kohn and Rostoker.[12]

Multiple-scattering theory is most conveniently formulated with Green functions instead of wavefunctions. Eq. (6) is replaced by

$$(-\nabla_{\mathbf{r}}^2 + V_{eff}(\mathbf{r}) - E)\,G(\mathbf{r},\mathbf{r}',E) = -\delta(\mathbf{r}-\mathbf{r}') \tag{7}$$

with the appropriate boundary condition that $G(\mathbf{r},\mathbf{r}',E)$ vanishes for $\mathbf{r}$ or $\mathbf{r}'$ going to infinity. Eq. (7) can be transformed into an integral equation, the Dyson equation

$$G(\mathbf{r},\mathbf{r}',E) = g(\mathbf{r},\mathbf{r}',E) + \int g(\mathbf{r},\mathbf{r}'',E)\,V_{eff}(\mathbf{r}'')\,G(\mathbf{r}'',\mathbf{r}',E)\,d\mathbf{r}'' \tag{8}$$

where the free space Green function $g(\mathbf{r},\mathbf{r}',E)$ is defined by (7) with vanishing potential $V_{eff}(\mathbf{r}) = 0$. The fundamental property of multiple-scattering theory is the fact that for numerical applications it is possible to solve (8) by a system of linear equations

$$G_{LL'}^{nn'}(E) = g_{LL'}^{nn'}(E) + \sum_{n''}\sum_{L''L'''} g_{LL''}^{nn''}(E)\,t_{L''L'''}^{n''}(E)\,G_{L'''L'}^{n''n'}(E) \tag{9}$$

where upper indices refer to atomic sites and lower indices to angular-momentum quantum numbers $L = (\ell,m)$. With the matrix elements $G_{LL'}^{nn'}(E)$ the Green function $G(\mathbf{r},\mathbf{r}',E)$ can be written in atom-centered coordinates as

$$G(\mathbf{r}+\mathbf{R}^n,\mathbf{r}'+\mathbf{R}^{n'},E) = \delta^{nn'}\,G_s^n(\mathbf{r},\mathbf{r}',E) + \sum_{LL'} R_L^n(\mathbf{r},E)\,G_{LL'}^{nn'}(E)\,R_{L'}^{n'}(\mathbf{r}',E) \tag{10}$$

where $G_s^n(\mathbf{r},\mathbf{r}',E)$ and $R_L^n(\mathbf{r},E)$ are single-site quantities, solely determined by the potential $V_{eff}(\mathbf{r})$ restricted to a cell surrounding atom n at position $\mathbf{R}^n$. Definitions of $G_s^n(\mathbf{r},\mathbf{r}',E)$ and $R_L^n(\mathbf{r},E)$ can be found in[2] and the cells are assumed to be non-overlapping and space-filling. The space-filling property may require to choose the expansion-centers $\mathbf{R}^n$ not only at the atomic sites but also elsewhere at empty sites. The scattering matrices

$$t^{n}_{LL'}(E) = \int_{\Omega^n} \mathbf{R}^{n}_{L'}(\mathbf{r},E)\, V_{eff}(\mathbf{r}+\mathbf{R}^n)\, J_{L'}(\mathbf{r},E)\, d\mathbf{r} \tag{11}$$

are also single-site quantities defined by integrals over the atomic cells with $J_L(\mathbf{r},E)$ given as products of spherical Bessel functions and spherical harmonics[2] while the free-space coefficients $g^{nn'}_{LL'}(E)$ in (9) are given as sums over products of spherical Hankel functions and spherical harmonics.

The choice of multiple-scattering theory instead of basis-set methods to solve the one-particle Schrödinger equation (6) leads to several important advantages. There are no ambiguities in choosing basis functions, the size of the matrices in the angular-momentum indices is most economical and the computations are decoupled into single-site properties like $t^{n}_{LL'}(E)$, $R^{n}_{L}(\mathbf{r},E)$ $G^{n}_{s}(\mathbf{r},\mathbf{r}',E)$, and their structural combination is described by a linear system of equations (9) which is simpler to solve than the determination of eigenvalues and eigenfunctions necessary in basis-set methods. The r^{-1} singularities of the potential at the nuclear positions are treated exactly. The same is true for the kinetic energy since the Laplacian of (6) and (7) does not appear in (8). Similar to plane-wave methods a single parameter governs the accuracy, here it is ℓ_{max} the highest angular momentum taken into account. One large disadvantage of multiple-scattering theory was for many years the restriction to muffin-tin potentials, i.e., to spherically symmetric potentials within non-overlapping spheres around the atoms and to a constant potential in the interstitial region between the spheres. It has now become clear that the muffin-tin approximation is not necessary and that multiple-scattering theory can be used for potentials of arbitrary form.[13]

The convergence of full-potential multiple-scattering theory with respect to the number of angular momenta taken into account, which determines the size of the matrices in (9) with respect to their L-indices, is still a somewhat controversial issue.[13,14] The form of the multiple-scattering equations presented above probably leads to the fastest convergence, at least this is the experience gained in Jülich where many full-potential calculations have been performed up to now. This experience shows that the inclusion of s, p, d and f angular momenta, leading in (9) to matrices of size 16 times N with N being the number of atoms, seems to be enough for accurate electronic-structure calculations.

The biggest advantage of multiple-scattering theory is the possibility to separate the calculations into several steps, since (8) does not only describe the connection between the Green function of the considered system and the Green function of free space, but also the connection between Green functions of two arbitrary systems if $V_{eff}(\mathbf{r})$ is understood as the difference between the two potentials. This means that one can first use Fourier transformation to compute (9) for ideal periodic bulk solids and then use the finite range of the screened potential perturbation to compute (9) directly in real space for a localized defect in the bulk. The intermediate step of constructing a surface as a two-dimensional horizontal periodic perturbation which is localized in vertical direction is also possible, thus allowing for density-functional electronic-structure calculations of adsorbate atoms.[15]

The efficiency of the Green-function multiple-scattering formalism for large-scale electronic-structure calculations has recently been demonstrated by calculations for iron-row transition-metal impurities in Pd.[16] Clusters of up to 1061 atoms were treated self-consistently with the proper embedding in bulk Pd guaranteed by the Green-function formalism. The results showed that the magnetization clouds around the "giant-moment" impurities are rather extended over a large number of atoms. While these calculations were done with spherical potentials, the use of full potentials is not much more time-consuming, since the question of spherical or full potentials only affects the single-site part of the calculations and thus requires an effort which scales only linearly with the number N of atoms. In contrast, the effort to solve (9) scales with N^3 and thus overwhelms the calculations for large N.

Full-Potential Calculations

The difficult problem in full-potential multiple-scattering theory is the accurate calculation of the single-site wavefunctions $R_L^n(\mathbf{r},E)$. Here they are expanded in spherical harmonics

$$R_L^n(\mathbf{r},E) = \sum_{L'} R_{L'L}^n(r,E)\, Y_{L'}(\hat{r}) \tag{12}$$

and a similar expansion is used for the single-cell potentials

$$V^n(\mathbf{r}) = \sum_{L} V_L^n(r)\, Y_L(\hat{r})\,. \tag{13}$$

The non-spherical components $V_L^n(r)$ for $\ell > 0$ lead to coupled radial equations for $R_{L'L}^n(r,E)$ which are conveniently solved in two steps. First the solution $\tilde{R}_\ell^n(r,E)\, Y_L(\hat{r})$ for the spherically symmetric potential $V_{\ell=0}^n(r)/\sqrt{4\pi}$ is calculated and then the radial integral equation

$$R_{L'L}^n(r,E) = \tilde{R}_\ell^n(r,E)\,\delta_{L'L} + \int \tilde{g}_{\ell'}(r,r',E) \sum_{L''} \Delta V_{L'L''}^n(r')\, R_{L''L}^n(r',E)\, r'^2\, dr' \tag{14}$$

is solved by iteration.[17] In (14) the radial Green function $\tilde{g}_{\ell'}^n(r,r',E)$ is the one corresponding to the spherically symmetric potential $V_{\ell=0}^n(r)/\sqrt{4\pi}$ and $\Delta V_{L'L''}^n(r)$ is the small and purely anisotropic part of the potential given by

$$\Delta V_{L'L''}^n(r) = \int_{4\pi} Y_{L'}(\hat{r})\, [V^n(\mathbf{r}) - V_{\ell=0}^n(r)/\sqrt{4\pi}\,]\, Y_{L''}(\hat{r})\, d\hat{r}\,. \tag{15}$$

Reliable charge densities and energies can usually be obtained if (14) is iterated twice. Since the potential $\Delta V_{L'L''}^n(r)$ vanishes near the atomic positions, it is enough if the integral in (14) is calculated in the outer region of the cell. This represents a considerable saving of computation time since near the atomic positions many mesh points are needed to describe the core electrons accurately.

In full-potential multiple-scattering theory accurate integrals over the atomic cells are needed. These cells are here chosen as the Wigner-Seitz cells of the Cu fcc lattice and have a complicated faceted geometry. The cell integration is done by introducing a Heavyside step function $\Theta(\mathbf{r})$, being equal to one in the cell and zero otherwise, and by introducing its angular-momentum expansion $\Theta(\mathbf{r}) = \sum_L \Theta_L(r)\, Y_L(\hat{r})$. This leads to

$$\int_{cell} f(\mathbf{r})\, d\mathbf{r} = \int f(\mathbf{r})\, \Theta(\mathbf{r})\, d\mathbf{r} = \sum_L \int_0^\infty f_L(r)\, \Theta_L(r)\, r^2\, dr \tag{16}$$

when $f(\mathbf{r})$ is expanded in spherical harmonics as $f(\mathbf{r}) = \sum_L f_L(r)\, Y_L(\hat{r})$ which is done for all relevant quantities like electron densities, potentials or their products. If the highest angular momentum in (9–10) is denoted by ℓ_{max}, the spherical-harmonics expansion for the density (defined in (17) below) contains spherical harmonics up to $2\,\ell_{max}$ and products of densities and potentials contain spherical harmonics up to $4\,\ell_{max}$. This gives a natural cutoff for the L sum in (16). This cutoff is therefore not affected by the slow convergence of the expansion of $\Theta(\mathbf{r})$ in spherical harmonics. The so-called shape-truncation functions $\Theta_L(r)$ can be calculated accurately and efficiently by the method described in.[18]

Special Calculational Problems for Free Clusters

The central quantity of density-functional theory, the density n(r), is easily determined from the Green function by

$$n(\mathbf{r}) = -\frac{2}{\pi}\,\mathrm{Im}\int_{-\infty}^{\mu} G(\mathbf{r},\mathbf{r},E)\,dE \tag{17}$$

where the integral is over the energy range of all occupied states with energy E less than the chemical potential μ. In infinite metallic solids the states around μ are continuous and (17) is most conveniently calculated by deforming the integration contour from the real-energy axis into the complex-energy plane where around 30 to 40 integration mesh points along the contour are enough to evaluate (17) very accurately.[19] Moreover, the chemical potential μ is easily determined from the condition that the ideal periodic solid must satisfy charge neutrality. For localized perturbations like impurities the chemical potential μ is fixed to its value in the unperturbed solid.

The situation in calculations for free clusters is much more difficult. The eigenstates are discrete and Im $G(\mathbf{r},\mathbf{r},E)$ consists of an assembly of delta functions similar to densities of states of finite systems. The problem in the Green-function formalism presented above is that these eigenstates are *not* calculated as in a Rayleigh-Ritz procedure. Instead of eigensystems of matrices only matrix inversions (solving linear systems of equations) are needed according to (9). This precludes the possibility of state counting as a method to determine the chemical potential μ. Instead of that μ can be determined by iteratively adjusting it to obtain charge neutrality. In principle, this adjustment could be achieved simultaneously with the iterative self-consistent determination of $n(\mathbf{r})$ resulting from $V_{eff}(\mathbf{r})$ by (7) and (17) and of $V_{eff}(\mathbf{r})$ resulting from $n(\mathbf{r})$ by the functional derivatives of (3) and (4) with respect to $n(\mathbf{r})$. This simultaneous iteration of μ, $n(\mathbf{r})$ and $V_{eff}(\mathbf{r})$ is not yet implemented in the computer programs available in Jülich, mainly because it is not yet investigated how the sophisticated generalized Anderson technique[16] used to accelerate the self-consistency iterations is affected by the simultaneous determination of μ.

Another problem of using (17) for free clusters is obviously the treatment of partially filled states. In local-spin-density-functional calculations, for instance, a Ni atom in configuration $3d^84s^2$ has a spin-down d state with 3/5 filling. Since μ is fixed at a delta-function peak in (17) such a partial filling cannot be achieved. A technique to avoid this problem is to apply density-functional theory in a finite-temperature version for the electronic system. By that the delta-function peaks are essentially broadened into Lorentzian peaks and can be partially filled.

The introduction of finite temperatures requires to replace (17) by

$$n(\mathbf{r}) = -\frac{2}{\pi}\,\mathrm{Im}\int_{-\infty}^{\infty} G(\mathbf{r},\mathbf{r},E)\,f_T(E)\,dE \tag{18}$$

where $f_T(E) = (1 + \exp\frac{E-\mu}{kT})^{-1}$ is the Fermi-Dirac distribution function and to minimize the grand-canonical functional

$$\Omega[n] = E[n] - \mu N - TS \tag{19}$$

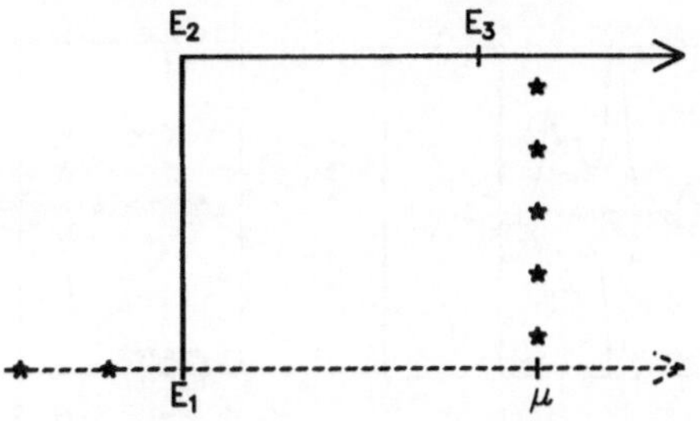

Figure 1. The contour in the complex energy plane. T = 800 K is used. For a description of E_1,E_2, E_3 see text.

instead of the energy functional E[n] given in (1). The grand-canonical functional is conveniently evaluated by

$$\Omega[n] = -\int_{-\infty}^{\infty} N(E)\, f_T(E)\, dE \tag{20}$$

where $N(E) = \int_{-\infty}^{E} n(E')\, dE'$ with $n(E) = -\frac{2}{\pi} \operatorname{Im} \int G(\mathbf{r},\mathbf{r},E)\, d\mathbf{r}$ is the integrated density of states. Equation (20) follows after some algebra from the definition of electron number

$$N = \int_{-\infty}^{\infty} n(E)\, f_T(E)\, dE \tag{21}$$

and electron entropy

$$S = -k \int_{-\infty}^{\infty} n(E)\, [f_T(E)\, \ell n\, f_T(E) + \bar{f}_T(E)\, \ell n\, \bar{f}_T(E)]\, dE \tag{22}$$

with $\bar{f}_T(E) = 1 - f_T(E)$.

The integrals (18) for the density n(**r**) and (20) for the grand-canonical potential Ω[n] can both be evaluated by the method of complex-energy integration.[19] The Green function G(**r**,**r**',E), the density of states n(E) and the integrated density of states N(E) are analytic functions of E in the entire complex-energy plane except for singularities on the real-energy axis which can be poles corresponding to exponentially decaying core states and branch cuts corresponding to bands of continuous states. As a matter of fact this means that for a proper definition, the energy E in the integrals (17), (18), (20–22) must contain an (infinitesimally) small positive imaginary part. Further singularities in these integrals are the poles of the Fermi function at energies $E_n = \mu + i\pi\,(2n - 1)\,kT$ for all integers n.

The complex contour for the numerical evaluation of (18) and (20) should fulfill two contradictory requirements. It should be far from the real axis so that G(**r**,**r**',E) and N(E) are very smooth functions of E and it should be short so that not too many mesh points are needed for the numerical integrations. Here the following procedure was chosen. The core states were treated separately as atomic states, thus allowing one to choose the starting point E_1 of the contour just below the 4s, 4p and 3d valence states of the iron-row transition metals. The contour shown in Fig. 1 then consists of three parts, the first going from E_1 to $E_2 = E_1 + 10\, i\, \pi\, kT$, the second going from E_2 to $E_3 = \mu - 30\, kT + 10\, i\, \pi\, kT$, and the third going from E_3 to $E_4 = \infty$. In addition the residues at five Fermi-function poles E_n =

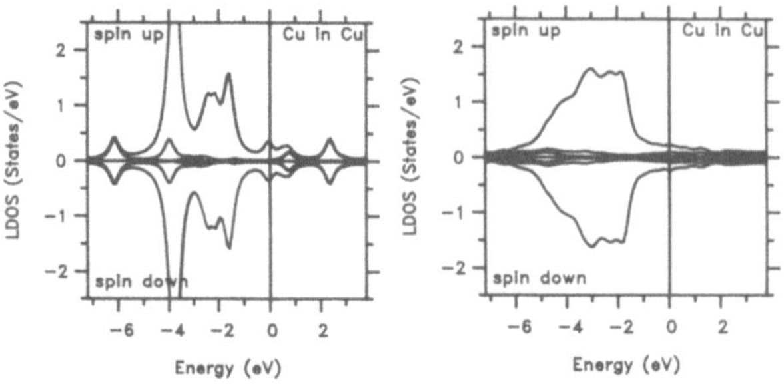

Figure 2. Local density of states for the central atom in the free cluster Cu_{13} (left) and for bulk Cu (right). The results for spin-down direction are plotted downwards on an inverted scale. For each spin direction three curves are given, the one nearest to the axis shows the s contribution, the next one shows the sum of s and p contributions, and the one farthest from the axis shows the sum of s, p, and d contributions.

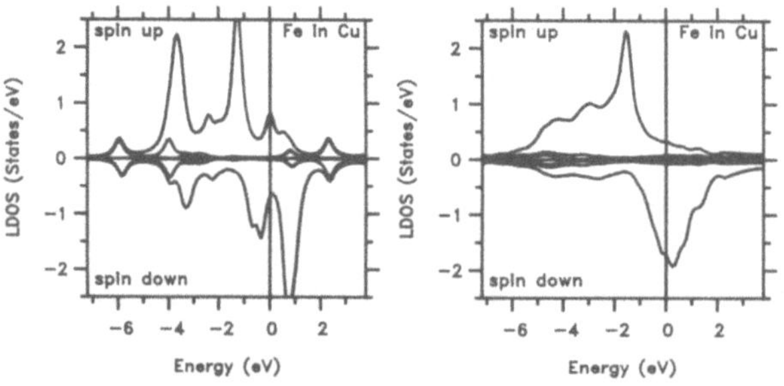

Figure 3. Local density of states for the Fe atom in the free cluster $FeCu_{12}$ and embedded in bulk Cu. Description of the curves as in Fig. 2.

$\mu + i\pi (2n - 1) kT$ with $n = 1.5$ must be taken into account. Between E_1 and E_2 and between E_2 and E_3 standard Gauss-Legendre integration rules and the replacement $f_T(E) = 1$ are used. For the integration between E_3 and E_4 special Gauss rules for the Fermi-Dirac weight function were developed according to the description given in.[20]

Numerical Details

All calculations were done with the exchange-correlation potential given in the Vosko-Wilk-Nusair parametrization[9] of the quantum-Monte-Carlo results of Ceperley and Alder[8] and with an angular-momentum cutoff at $\ell_{max} = 4$ in (9–11). The electronic temperature to facilitate the energy integrals was chosen as T = 800 K which essentially leads to broadened density-of-states peaks with 32 mRy full width at half maximum. The energy mesh used consists of 33 points: 5 points at Fermi-function poles, 8 points between E_1 and E_2, 16 points between E_2 and E_3, and 4 points between E_3 and E_4. The lattice constant of bulk Cu was taken as 6.8314 atomic units which is the experimental value. The positions of the atoms in the free clusters as well as in the embedded clusters were chosen to coincide with the positions of the atoms in the Cu fcc lattice. No relaxations of the atoms were taken into account because they would come out different in free and embedded clusters, thus making their comparison more difficult. All systems considered obeyed the cubic O_h point-group symmetry which was exploited as described in.[16]

The embedded clusters consisted of 13 atoms, i.e. one impurity and 12 Cu neighbor atoms, for which the potentials were determined self-consistently. Potential perturbations on all other surrounding Cu atoms (being infinitely many) were neglected. From the experience gained over the years in Jülich, this neglect is well justified in calculations for local magnetic properties of impurities like magnetic moments, spin-polarized densities of states, and hyperfine fields. (See also the discussions in [16].) The potential of bulk Cu was determined self-consistently with an accuracy of 10^{-6} Ry defined as the average difference between input and output potentials of the last self-consistency iteration. In these iterations and in the final calculation of the Green-function matrix elements $g^{nn'}_{LL'}(E)$ of the Cu host crystal for use in (9), the Fourier back transformations, i.e. the Brillouin-zone integrations, were done with 6144 **k** points in the 1/48th irreducible zone using a modified tetrahedron integration scheme.[21–22]

The free clusters also consisted of 13 atoms. Additionally, 66 empty cells surrounding the clusters were taken into account in the self-consistent calculations. The whole arrangement satisfied O_h point-group symmetry which was exploited in the calculations. Electron densities and potentials outside of these 79 sites were neglected. The approximation is justified because in density-functional theory the densities decay exponentially. The charge on an outermost site was always less than 0.005 e. Some preliminary calculations with a fixed chemical potential $\mu = -0.296$ Ry were also done for larger clusters consisting of 225 sites, i.e. one impurity atom, 78 surrounding Cu atoms, and 146 empty sites. For all clusters, embedded or free, self-consistency was assumed if in the last iteration input and output potentials agreed better than 10^{-6} Ry on the average. Radial integrals as in (3, 11, and 14) were done on an exponentially spaced mesh within muffin-tin spheres and, in the outer-cell regions, on linear meshes properly adjusted to the kinks of the space-truncation functions. Full potentials and densities were taken into account at the 148 outer mesh points of a total of 424 radial mesh points. The density-of-states pictures shown below have been calculated with a Lorentzian broadening with 32 mRy full width at half maximum which corresponds to the temperature T = 800 K used during the self-consistency iterations.

III. RESULTS

A common way of comparing the electronic structure of bulk and free-cluster calculations is a comparison of the density of states. Figs. 2-5 show the local densities of states for the central atom of the free clusters $CuCu_{12}$, $FeCu_{12}$, $MnCu_{12}$, and $CrCu_{12}$ in comparison with the local densities of states of the embedded clusters in bulk Cu. The local densities of states for each spin direction are obtained by the formula $n_{loc}(E) = -\frac{1}{\pi} \mathrm{Im} \int_{cell} G(\mathbf{r},\mathbf{r},E)\, d\mathbf{r}$ where the integration is over the Wigner-Seitz cell at the impurity site and the appropriate Green function for each spin direction is used.

The pictures for the 13-atom clusters all show bonding s peaks at about -5.5 eV, bonding p peaks at about -3.5 eV, anti-bonding s peaks at about 1 eV, and anti-bonding p peaks at about 2.5 eV with zero energy chosen at the chemical potential μ. These peaks eventually broaden into the sp bands of the solid. The main contributions to the local densities of states come from the d electrons and show a clear spin splitting, at least for $FeCu_{12}$ and $MnCu_{12}$. The results for $FeCu_{12}$ in Fig. 3 are in good agreement with the local densities of states for $FeCu_{18}$ as calculated by Blaha and Callaway.[5] Comparing the results for the 13-atom clusters with the impurities in bulk, the main differences occur for the s and p contributions, although the d contributions of $MnCu_{12}$ and $CrCu_{12}$ also considerably differ from the ones of Mn and Cr impurities in bulk Cu. This latter difference, in particular with respect to the spin splitting is also reflected in the calculated local magnetic moments which are 2.65 μ_B , 3.52 μ_B, 3.03 μ_B for Fe, Mn, Cr impurities in bulk Cu and 2.15 μ_B, 1.62 μ_B, 0.86 μ_B for Fe, Mn, Cr in free clusters surrounded by 12 Cu atoms. Whereas the moments for Fe agree reasonably in both systems, the moments for Mn and Cr show drastic differences. Of course, it cannot be ruled out that more than one magnetic configuration can

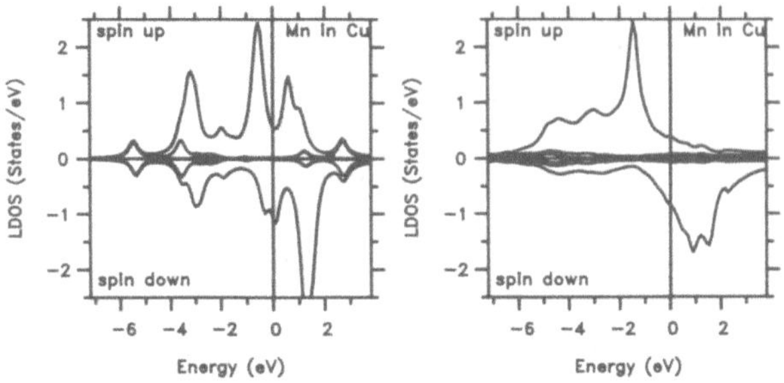

Figure 4. Local density of states for the Mn atom in the free cluster $MnCu_{12}$ and embedded in bulk Cu. Description of the curves as in Fig. 2.

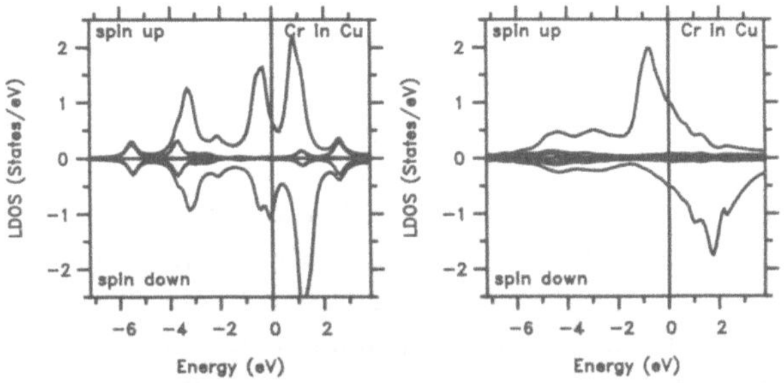

Figure 5. Local density of states for the Cr atom in the free cluster $CrCu_{12}$ and embedded in bulk Cu. Description of the curves as in Fig. 2.

be found for these clusters as recently has been demonstrated for free V and Cr clusters by Lee and Callaway.[23] Thus further investigations of the free clusters $MnCu_{12}$ and $CrCu_{12}$ with respect to this issue are desirable.

It is clear that very local properties like hyperfine fields of impurities in bulk Cu cannot be described by the 13-atom clusters, both because of the different magnetic moments and because of the poor description of 4s valence states which considerably contribute to the hyperfine fields. The calculated Fermi contact hyperfine fields for $FeCu_{12}$, $MnCu_{12}$, $CrCu_{12}$ are -12 T, -8.4 T, -4 T in disagreement with the results for impurities in bulk calculated as -11.1 T, -13.8 T, -11.2 T.

The pictures for the 79-atom clusters (not shown) are in much closer correspondence to the ones for the impurities in bulk Cu, at least as far as p and d contributions are concerned. The s contributions are rather strange with a simple peak located at about 2 eV above the chemical potential. Most probably this is an artifact of the present calculations which do not yet use the correct chemical potential μ. The present choice $\mu = -0.296$ Ry leads to a surplus charge of about 0.3 e for the 79-atom clusters considered and no local occupied s states are found in the calculations at the central sites. Nevertheless, it seems that the local densities of states of the central atoms in the 79-atom clusters are in fair agreement with the ones of the impurities in bulk Cu. The same is true for the local magnetic moments which are calculated as 2.39 μ_B, 3.21 μ_B and 2.70 μ_B for Fe, Mn, Cr surrounded by 78 Cu atoms. Because of the missing occupied s states a comparison of hyperfine fields makes no sense.

IV. CONCLUSIONS

The aim of the present work was to show that full-potential multiple-scattering theory can be used to solve the density-functional one-particle equations for free and eigenvectors are needed as in basis-set methods based on the Rayleigh-Ritz variational principle. Compared to systems like bulk solids, impurities in the bulk or at the surface, where states with energies near the chemical potential are continuous, in free clusters the states are discrete and partially filled and require special techniques like the use of finite temperatures and integration contours in the complex-energy plane.

The calculated results show that free clusters of 13 atoms are too small to describe the electronic properties of impurities in solids, even energy-integrated quantities like local magnetic moments may turn out completely different. The situation is much more favorable for free cluster of 79 atoms, a fact obvious even from the preliminary calculations done so far. A more complete comparison between free and embedded clusters is planned for the future and requires further investigations. In particular, the temperature dependence and the possibility of several stable configurations and their energy differences should be studied.

REFERENCES

1. P.J. Braspenning, R. Zeller, A. Lodder, and P.H. Dederichs, Phys. Rev. B **29**, 703 (1984).
2. R. Zeller, J. Phys. C **20**, 2347 (1987).
3. K.H. Johnson, D.D. Vvedensky, and R.P. Messmer, Phys. Rev. B **19**, 1519 (1979).
4. R. Zeller, R. Podloucky, and P.H. Dederichs, Z. Physik B **38**, 165 (1980).
5. P. Blaha and J. Callaway, Phys. Rev. B **33**, 1706 (1986).
6. D.D. Vvedensky, M.E. Eberhart, and M.E. McHenry, Phys. Rev. B **35**, 2061 (1987).
7. P.C. Hohenberg and W. Kohn, Phys. Rev. **136**, B864 (1964).
8. D.M. Ceperley and B.J. Alder, Phys. Rev. Lett. **49**, 566 (1980).
9. S.H. Vosko, L. Wilk, and M. Nusair, Can J. Phys. **58**, 1200 (1980).
10. Lord Rayleigh, Philos. Mag. **34**, 481 (1892).
11. J. Korringa, Physica **13**, 392 (1947).
12. W. Kohn and N. Rostoker, Phys. Rev. **94**, 1111 (1954).
13. W.H. Butler, A. Gonis, and X.-G. Zhang, Phys. Rev. B **45**, 11527 (1992).
14. W.H. Butler, A. Gonis, and X.-G. Zhang, Phys. Rev. B **48**, 2118 (1993).
15. P. Lang, V.S. Stepanyuk, K. Wildberger, R. Zeller, and P.H. Dederichs, Sol. State Commun. **92**, 755 (1994).
16. R. Zeller, Modelling Simul. Mater. Sci. Eng. **1**, 553 (1993).
17. B. Drittler, Diss. TH Aachen, Berichte des Forschungszentrums Jülich, Jül-2445 (1991).
18. N. Stefanou, H. Akai, and R. Zeller, Comput. Phys. Commun. **60**, 231 (1990).
19. R. Zeller, J. Deutz, and P.H. Dederichs, Sol. State Commun. **44**, 993 (1982).
20. R.P. Sagar, Comput. Phys. Commun. **66**, 271 (1991).
21. J.E. Müller and J.W. Wilkins, Phys. Rev. B **29**, 4331 (1984).
22. R. Zeller, in *Metallic Alloys: Experimental and Theoretical Perspectives - 1993 Dearfield Beach*, Proceedings of the NATO Advanced Research Workshop, edited by J.S. Faulkner and R.G. Jordan, NATO ASI Series E: Vol. 256 (Kluwer Academic Publishers, Dordrecht, 1994).
23. K. Lee and J. Callaway, Phys. Rev. B **49**, 13906 (1994).

EMBEDDED CLUSTER THEORY: REACTIONS ON METAL AND SEMICONDUCTOR SURFACES

J.L. Whitten

Department of Chemistry
North Carolina State University
Raleigh, NC 27695

I. INTRODUCTION

The goal of this research is the development and application of theoretical techniques that will provide a molecular level understanding of surface processes, especially reaction mechanisms, energetics and adsorbate structure. Electronic structures are described by an *ab initio* embedding formalism that permits an accurate determination of reaction energetics and adsorbate geometry for both metal and semiconductor surfaces. This work relates to two different subject areas: catalytic processes on metals and properties of electronic materials. An important objective of theoretical studies is to delineate effects related to surface sites, adsorbate structure, impurities, coverage, mobility, and reaction energetics. Topics to be addressed include:

- adsorbate bonding to surfaces (general principles and mechanistic issues)
- adsorbate structure and spectroscopy as a function of surface site
- chemisorption energetics of molecules and molecular fragments
- interaction of coadsorbed species
- surface reactions (heats of reaction and activation energies)
- nonequilibrium geometries, mechanistic issues and potential surfaces

In order to describe the energetics of such surface reactions accurately, a first-principles embedding theory is needed and the main focus of our research has been the development of techniques for treating electronic interactions on metals and electronic materials by *ab initio* many-electron theory. Embedding is defined via a localized orbital definition of an electronic subspace that encompasses the adsorbate and neighboring surface atoms. In this way, it is possible to carry out energy variational calculations of molecular quality on a surface portion of a system that is coupled to a more approximately described bulk lattice.

Detailed computations have helped classify the reactivity of surface sites for different types of adsorbates and, in a few cases, reactions of coadsorbed species on transition metal surfaces have been treated. Numerous investigations have been published.[1-14] In this report, the embedding theory is described and a few examples are provided to illustrate its applicability to metals and non-metals.

II. THEORETICAL APPROACH

In order to describe reactions at a surface involving bond formation or dissociation, sophisticated wavefunctions are required to account for changes in polarization and electron correlation. *Ab initio* configuration interaction theory provides an attractive way to proceed providing systems are of manageable size. The purpose of the embedding scheme is to organize the theoretical approach in a manner that will permit an accurate many-electron treatment of the adsorbate/surface portion of a system while coupling this region with the bulk.[2, 6, 11]

Calculations are carried out for the full electrostatic Hamiltonian of the system (except for core electron pseudopotentials), with wavefunctions constructed by self-consistent-field (SCF) and configuration interaction (CI) expansions. For N electrons, Q nuclei

$$H = \Sigma_i^N\left(-\tfrac{1}{2}\nabla_i^2\right) + \Sigma_i^N\Sigma_k^Q\left(-Z_k/r_{ki}\right) + \Sigma_{i<j}^N 1/r_{ij} \tag{1}$$

CI wavefunctions are generated from an initial SCF configuration, or from multiple parent configurations in the case of stretched or broken bonds, by one- and two-particle excitations from the main components of the wavefunction to give expansions of the form

$$\Psi = \Sigma_k \lambda_k A\left(\chi_1^k \chi_2^k \ldots \chi_n^k\right). \tag{2}$$

(A = antisymmetrizer). Perturbation theory is used to incorporate numerous small contributions not explicitly included in the CI expansion.

Surfaces are modeled initially by a large cluster of atoms and this cluster is subsequently reduced to a smaller effective size by the embedding procedure. Substrate structures are initially taken from experiment or idealized models and surface and local region geometries are subsequently optimized. In the treatment of transition metals, the first stage of the embedding procedure deals only with the most delocalized part of the electronic system, the s,p band; d functions on surface atoms are added after a surface s,p electronic subspace is defined.

Self-consistent-field calculations are performed on the s,p band and occupied orbitals are then localized by a unitary transformation to obtain a set of orbitals localized around the periphery of the cluster (the surface of interest is not part of the periphery). Letting φ and φ' denote SCF and localized orbitals, respectively, the total wavefunction is invariant under the transformation

$$\text{delocalized} \quad \Phi' = U\Phi \quad \text{localized} \tag{3}$$

$$\Psi = A(\varphi_1\varphi_2\varphi_3\ldots\varphi_N) = A(\varphi_1'\varphi_2'\varphi_3'\ldots\varphi_N'). \tag{4}$$

Electron exchange with atomic orbitals, $\{s_k\}$, of periphery atoms is used as the criterion to define the localization transformation. Maximization of the positive definite exchange integral, $\gamma = \langle \varphi'(1)\varphi'(2)|r_{12}^{-1}|s_k(1)s_k(2)\rangle$, summed over periphery atoms k, produces an eigenvalue spectrum

$$\gamma_1 \geq \gamma_2 \geq \gamma_3 \geq \ldots \geq \gamma_p \geq \ldots \geq \gamma_N \tag{5}$$

with corresponding eigenfunctions, $\{\varphi_k'\}$. The localized orbitals, $\varphi_1', \varphi_2', \varphi_3' \ldots \varphi_p'$, primarily

Table 1. Fe_{104} localized orbitals. Localization about the 52 periphery atoms of the 104 atom cluster. Orbitals 27 - 32 are indicated in parentheses. Exchange eigenvalues, in a.u., and overlap integrals for each localized orbital φ' with an orbital φ'' obtained by removing basis functions not on the periphery followed by renormalization are listed.

Exch.	<φ'φ">	Exch.	<φ'φ">	Exch.	<φ'φ">
.91	.992	.35	.905	.09	.026
.91	.994	.34	.960	.08	.129
.87	.998	.32	.986	.07	.054
.79	.992	.29	.892	.07	-.062
.79	.986	.28	.928	.07	.053
.70	.995	.26	.900	.06	-.025
.66	.995	.25	.948	.06	.023
.55	.999	.25	.935	.05	-.062
.54	.981	.25	.929	.04	-.034
.49	.983	.25	.972 (27)	.04	.021
.48	.982	.19	.885 (28)	.04	-.004
.45	.980	.18	.741 (29)	.04	-.031
.44	.983	.17	.634 (30)	.02	.026
.43	.967	.17	.480 (31)	.01	.023
.42	.981	.11	.252 (32)	.01	-.010
.41	.978	.10	.333	.01	.008
.36	.949	.10	.115	.01	.025
				.001	.022

reside on the periphery atoms (with proper choice of p), but tails extend into the interior of the cluster and toward the surface. A convenient way to analyze the degree of localization is to truncate orbital expansions by removing basis functions not on the periphery, renormalize and then compute the overlap with the original orbital. Boundary potentials are derived from the set of periphery orbitals by analyzing their penetration into the interior of the cluster. Localized electron densities corresponding to these orbital tails give rise to Coulombic and exchange potentials, $|r_{12}^{-1}\rho(2)\rangle$ and $|r_{12}^{-1}\gamma(1,2)\rangle$, and projectors, $\Sigma_m \varepsilon_m |q_m\rangle\langle q_m|$. Table 1 provides an illustration of the localization analysis for a symmetric, three-layer Fe_{104} (37, 30, 37) cluster model of the (110) surface in which SCF orbitals are localized about the 52 atoms on the periphery. The table shows that 30 doubly occupied orbitals (thus, eight electrons from the interior) strongly involve the periphery. The third layer of the cluster is not modeled in this illustration. See Ref. 6 for details.

The treatment of the adsorbate-surface system is then carried out by augmenting the basis in the region around the surface site of interest to include functions describing polarization and correlation contributions. Transition metal d functions on surface atoms are introduced explicitly at this stage. Final electronic wavefunctions including the adsorbate are constructed by configuration interaction

$$\Psi = \Sigma_k \lambda_k A(\chi_1^k \chi_2^k \ldots \chi_n^k \varphi_1' \ldots \varphi_p') \equiv \Sigma_k \lambda_k A(\chi_1^k \chi_2^k \ldots \chi_n^k), \tag{6}$$

and the coupling of the local electronic subspace and adsorbate to the bulk lattice electrons, $\{\varphi'_j, j = 1, p\}$, is represented by the modified Hamiltonian,

$$H = \Sigma_i^n - \tfrac{1}{2}\nabla_i^2 + \Sigma_i^n \Sigma_k^Q - Z_k/r_{ki} + \Sigma_{i<j}^n 1/r_{ij}$$

$$+\Sigma_m \left\{ |r_{12}^{-1}\rho_m(2)> - |r_{12}^{-1}\gamma_m(1,2)> \right\} + \Sigma_m \varepsilon \, |q_m><q_m| \qquad (7)$$

where ρ_m, γ_m and q_m denote atomic functions or densities derived from $\{\varphi'_j, j = 1, p\}$, and the dimensionality of the electronic system is reduced from N to n.

The level of computational difficulty in the final model is that of a cluster (typically 20 - 30 atoms) in the presence of electrostatic potentials and projectors describing the embedding. Note that atoms in the surface region and their nearest neighbors do not have effective potentials (except for core electrons). The effective atomic potentials begin at the next shell of atoms and functions are also present in this shell to allow a polarization of electrons on reaction with adsorbates.

III. DISSOCIATION OF METHANE ON AN FE/NI(111) ALLOY SURFACE

Recent experimental and theoretical studies have shown that the activation barrier for the dissociation of CH_4 on nickel is in the range 12 to 17 kcal/mol depending on whether tunneling contributions are included. The underlying idea of the present work is to increase the reactivity of the nickel surface by doping it with iron. The objective is to reduce the activation energy barrier for the dissociation of methane but not to create a surface that is so reactive that the products are permanently pinned to the dissociation site. The present study which is described in detail in Ref. 10 concerns the energetics of the dissociation reaction, $CH_4 = CH_3 + H$ on a Ni(111) surface in which a single Ni atom is replaced by an Fe atom.

The cluster geometry and surface region of the Fe/Ni(111) surface are shown in Fig. 1. Embedding theory is used to reduce the Ni_{88} cluster to a 41-atom model depicted as shaded atoms: the surface layer of 19 atoms, a second layer of 14 atoms, and a third layer of 8 atoms. An effective [1*s*-2*p*] core potential is used for the central Fe atom and 3s, 3p, double zeta 3d, 4s and 4p orbitals are explicitly included. The six Ni atoms surrounding the central Fe atom have an effective [1*s*-3*p*] core potential and valence 3d, 4s and 4p orbitals. In the atomic configuration of Fe, $(3s^2)(3p^6)(3d^7)(4s^1)$, the $3d_{xy}$, $3d_{xz}$ and $3d_{yz}$ orbitals are singly occupied in the initial SCF calculations. For the six surface Ni atoms, the $3d_{z^2}$ is singly occupied in the valence configuration of $(3d^9)(4s^1)$. Other Ni atoms are described by an effective core potential for [1*s*-3*d*] electrons, and a single 4s orbital. For all boundary atoms, and those in the third layer, the core potential is further modified to account for bonding to the bulk as defined by the embedding procedure. The same cluster, basis and core potentials are used in all subsequent calculations on the Fe/Ni(111) surface for CH_4 adsorption and for the $CH_4 = CH_3 + H$ dissociation.

Dissociation is considered atop iron with CH_3 and H moving to three-fold product sites on either side of the iron atom. Figure 2 depicts the dissociation pathway. For CH_4 adsorbed at an atop Ni site on the Ni(111) surface, the transition state of $H_3C...H$ corresponds to a C-H bond stretch from 1.08 Å to 1.53 Å and a C-Ni distance of 2.41 Å. The activation energy is 16.7 kcal/mol relative to CH_4 at infinite separation from the Ni surface.[7] Figure 2 shows four intermediate $H_3C...H$ geometries on the substitutional Fe/Ni(111) surface with the same stretched C-H bond of 1.53 Å at a C to surface distance of 2.43 Å. The stretched $H_3C...H$ with C directly above Fe, Fig. 2d, corresponds to the transition state for which the total energy is 5.7 kcal/mol higher than that of CH_4 at infinite

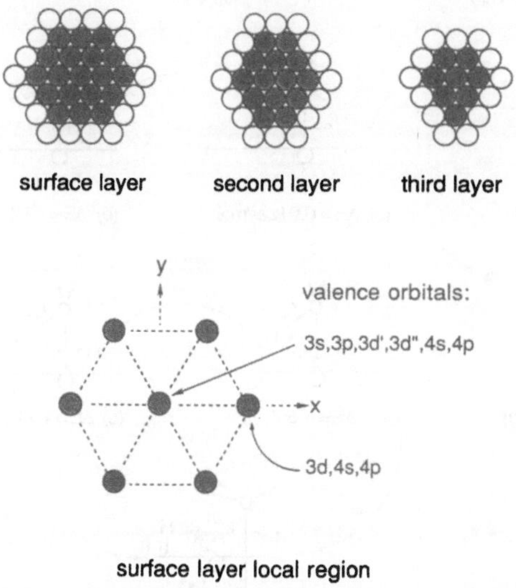

Figure 1. Cluster geometry and local region of the nickel cluster used to model the (111) crystal face of nickel. The three layer, 88-atom cluster, consists of a surface layer of 37 atoms, a second layer of 30 atoms and a third layer of 21 atoms. Embedding theory is used to reduce the Ni_{88} cluster to a 41 atom model depicted as shaded atoms. Atoms surrounding the seven local region atoms in the surface layer, those surrounding the four central atoms in the second layer, and shaded atoms in the third layer are described by effective potentials for $(1s^2) \ldots (3p^6)(3d^9)(4s^a)$ configurations with a = 0.5, 0.3 and 0.75, respectively. Unshaded atoms have neutral atom $(1s^2) \ldots (3p^6)(3d^9)(4s^1)$ potentials. All atoms have Phillips-Kleinman projectors $\Sigma_m \varepsilon_m | q_m \rangle\langle q_m |$ for the fixed electronic distribution. The nearest neighbor Ni-Ni distance is 2.48 A. In the Fe/Ni(111) study, the central Ni atom in the surface layer is replaced by an Fe atom.

separation from the surface. The CH_3 and H products are bound to the surface with about the same adsorption energy as for Ni(111). Thus, the iroy/nickel alloy surface has reduced the activation barrier but has left the dissociation products as reactive as on pure nickel.

IV. DIAMOND NUCLEATION ON NICKEL

One of the objectives of diamond film research is the growth of single crystals of sufficient size and quality to be useful for electronic devices and other applications. Both covalent materials and metals have been used as substrate surfaces. Although attractive because of their close lattice match to diamond, metals such as Ni and Cu have had only limited success as substrates. One of the attractive features of a metal like Ni, as well as a drawback, is its catalytic activity. The reactivity of the nickel surface leads to dissociative reaction of many molecular species, including hydrocarbons, and the strong bonding of the resulting carbon fragments to the surface could in principle provide a good starting point for carbon film growth. On the other hand, it is well known that graphite overlayers form on nickel and are poisons to further catalytic activity as well as inhibitors to the growth of an overlayer of tetrahedrally bonded carbon.

In this study, we consider Na, H and C atoms implanted interstitially below the surface layer and probe changes in the reactivity of the surface. Benzene is used as a probe molecule. In its planar equilibrium geometry, benzene bonds to the surface through its pi-electron system as does graphite while in a puckered, tetrahedral geometry, it represents a subunit of the diamond lattice. Analysis of the adsorption energetics of planar and non-planar structures of C_6H_6 measures the effect of Na, H and C interstitial species. A full

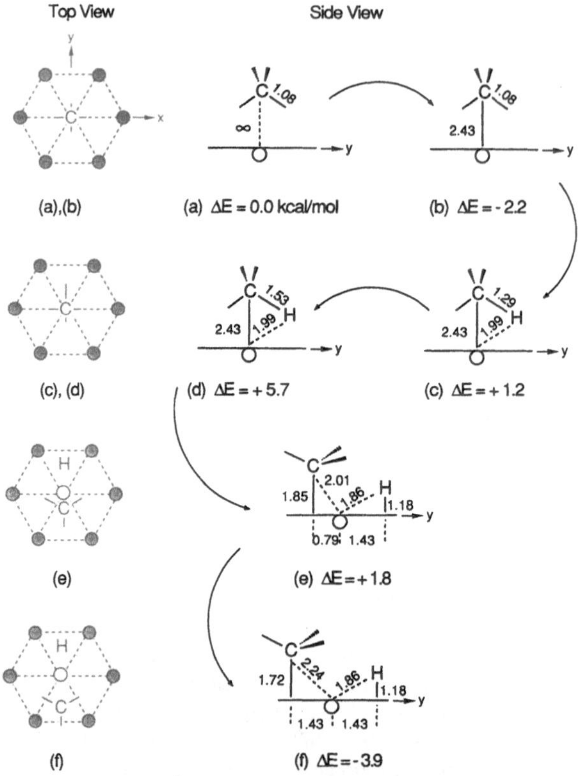

Figure 2. Dissociation of CH_4 on Fe/Ni(111). The minimum energy dissociation pathway shows a transition state 5.7 kcal/mol above the energy of the desorbed state.

account of this work including comparisons with adsorbed CH_3 in planar and nonplanar geometries is reported in Ref. 12.

The 41-atom embedded cluster used to model the (111) crystal face of nickel consists of three layers of atoms, 19 atoms in the surface layer, 14 in the second and 8 in the third layer as depicted in Fig. 1. The region of the surface of primary interest consists of seven atoms and on these atoms, 3*d*, 4*s* and 4*p* orbitals are explicitly included in the valence basis and the remaining electrons are described by a [1*s*-3*p*] Ni core potential.

Table 2. C_6H_6 adsorbed at a hollow 3-fold site on Ni(111) with subsurface Na, H and C interstitial atoms. Adsorption energies, E_{ads}, and distances from C to the surface (R_e for planar C_6H_6 and r_e for the lower three C atoms in the puckered structure) are given. ΔE is the energy required to create the puckered structure relative to gas phase planar benzene. Results are from configuration interaction calculations and are corrected for basis superposition effects of approximately 3 to 5 kcal/mol.

	planar		puckered		
	E_{ads} (kcal/mol)	R_e (Å)	E_{ads} (kcal/mol)	r_e (Å)	ΔE (kcal/mol)
gas phase	---	---	---	---	198
Ni(111)	18	2.47	-81	1.80	99
Na interst.	10	2.51	-59	1.75	69
H interst.	19	2.47	-64	1.75	83
C interst.	44	2.45	-90	1.85	134

Adsorption energies relative to the gas phase puckered benzene structure are exothermic, 99, 129, 115, and 64 kcal/mol for Ni(111), Na, H and C interstitial cases, respectively.

Based on earlier theoretical and experimental studies, we consider adsorption at the hollow 3-fold surface site. Interstitial atoms are positioned below this surface site and midway between the first and second layers, *i.e.*, at the center of an octahedral hole in the lattice.

Calculated adsorption energies for planar benzene on the (111) surface in the presence of Na, H and C interstitials are reported in Table 2. The adsorption energy, E_{ads}, changes dramatically in the presence of Na and C interstitials compared to the Ni(111) surface without interstitial species. The 44 kcal/mol value for the C interstitial is twice that for Ni(111) while the Na interstitial significantly weakens the C_6H_6-surface bond to a value of 10 kcal/mol, about one-half that for Ni(111). The optimized distances from the ring to the surface remain about the same, 2.47 ± 0.04 Å, in the presence of Na, H and C interstitials as for Ni(111).

Adsorption of the puckered ring geometry is now considered where bond orientations around each carbon are tetrahedral (sp^3 hybridization). The C-C bond length in diamond, 1.54 Å, is used and C-H bond lengths are fixed at their normal value in hydrocarbons, 1.08 Å. The geometry is depicted in Fig. 3. In this figure, the three carbon atoms involved in bonding with the surface are equidistant from the surface and their bonds with the surface are represented by dashed lines. If the tetrahedral orientation is maintained, these bonds would point almost exactly at the centers of the 3-fold surface sites (the displacement from the center is 0.02 Å). In the adsorption energy calculations, the distance from the ring to the surface is optimized without distorting the tetrahedral structure of the ring.

Adsorption energies and calculated distances are reported in Table 2 for the adsorption of the puckered ring structure at the 3-fold hollow site in the presence and absence of Na, H and C interstitial atoms. Adsorption energies are endothermic relative to planar C_6H_6 at infinite separation ranging from -90 kcal/mol for the C interstitial to -59 kcal/mol for the Na interstitial. Relative to the gas phase puckered benzene structure, the adsorption energies are exothermic, 99, 129, 115, and 64 kcal/mol for Ni(111), Na, H and C interstitial cases, respectively. Adsorption of the puckered structure corresponds to the

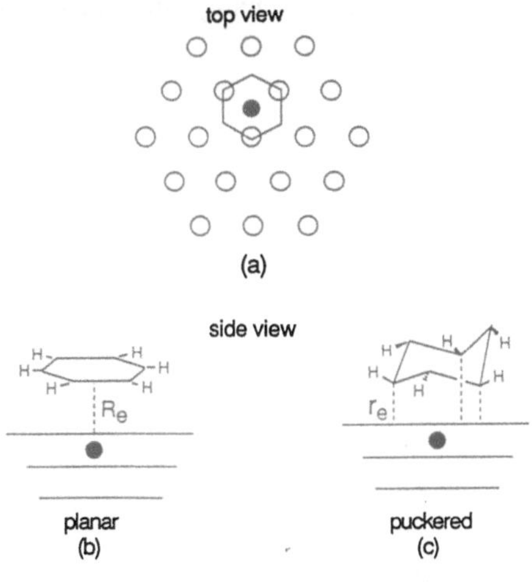

Figure 3. Top view of adsorbed C_6H_6 at a hollow 3-fold site on Ni(111), (a); planar benzene (b) and puckered diamond subunit (c). The dark circle represents a Na, H or C interstitial atom which is below a hollow 3-fold site, midway between the first and second layers, *i.e.*, at the center of an octahedral hole in the lattice. The vertical distance from the nucleus of the interstitial atom to the first and second layers is 1.01 Å.

formation of three bonds with the surface. Perpendicular distances from the lower layer of C atoms to the surface are between 1.75 Å and 1.85 Å for the different cases.

Comparing these values with the planar benzene adsorption energies leads to the following conclusions.

(1) Distortion of benzene in the gas phase to the puckered structure is a very energetic process, 198 kcal/mol endothermic. On the Ni(111) surface, much less energy, 99 kcal/mol, is required for this distortion.

(2) On the surface, the planar structure is more stable than the puckered form even in the presence of interstitial species.

(3) The presence of Na as an interstitial atom, and to a lesser extent H, preferentially stabilizes the puckered ring structure. Interstitial C favors the planar structure.

(4) The energy required for the planar to puckered distortion is 69 kcal/mol in the presence of Na, 83 kcal/mol with H, and 134 kcal/mol with C.

Although there are complicating factors, such as changes in surface structure, the present results suggest that electropositive implants such as Na may be useful in stabilizing tetrahedral carbon on a nickel surface thereby promoting the nucleation of diamond.

V. ADSORPTION OF CH_2 ON C(100)

The C(100) surface is potentially attractive for epitaxial growth of diamond films of high quality. During film growth, hydrogen is used to maintain a tetrahedral geometry of carbon. Hydrogens are subsequently stripped from the surface by reaction with atomic H from plasmas or other sources, carbon is added to the surface and growth continues. One of the important questions concerning the dimerized C(100)-2x1 surface is the difficulty of breaking surface dimer bonds by CH_x species and H. There is evidence that the 2x1 surface is resistant to attack by atomic hydrogen, and presumably also CH_x. The present study explores some of the basic issues and concludes that CH_2 is able to react with C(100) with absolutely no energy barrier.

The cluster, $C_{12}H_{20}$, shown in Fig. 4, is used to model the C(100) surface and the dimerization reconstruction. On adsorption, CH_2 forms bonds with atoms $C_1$and C_2. Multi-parent CI calculations are carried out on an electronic subspace defined by a localization of occupied and virtual orbitals, based on exchange maximization with basis functions on atoms C_1 and C_2 and CH_2.

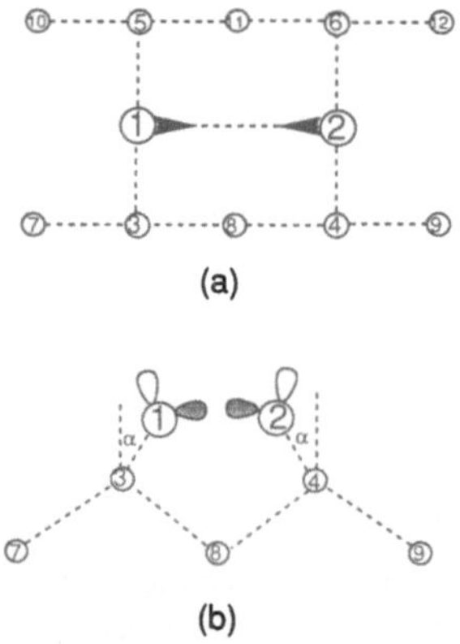

Figure 4. The $C_{12}H_{20}$ cluster used to model the C(100)-2x1 surface. On adsorption, CH_2 bonds with atoms C_1 and C_2 and the surface returns to a nearly ideal, undimerized, geometry.

We first calculate the total energy of the system as a function of the distance of CH_2 from the reconstructed C(100)-2x1 surface. The surface atoms are then relaxed to their equilibrium geometry. As CH_2 approaches the surface, as shown in Fig. 5, the total energy decreases monotonically to a value 4.7 eV lower than that for the desorbed state. These results which are for a fixed C_1-C_2 distance of 1.4 Å show that there is no barrier to the chemisorption of CH_2 on the C(100)-2x1 reconstructed surface.

Figure 5 also shows the energy of the $CH_2/C_{12}H_{20}$ system as a function of the CH_2-surface distance and α, the angle of rotation of the surface atoms C_1 and C_2 about stationary atoms in the second layer. The energy decreases considerably as α changes from 35°, to 20° to 10° showing that the surface is de-reconstructed by CH_2. The lowest energy (for $\alpha = 9.2°$) is 6.2 eV exothermic with respect to the desorbed state $CH_2 + C_{12}H_{20}$(2x1). This rotation corresponds to a dimer bond length of 2.23 Å. The energy increases slightly when α decreases to 0°, the ideal C(100)-1x1 surface. In the absence of adjacent CH_2 adsorbates, it is the singlet coupling of the dangling bonds of C_1 and C_2 that prevents complete return to the 1x1 geometry.

The implication of these results is that C and CH would also likely react with essentially no barrier, and it would follow that difficulties that have been encountered in the growth of C(100) are not due to reduced reactivity of the surface to CH_x species.

VI. SUMMARY

A large number of reactions on metal and semiconductor surfaces have now been carried out using the embedding procedure described in this report and earlier publications. The report summarizes several of these studies and provides an overview of the main features of the embedding procedure.

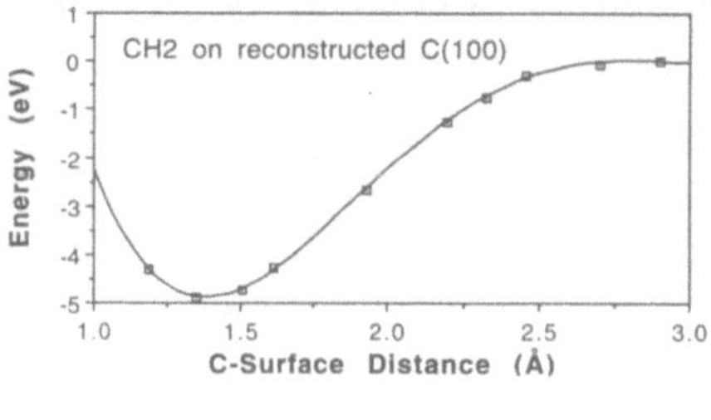

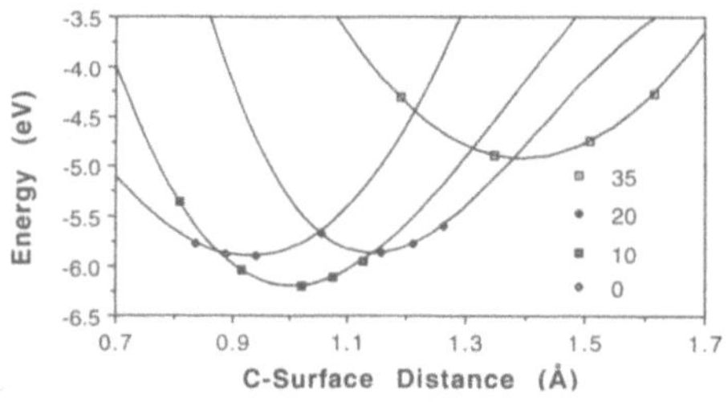

Figure 5. Reaction of CH_2 with a cluster model of the C(100) surface. The upper figure is for reaction with the reconstructed (dimerized) surface. The lower figure shows the variation in adsorption energy for different angles of rotation of C_1 and C_2 about stationary atoms in the second layer. The angles 35° and 0° correspond to the dimerized and ideal surfaces, respectively.

ACKNOWLEDGMENTS

This work was supported by the U.S. Department of Energy, grant # DE-FG05-90ER45436. The nickel interstitial work and the study of CH_2 adsorption on C(100) was supported by the Strategic Defense Initiative Organization-Innovative Science and Technology Office through the Office of Naval Research and a subcontract from the Research Triangle Institute to North Carolina State University. The adsorption studies on nickel and iron surfaces were supported by the U. S. Department of Energy. For applications described in this report, I have drawn on published and unpublished calculations by my colleagues Dr. Hong Yang and Dr. Ze Jing, to whom I am indebted for their work and stimulating discussions.

REFERENCES

1. There are numerous references to experiments and other theoretical work related to the applications discussed in this report. These are found in the primary publication cited.
2. J. L. Whitten and T. A. Pakkanen, Phys. Rev. B**21**, 4357 (1980).
3. H. Yang and J. L. Whitten, Surface Science, **255**, 193 (1991), and references therein.
4. Z. Jing and J. L. Whitten, Surface Science, **250**, 147 (1991).
5. Z. Jing and J. L. Whitten, Phys. Rev. B**44**, 1741 (1991).
6. J. L. Whitten, in *Cluster Models for Surface and Bulk Phenomena*, p. 375, G. Pacchioni and P. S. Bagus, eds., Plenum, (1992).
7. H. Yang and J. L. Whitten, J. Chem. Phys. **96**, 5529 (1992).
8. H. Yang and J. L. Whitten, J. Chem. Phys., **98**, 5039-5049 (1993).
9. Z. Jing and J. L. Whitten, J. Chem. Phys., **98**, 7466-7470 (1993). Si(100), and references to other calculations on silicon contained therein.
10. H. Yang and J. L. Whitten, Surface Science, **289**, 30-38 (1993).
11. J. L. Whitten, Chem. Phys., **177**, 387-397 (1993).
12. H. Yang, J. L. Whitten and R. Markunas, Applied Surf. Sci., **75**, 12-20 (1994).
13. J. L. Whitten, P. Cremaschi and R. Markunas, Applied Surf. Sci., **75**, 45-50 (1994).
14. Z. Jing and J. L. Whitten, Materials Research Society Meeting (1994); Phys. Rev. B**50**, in press, (1994).

CLUSTER STUDIES OF La_2CuO_4 GEOMETRIC DISTORTIONS ACCOMPANYING DOPING

Richard L. Martin

Theoretical Division, MS B268
Los Alamos National Laboratory
Los Alamos, NM 87545

I. INTRODUCTION

It is well established[1] that La_2CuO_4, the parent material of the superconducting cuprates, is an anti-ferromagnet with a gap to charge excitations of 2eV. This stoichiometric composition is described well by a model in which the layered planes of CuO are replaced by a 2-dimensional lattice of *effective* Cu sites, each of which is occupied by one electron. The low-lying excited states observed experimentally (spin waves) are quantitatively described by a Heisenberg spin Hamiltonian, in which each electron is assumed to be localized on a single site and only its spin degrees of freedom are relevant. Both neutron scattering and Raman experiments yield an effective near-neighbor anti-ferromagnetic coupling constant of ~0.125eV. The parent material can therefore be envisioned as a system in which the electrons are highly localized.

Upon replacement of La^{3+} with Sr^{2+} ($La_{2-x}Sr_xCuO_4$) the picture changes. The ion Sr^{2+} requires an additional electron relative to La^{3+}, and the substitution therefore has the effect of removing electrons from the CuO planes, or alternatively, introducing *holes* into the lattice. The long-range anti-ferromagnetic order is quickly lost, and by x~.05 the material begins to exhibit superconductivity at low temperatures. In the *normal state*, i.e. at temperatures above the superconducting transition temperature, the electrons behave as if they were delocalized, i.e. as a metal, albeit a very unusual one. Evidence for spin fluctuations in the normal state are found in neutron scattering experiments even for doping concentrations of ~0.15, near the maximum in the plot of T_c vs. Sr concentration. A precise understanding of the properties of this transition from localized to itinerant behavior, from insulator to metal, is therefore of great interest.[1]

Cluster calculations, both first-principles and those based on model Hamiltonians, have played an important role in the development of our understanding of this problem. First principles calculations, such as those discussed in this article, suffer from the significant disadvantage that they cannot be performed on clusters large enough to observe the pairing interaction directly. They can be used, however, to examine experiments which probe local interactions, to generate parameters for simpler models which can be extended to larger clusters, and to provide some insight into the nature of the local electronic structure.

In this article, I shall focus on the character of the first few holes doped into La_2CuO_4. Are they localized to a specific site or a small unit of the lattice? This question has been difficult to pin down experimentally.[2] Many researchers conclude that the addition of a hole into the antiferromagnetic background induces structural deformations which have a pronounced impact on those experiments which probe the local geometry about a site. This implies that the holes are initially localized about one or a few sites; that they behave as polarons. In this contribution, the local geometric distortions associated with the various charge states of small clusters are studied with ab initio quantum chemistry techniques. In these investigations I initially beg the question of whether the holes in the actual material are localized. By confining them to small clusters I simply examine what would be expected to happen if they were. This information may be helpful in interpreting experiments which attempt to answer such questions. The second issue, whether and why the holes would tend to localize in the infinite lattice, is addressed here only qualitatively at the end of the discussion.

II. COMPUTATIONAL DETAILS

This research builds upon previous work described in detail elsewhere[3-5] and therefore only a brief summary of the approach will be provided. The clusters to be examined are CuO_6 and Cu_2O_{11}. In order to provide a proper electrostatic environment for the cluster, the primary CuO_6 and Cu_2O_{11} units are imbedded in a Madelung/Pauli background potential as shown in Fig. 1. In brief, several neighbor shells about the primary cluster (~ 1000 point charges) are included explicitly using formal charges of 3+, 2+, and 2- for La, Cu, and O, respectively. The positions are taken from an experimental study of $La_{2-x}Sr_xCuO_4$, with x=0.15, in the tetragonal phase.[6] Near the periphery of this point charge field, the positions and magnitudes of additional charges were determined by a least squares fit so as to reproduce the exact Madelung potential at the sites in the primary cluster. If the embedding potential contains only this Madelung field, it is not possible to determine reliable geometries for the cluster. For example, imagine computing the total energy of the CuO_6 cluster as a function of the equatorial O position. Since the O is negatively charged, it will be attracted to the neighboring 2+ point charge representing a background Cu ion. This is as it should be, but what is left out of the description is the fact that at short distances the

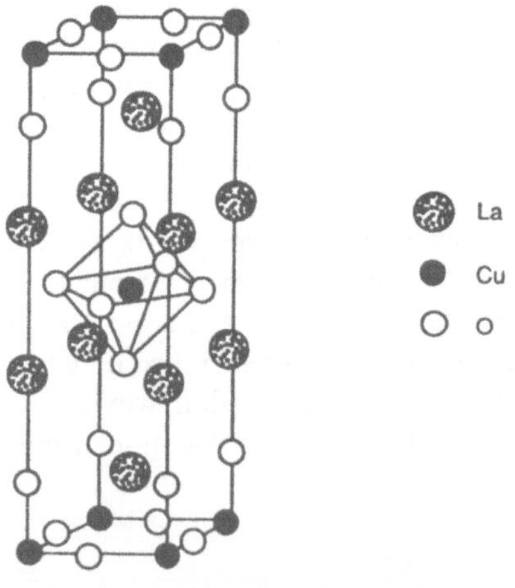

Figure 1. The basic cell used to generate the background potential for CuO_6.

electrons of the O feel a repulsion from the core electrons of the Cu ion. This effect is not represented by the Madelung potential alone; the Pauli repulsion associated with the ion core of the immediate neighbors of the O must be included as well.[7] Therefore, the cationic nearest-neighbors of the atoms in the primary cluster are represented by effective ion core potentials. These effective potentials were generated for the free ion, i.e. Cu^{2+} and La^{3+}, and have a leading term which is identical to that provided by the appropriate formal charge. Additional terms in the effective potential enforce the Pauli repulsion of the electrons in the primary cluster with the core electrons of the neighbor. This improvement is critical for the proper description of cluster geometries.

The basis sets for the Cu and O atoms are the standard atomic expansions used and tested previously on CuO_6, Cu_2O_7, and Cu_2O_{11}. As before, the [Ne] core of the Cu atom is replaced by the effective core potential of Hay and Wadt.[8] Hartree-Fock and configuration-interaction (CI) calculations were performed using the MESA suite of electronic structure programs.[9]

III. RESULTS

CuO_6

There are three charge states of the CuO_6 cluster shown in Fig. 1 which are of interest. The first is CuO_6^{11-}. The electron count in this cluster corresponds to closed-shell $Cu^{1+}(d^{10})$ and filled $O^{2-}(p^6)$ ligands. It therefore models the situation in which an additional electron is added to La_2CuO_4. The model for the undoped, *neutral*, cluster is CuO_6^{10-} (formally Cu^{2+}). Addition of a hole to the undoped model yields $CuO_6^{9-}(Cu^{3+})$.

Previous SCF and CI calculations,[3,4] as well as cluster calculations[10,11] using parameters extracted from constrained density-functional-theory(DFT)[12-14] or the ZINDO semi-empirical method[15] agree that the ground states of these clusters can be well understood in terms of occupying an orbital of $b_{1g}(d_{x^2-y^2})$ symmetry with $0(CuO_6^{9-})$, $1(CuO_6^{10-})$, or $2(CuO_6^{11-})$ electrons. The general character of the effective orbital is an anti-bonding combination of the Cu $3d_{x^2-y^2}$ orbital and the b_{1g} linear combination of the $O2p_\sigma$ orbitals. The precise nature is a function of the charge state. In CuO_6^{11-} (d^{10}), it is ~60% $d_{x^2-y^2}$ and ~40% $O2p_\sigma$ In the undoped cluster, the effective orbital is rehybridized and becomes more ionic. SCF calculations suggest the unpaired electron is ~90% Cu $3d_{x^2-y^2}$ in character. CI calculations induce more hybridization, leading to an unpaired spin density which is ~80% Cu $3d_{x^2-y^2}$. Upon adding a second hole, the orbital rehybridizes again, becoming an ~50:50 mix of $3d_{x^2-y^2}$ and $O2p_\sigma$. This rehybridization is accomplished via a strong charge-transfer relaxation accompanying the addition of the second hole and leads to a prominent satellite in photo-emission. The position and intensity of this satellite is fairly well described by CI calculations reported earlier.[16]

Because the ground state of the neutral cluster is relatively ionic, with a single unpaired electron only slightly delocalized (~20%) onto the ligands, the low-lying excited states of the neutral are found to be crystal-field excitations. CI calculations predict the $d_{x^2-y^2} \rightarrow d_{xy}$ excitation to occur at 1.82eV, $d_{z^2} \rightarrow d_{x^2-y^2}$ at 1.89eV, and $d_{xy,yz} \rightarrow d_{x^2-y^2}$ at 2.28eV. These are of interest because the lowest of them fall within the insulating gap (2eV) of the anti-ferromagnet. Recently Liu et al.[17,18] have observed the $d_{xy} \rightarrow d_{x^2-y^2}$ exciton in La_2CuO_4 by resonance Raman experiments. It occurs at 1.7eV, in good agreement with the CI calculation (1.8eV). This feature is accompanied by two weaker sidebands at higher energy which have been assigned to vibrational and magnon excitations accompanying the transition.[19]

The next set of excited states in the neutral cluster correspond to $O2p \rightarrow Cu3d_{x^2-y^2}$ charge transfer excitations. These, according to the Zaanen, Sawatzky, Allen interpretation[20,21] correspond to the optical conductivity gap at 2eV in La_2CuO_4, and therefore this material is more precisely described as a charge-transfer insulator as opposed to a traditional Mott-Hubbard insulator. This interpretation is supported by the cluster calculations discussed above, but it appears that they place the optical gap at an energy higher than the 2eV observed experimentally. First principles cluster studies[3] of CuO_6 overestimate the gap dramatically, even when corrected for estimates of the increased $O2p$ bandwidth and additional polarization stabilization which would develop if larger clusters were studied. An estimate of ~5eV for the charge-transfer gap was made earlier.[3] This estimate is rather crude, as it did not come from a direct calculation of the lowest lying charge transfer state, but rather from a model based on parameters extracted from calculations on other states. A direct calculation on Cu_2O_{11}, e.g., might lead to a smaller gap. Calculations[10] on larger clusters based on parameters extracted from DFT are much more in line with experiment, but also place the gap too high at ~2.4eV. One suggestion[3] for this behavior is that all the calculations to date have addressed the vertical excitation energy, in which the geometry of the excited state is identical to that of the ground state. This is usually a good approximation for those systems where the excitation is delocalized over the entire lattice. If the charge-transfer excitation were significantly localized, as in the Zaanen-Sawatzky-Allen model, one might expect to observe the adiabatic charge transfer edge, i.e. the 0-0 transition from the ground state minimum to the excited state minimum. This possibility, as well as other indications of electron-phonon coupling,[2] motivated the calculations described next.

Table 1 presents axial and equatorial bond lengths optimized for the CuO_6^{11-}, CuO_6^{10-}, and CuO_6^{9-} cluster ground states. These clusters represent formal Cu populations of d^{10}, d^9, and d^8, respectively, corresponding to the electron count associated with n-doped, undoped, and p-doped clusters. The column labeled SCF reports self-consistent-field calculations, while the others refer to configuration-interaction calculations with single and double substitutions (CISD), and CISD with a correction for quadruple excitations using the Davidson approximation (CISD(Q)).[22] Because the optimum geometries differ by only ~0.01Å between the SCF and CISD calculations, I will focus in what follows on the SCF results. First of all, note that the axial bond distance (2.40Å) calculated for the undoped model is in good agreement with experiment (2.41Å). The error in the equatorial bond length (1.85Å computed vs. 1.89Å experiment) is larger. Upon n-doping the calculated equatorial bond length increases by 0.07Å, and the axial bond length becomes 0.15Å longer. An increase in the equatorial bond length is expected since the additional electron is entering an anti-bonding orbital. The large change in the axial CuO distance is somewhat

Table 1. Optimized CuO_6 Bond Lengths(Å)

State	SCF		CISD		CISD(Q)		Expt	
	R_{ax}	R_{eq}	R_{ax}	R_{eq}	R_{ax}	R_{eq}	R_{ax}	R_{eq}
$CuO_6^{11-}(d^{10})$	2.55	1.92	2.55	1.92	2.55	1.92	-----	-----
$CuO_6^{10-}(d^9)$	2.40	1.85	2.41	1.85	2.41	1.85	2.41	1.89
$CuO_6^{9-}(d^8)$	2.32	1.77	2.33	1.78	2.33	1.78	-----	-----

more surprising, and is probably associated with the rather large amount of Cu4s character which is mixed into the d_{z^2} orbital in this nominally d^{10} state.[4] The geometry relaxation stabilizes the energy by ~0.5eV. Upon p-doping the CuO bond lengths contract, consistent with removing an electron from the anti-bonding orbital. Both the equatorial and axial bond lengths decrease by 0.08Å, leading to a stabilization energy of ~1.8eV. Note that this is significantly larger than that associated with n-doping. As expected from the character of the anti-bonding orbital, most of the stabilization energy is attributable to motion of the equatorial oxygens.

The good agreement between theory and experiment for the CuO bond lengths in the undoped case suggests that the Madelung/Pauli embedding potential is a reasonable description of the background. It is possible to get another handle on the appropriateness of the background potential by determining the force constants for the equatorial CuO stretch. The potential is found to be reasonably harmonic and somewhat too hard. For example, if an equatorial O is moved toward the Cu atom in the undoped cluster, a force constant of ~48eV/Å^2 is found at the SCF level of approximation. If, on the other hand, the O is moved away from the central Cu and toward the background Cu^{2+} effective ion potential, a force constant of 49eV/Å^2(SCF) results. In the simplest model, uncoupled springs for each CuO bond, this force constant leads to a frequency for the CuO symmetric stretch (the breathing mode) of ~900cm^{-1}. Experimentally this mode is found in the vicinity of 650cm^{-1}.[23] This suggests that the embedding potential provides a somewhat too *stiff* environment in the plane, consistent with the theoretical equatorial bond length being somewhat too small. In the calculations on Cu_2O_{11} to be described later, it is possible to extract an SCF force constant for the motion of an O which is not a direct neighbor of a background potential site. In that case a smaller frequency, ~800cm^{-1}, results. This is not unreasonable given the simplicity of the model and the tendency of the SCF approximation to overestimate vibrational frequenncies in small molecules by 10-15%. In contrast, the force constant for motion of the axial O is weaker and quite anharmonic. Motion toward the Cu yields k=29eV/Å^2, while motion away from the Cu gives 7eV/Å^2.

Cu_2O_{11}

Now consider the Cu_2O_{11} cluster. It is composed of two of the CuO_6 clusters shown in Fig. 1 sharing an equatorial oxygen. In previous work, it was found that the electronic states of this cluster could be described fairly well with a single-band Pariser-Parr-Pople (PPP) model (also known as the single-band extended-Hubbard model). In this single band approximation, the electronic degrees of freedom associated with the oxygen and copper orbitals are contracted into a single *effective* orbital of $d_{x^2-y^2}$ symmetry on each CuO_6 unit. Cu_2O_{11} then becomes isoelectronic with H_2 leading to the states shown schematically in Fig. 2. The two effective orbitals give rise to two molecular orbitals, σ_g and σ_u, which can accommodate from n=0 to n=4 electrons. Within this model, the energies of the available states are given by:

$$E_0 = 0 \tag{1}$$

$$E_{C0} = \varepsilon + t \tag{2}$$

$$E_{C1} = \varepsilon - t \tag{3}$$

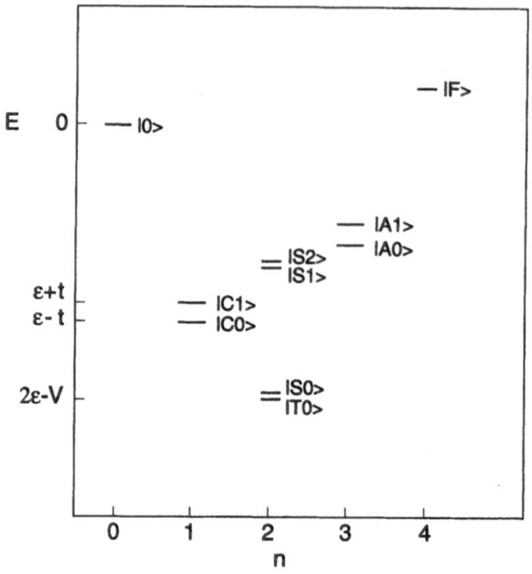

Figure 2 A diagram of the states allowed for a simple two-site model as a function of the number of electrons.

$$E_{T1} = 2\varepsilon + V \tag{4}$$

$$E_{S0} \cong E_{T1} - \frac{(2t)^2}{U-V} \tag{5}$$

$$E_{S1} = 2\varepsilon + U \tag{6}$$

$$E_{S2} \cong E_{S1} + \frac{(2t)^2}{U-V} \tag{7}$$

$$E_{A0} = 3\varepsilon + U + 2V + t \tag{8}$$

$$E_{A1} = 3\varepsilon + U + 2V - t \tag{9}$$

$$E_F = 4\varepsilon + 2U + 4V \tag{10}$$

where ε is the site energy, t is the hopping integral, and U and V are the on-site and nearest-neighbor Coulomb repulsion energies, respectively. One energy difference in the figure can be compared directly with experiment, and that is the splitting between the lowest singlet and triplet states of the n=2 species. This difference is computed[5] to be ~90meV in the CISD(Q) approximation, in fair agreement with the experimentally observed Heisenberg coupling constant of ~125meV. Additional CISD calculations for the states in Fig. 2 were performed and the results fit to Eqs. 1-10 to determine the four parameters of the PPP model.[5] In general, the relative energies of the various states could be reproduced by this simple model to within ~0.2eV.

Given the complexity of the CI calculations, which treat all 110 electrons arising from the ligands as well as 18 electrons associated with the metal centers, it is perhaps surprising that the results for these states can be so accurately mapped onto the PPP model.

There are, of course, aspects of the calculations which are not captured. The low-lying crystal-field excitons in the undoped material, i.e. $d_{z^2} \rightarrow d_{x^2-y^2}$, are missing in a single-band model, as are interesting correlations among the O2*p* and Cu3*d* electrons which are represented in the PPP model as *renormalized* parameter values. In the remainder of this paper, I will focus on one very interesting aspect of the electronic structure which depends critically on correlation among the motion of the O2*p* and Cu3*d* electrons but which shows up in a single-band model as a renormalized site-energy and an *on-site* electron-phonon coupling.

Consider the state |C0> in Fig. 2. This is the ground state of the cluster when a single electron occupies the band. In hole notation, it corresponds to adding a hole to the *neutral* antiferromagnetic state. If symmetry-restricted SCF calculations are performed for |C0>, the singly occupied orbital is delocalized equally over both Cu centers. The imposition of symmetry on the molecular orbitals, however, is a restriction, and the calculation can also be performed in a reduced symmetry group in which the mirror plane bisecting the Cu-O-Cu axis is not enforced. *Such a calculation gives rise to a new, broken-symmetry, SCF solution which lies 1.9eV lower in energy than the symmetry-restricted one.* In this solution, the additional hole is strongly localized on one of the CuO_6 units, say the left one, while the $d_{x^2-y^2}$ electron is localized on the other. It is degenerate with its mirror image in which the hole is localized on the right.

The origin of this behavior is the relatively small *bandwidth* of the $d_{x^2-y^2}$ derived molecular orbitals and the large polarizability of O^{2-}. If the hole localizes, the O2*p* electrons can polarize to stabilize it. If this polarization energy is larger than the kinetic energy lost by localization ($\sim t = 1/2(E_{C0}-E_{C1})$), then the broken-symmetry solution is lower in energy than the delocalized solution and will be found by the variational SCF approach. The important point is that the broken-symmetry solutions are alerting us to an important left-right correlation between the motion of the d-electrons in the band and the O2*p* electrons. When the d-electron *hops* from the left side to the right, the O2*p* electrons polarize in the opposite direction in order to get out of its way. Alternatively, when the hole moves to the left, the O2*p* electrons polarize to the left side as well. These broken-symmetry solutions are not acceptable eigenstates of the full Hamiltonian, which has mirror symmetry and parity as a good quantum number. Appropriate eigenstates can be formed as linear combinations of the two broken-symmetry SCF states. In essence, the broken symmetry solutions are saying that the better zeroth-order description of the electronic structure is to begin with an *electronic polaron* which is then allowed to delocalize to form the band.

Note that this symmetry-breaking is recoverable starting from the symmetry-restricted molecular orbitals when the important correlations are included in a CI calculation. However, this interesting behavior becomes a non-event in the mapping to a single-band model, since it simply lowers the energies of the states |C0> and |Cl> by comparable amounts (Fig. 2). In terms of the PPP model, it simply renormalizes the site energy ε.[24] The direct effects of the dynamical correlations responsible for the broken-symmetry have been folded out of the model. They do survive in an indirect sense in the variation of the site energy with oxygen motion. Fig. 3 displays the energy of the two broken-symmetry states as a function of the position of the central O. For the determinant in which the hole is localized on the left side, the central O finds a minimum displaced some 0.12Å toward the side with the hole. The slope of the curve at the symmetric point, dε/dq, is 4eV/Å. The well depth is 0.3eV.

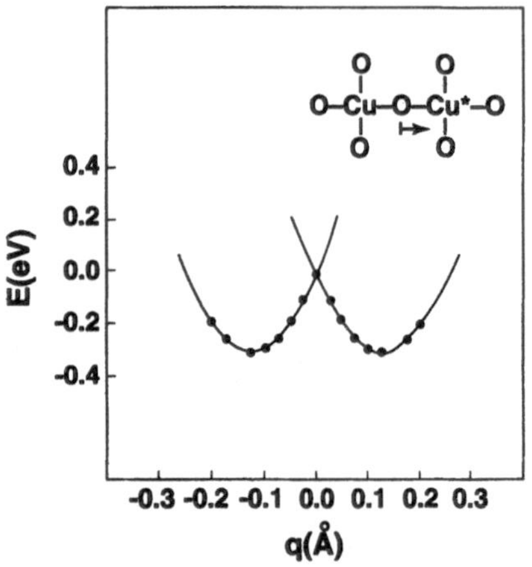

Figure 3. The *diabatic* potential energy curves for the broken-symmetry wavefunctions as a function of the motion of the bridging oxygen (q). The minimum in each curve corresponds to O motion toward the site with the localized hole. The slope of the curve at q=0 is 4eV/Å. The minimum is displaced by 0.12Å from the symmetric position and the well depth is ~0.3eV.

This information is used to augment the PPP Hamiltonian with an on-site electron-phonon coupling (Table 2). In addition to dε/dq, a force constant for the CuO stretch is needed. The constant in Table II was extracted from an SCF calculation on the triplet state, |T0>. The model Hamiltonian is then

$$H = H_{PPP} - \frac{\partial \varepsilon}{\partial q}\sum_{i,\sigma}(1 - n_{i,\sigma})q_i + \frac{1}{2}\sum_i (M\dot{q}_i^2 + Kq_i^2) \tag{11}$$

where i,σ label site and spin indices, respectively, $n_{i\sigma}$ is the number operator, M the oxygen mass, q_i the O displacement, and K the force constant.[25]

The double-well potential curves of Fig. 3 correspond to *diabatic* electronic states. The hopping integral between the two states must be considered to determine if the actual system will break symmetry. In the case of the Cu_2O_{11} cluster, if all four equatorial O atoms about the localized hole are allowed to move, the localized state can stabilize itself by ~1.2eV. This is larger than the hopping integral, ~0.65eV, and so we expect the adiabatic curve to show a double well; a polaron will form. In the infinite lattice, it is not clear what will happen. In this case the appropriate kinetic energy lost by localizing the hole is of

Table 2. Suggested PPP Holstein Parameters(eV)

ε	-5.5
t	-0.65
U	11
V	2.5
K_{CuO}	40eV/Å^2
dε/dq	4eV/Å
(U-V)/t	13

the order of half the bandwidth, or 4t=2.5eV. This is larger than the 1.2eV to be gained by the lattice relaxation, and so one would at first not expect an on-site coupling of this magnitude to form a polaron. In making this estimate, however, the bare hopping integral was employed for the kinetic energy. When the Coulomb interactions are considered a gap appears between upper and lower Hubbard bands and the effective hopping integrals are strongly reduced.[26] Near half-filling this reduction can be of the order of t/U , or ~0.1. By way of illustration, if $t_{eff} = 0.1t$, the loss of kinetic energy through localization is then only ~0.25eV, and the 1.2eV of lattice stabilization to be gained by forming a polaron begins to look attractive. A possible scenario, given the parameters found in this work, is that calculations on a larger lattice will find that the first few holes doped into the plane form polarons. Monte Carlo simulations suggest that t_{eff} increases as x.[26] Thus as more holes are doped into the plane, the polaron may become progressively larger, ultimately approaching the quasi-free limit.[27]

IV. CONCLUSIONS

First principles cluster calculations , such as those discussed here, can and have been used to address some issues important in high-temperature superconductivity. In these studies, one grapples with the usual problems governing accuracy in small molecules: the completeness of the one- and many-electron basis sets. In addition, the appropriateness of the background potential and the convergence of the results as a function of cluster size must be concerns as well. Certain questions, such as the magnitude of the spin exchange constant, the positions of the crystal-field excitations, and the interpretation of the photoemission spectrum, can be addressed semi-quantitatively with first principles quantum chemistry methods applied to quite small clusters. Other properties (e.g. transport, pairing susceptibilities, etc.) require a more extended lattice, and therefore attempts to generate parameters for models which can be extended to larger clusters have also been the focus of SCF/CI and DFT studies. This bootstrap approach is pragmatic, but unsettling. There is always the worry that in mapping the results onto a simpler model, one throws out the baby with the bathwater. In the current context, I have discussed the origin and magnitude of an on-site electron-phonon coupling in single-band models of the cuprates. The correlations responsible for the broken-symmetry "electronic polaron" solutions are, however, not explicitly included in a single-band model; they require retention of both the O2*p* and Cu3*d* degrees of freedom. Are these important for the mechanism of superconductivity, or is it permissible to think of them as simply renormalizing the effective parameters of a single-band model? It is clear that there are many such unanswered questions in this field, and I believe cluster calculations will continue to contribute to their solution.

ACKNOWLEDGMENTS

This work was supported by the Department of Energy through the LDRD program of Los Alamos National Laboratory.

REFERENCES

1. See, for example, "High Temperature Superconductivity", edited by K.S. Bedell, D. Coffey, D.E. Meltzer, D. Pines and J.R. Schrieffer (Addison-Wesley, Reading, MA, 1990).

2. See, for example, "Lattice Effects in High-T_C Superconductivity", edited by Y. Bar-Yam, T.Egami, J. Mustre-de Leon and A.R. Bishop (World-Scientific, Singapore,1992).
3. R. L. Martin in "Cluster Models for Surface and Bulk Phenomena", edited by G. Pacchioni, P.S. Bagus and F. Parmigiani (Plenum Press, New York, 1992).
4. R.L. Martin and P.J. Hay, J. Chem. Phys. **98**, 8680(1993).
5. R.L. Martin, J. Chem. Phys. **98**, 8691(1993).
6. R.J. Cava, A. Santoro, D.W. Johnson, Jr., and W.W. Rhodes, Phys. Rev. B **35**, 6716(1987).
7. N.W. Winter, R.M. Pitzer, and D.K. Temple, J.Chem. Phys. **86**, 3549(1987).
8. P.J. Hay and W.R. Wadt, J. Chem. Phys. **82**, 270(1985); W.R. Wadt and P.J. Hay, *ibid*, **82**, 284(1985); P.J. Hay and W.R. Wadt, *ibid*, **82**, 299(1985).
9. P.W. Saxe, B.H. Lengsfield III, R.L. Martin and M. Page, MESA.
10. M.S. Hybertsen, E.B. Stechel, M. Schluter and D.R. Jennison, Phys. Rev. B **41**, 11068(1990).
11. A.K. McMahan, J.F. Annett and R.M. Martin, Phys. Rev. B **42**, 6268(1990).
12. A.K. McMahan, R.M. Martin and S. Satpathy, Phys. Rev. B **38**, 6650(1988).
13. M.S. Hyberstsen, M. Schluter, and N.E. Christensen, Phys. Rev. B **39**, 9028(1989).
14. J.F. Annett, R.M. Martin, A.K McMahan and S. Satpathy, Phys. Rev. B **40**, 2620(1989).
15. Y.J. Wang, M.D. Newton and J.W. Davenport, Phys. Rev. B **46**, 11935(1992).
16. R.L. Martin, Physica B 1**63**, 583(1990).
17. R. Liu, M.V. Klein, D. Salamon, S.L. Cooper, W.C. Lee, S. -W. Cheong and D.M. Ginsberg, J. Phys. Chem. Solids **54**, 1347(1993).
18. R. Liu, D. Salamon, M.V. Klein, S.L. Cooper, W.C.Lee, S.-W. Cheong and D.M. Ginsberg, Phys. Rev. Lett **71**, 3709(1993).
19. D. Salamon, R. Liu, M.V. Klein, S.L. Cooper, W.C. Lee, S.-W. Cheong and D.M. Ginsberg, preprint (University of Illinois, P-94-87-057).
20. J. Zaanen, G.A. Sawatzky and J.W. Allen, Phys. Rev. Lett. **55**, 418(1985).
21. H. Eskes, G.A. Sawatzky and L.F. Feiner, Phys. Status Solidi C **160**, 424(1988). See also, G.A. Sawatzky in "Earlier and Recent Aspects of Superconductivity", edited by J.G. Bednorz and K.A. Muller (Springer, New York, 1990).
22. S.R. Langhoff and E.R. Davidson, Int. J. Quantum Chem., **8**, 61(1974).
23. See Krakauer, Pickett and Cohen, Ref. 2, p. 229, and references therein.
24. This redefinition of ε propagates an influence on the effective magnitudes of U and V as well (see Fig. 2).
25. While this model captures the spirit of what is observed in the clusters, it is deficient in that it ignores the axial oxygens, and also retains the particle-hole symmetry of the PPP model. The lattice stabilization energy computed for p-doped CuO_6 is much greater than that obtained for n-doping (1.8eV vs. 0.5eV), whereas the model predicts them both to be ~1.2eV.
26. See comments by D. Scalapino in Ref. 1, p. 118.
27. Whether or not this scenario leads continuously or discontinually to quasi-free behavior is not as straightfroward as I have described it. See the discussion between Emin and Alexandrov and Ranninger in Ref. 2.

NEUTRON MAGNETIC FORM FACTOR IN INSULATING TRANSITION METAL COMPOUNDS VIA CLUSTER CALCULATIONS

T.A. Kaplan,[1,3] Hyunju Chang,[1,3] S.D. Mahanti,[1,3] and J.F. Harrison[2,3]

[1]Department of Physics and Astronomy
[2]Department of Chemistry
[3]Center for Fundamental Materials Research
Michigan State University
East Lansing, MI 48824

I. INTRODUCTION

Our motivation for calculating the magnetic form factor in insulating transition metal oxides was originally to solve a puzzle[1,2] that arose in connection with the low-temperature magnetic neutron Bragg scattering experiments[3-6] on insulating parents of high-T_c superconductors, namely, La_2CuO_4 and $YBa_2Cu_3O_6$, (which are antiferromagnetic). The puzzle is the following: The ordered antiferromagnetic (AF) moment was reported[3-6] to be about 0.64 Bohr magnetons (μ_B). As noted by several authors,[1-3] this number agrees with the Heisenberg model (generally considered to be appropriate for these materials at low temperatures) via spin wave theory.[7] In this model the large reduction from the nominal value for Cu^{+2} ($3d^9$) is due to quantum spin fluctuations. But this agreement leaves no room for a reduction of the moment by covalence effects, a reduction whose existence had been discussed much earlier.[8,9] And it was concluded by some[3] that such covalence was essentially ruled out by these facts. However, these materials were expected on other grounds to show a *large* degree of covalence,[1] a fact demanding the question raised in references 1 and 2, *Where is the covalence?!*.

The interpretation of the neutron data[3-6] involved usage of the known form factor $f(\vec{q})$ of K_2CuF_4 which is isostructural to La_2CuO_4 (in the absence of the form factor for the materials being considered). For this purpose only the value at the smallest-$|\vec{q}|$ reflection was needed. We pointed out[2] that because K_2CuF_4 is a *ferro*magnet, it does not contain the covalent reduction of the moment, and therefore application of its form factor to an antiferromagnet is not appropriate. We also noted an improved procedure,[2,10] and carried it out as far as possible given the available theoretical information regarding K_2CuF_4.[11] While this gave improvement vis à vis the puzzle, it seemed that the degree of improvement was not nearly enough, leaving the puzzle unsolved.[2] For its solution it was clearly necessary to have a reliable calculation of the form factor. And we felt that an important test of any such calculation was to predict correctly the *shape* of the form factor as a function of $\vec{q}$.

Neutron form factor data (at more than one value of $\vec{q}$) for a stoichiometric sample of La_2CuO_4 were not (and still are not) available, a fact that led us to focus initially on the

Electronic Properties of Solids Using Cluster Methods
Edited by T.A. Kaplan and S.D. Mahanti, Plenum Press, New York, 1995

closely related La_2NiO_4, for which there is apparently good data at many different $\vec{q}$ values.[12] The first attempt at calculating the form factor was phenomenological, namely an application[12] of HM theory[8,9] using linear combinations of Hartree-Fock ionic orbitals for Ni and O (Ni^{+2} and constrained O^{-2} orbitals). Assuming the two unpaired electrons in Ni are in an orbital taken as the spherical part of the Ni 3d states (and their spins are parallel), adjustment of the few parameters in the model (the covalent mixing parameters[8,9]), gave results in excellent agreement with experiment, as seen in Fig. 1.[12] This agreement suggested strongly that the large deviations of the form factor from the free ion function, in particular the large reduction and flattening at small $\vec{q}$, were due to covalence. However, the excellent agreement was spoiled when the model was made more realistic by putting the unpaired electrons in the two (orthogonal) e_g states expected to be occupied in the presence of the crystal field![12,13] (The former procedure, i.e. the one that gave good agreement with experiment, is of course incorrect: it violates the Pauli exclusion principle.)

At this point we decided to consider *ab initio* calculations of the form factor. A calculation[12] of the form factor of La_2NiO_4 based on band theory in the local density approximation (LDA) lead to spin density values far smaller than observed, and for La_2CuO_4 this type of error is even worse (*no* magnetic ordering is found).[14] We then looked into the well-known and much-studied antiferromagnets $KNiF_3$ and NiO, for which experimental results exist.[15,16] Although several *ab initio* cluster calculations have been carried out for $KNiF_3$[17,18] and NiO,[19,20] the interest there was in excitation energies rather than the ground state wave function, which is necessary to determine the form factor. The only cluster calculation of the form factor was in $KNiF_3$[17] more than 25 years ago, and was considered by the authors as too crude to even compare with experiment. Other attempts to get the theoretical form factor from the free ion Ni^{+2} including orbital effects gave poor agreement with experiment.[21] This situation motivated us to first carry out cluster calculations on these well-known materials using the techniques of *ab initio* quantum chemistry, which have developed rapidly during the past two decades. We followed a standard procedure, namely the basic cluster was chosen to contain one Ni^{+2} ion and the 6 nearest ligands (F^- or O^{-2}), and the rest of the lattice was treated using the point charge model. Corrections to the point charge model in the form of limited Pauli repulsion were also considered.

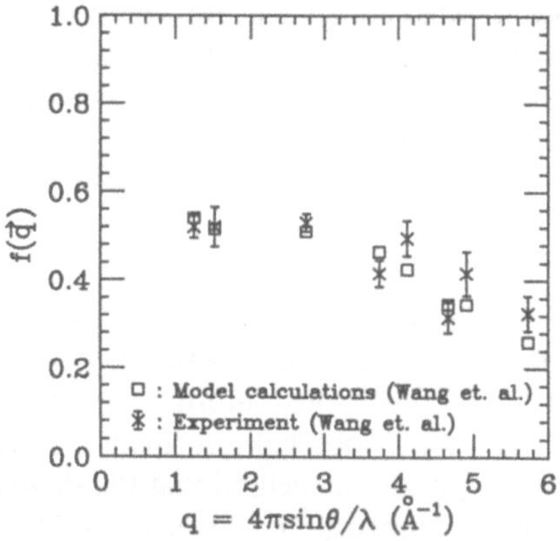

Figure 1. Form factor for La_2NiO_4. Experiment X ; phenomenological theory □ in [12]. Plotted are the values of $m(\vec{q}_a)$ divided by 2.

Our calculations on $KNiF_3$, NiO and La_2NiO_4, which were recently reported,[22] are reviewed in this paper, and we present our more recent work on the copper compounds, La_2CuO_4 and $YBa_2Cu_3O_6$ (YBCO). Regarding the results for the first three materials: we obtain excellent agreement with experiment for $KNiF_3$ and NiO, but the identical method applied to La_2NiO_4 leads to serious *dis*agreement with experiment. This suggests that another physical mechanism is at work in the latter case; indeed, theoretical arguments are given that a particular new mechanism, which leads to a novel type of magnetic ordering, may be at the root of the discrepancy. The shapes of the form factors for the cuprates agree rather well with the experiments; but the overall scale factor appears to be seriously wrong, a problem we argue is probably experimental in origin.

Section II contains an account of the background needed to interpret the theoretical results vis à vis the Bragg scattering experiments, including our calculation[2] of the spin density in the strong coupling Hubbard model. In Sec. III we review our results on the Ni compounds, $KNiF_3$, NiO, and La_2NiO_4, obtained in the cluster Hartree-Fock approximation. The similarly obtained results on La_2CuO_4 and $YBa_2Cu_3O_6$ are presented in Sec. IV. Correlation corrections are discussed in Sec. V. In Sec. VI we discuss the new mechanism, single-site singlet-triplet mixing, and its resulting novel kind of spin ordering which we argue might account for the major discrepancy between theory and experiment for La_2NiO_4. Section VII is a discussion. In the Appendix an interesting failure of Mulliken charge population analysis is noted and explained.

II. THEORY OF SPIN DENSITY INCLUDING COVALENCE AND QUANTUM SPIN FLUCTUATIONS

We review here the considerations of Ref. 2 which were motivated by the following situation. Prior to that work, theories of quantum or zero-point spin fluctuations[2,23,24] were based on the Heisenberg or other spin models, and therefore contained no covalence effects on the AF ordered moment; while the theory of covalence effects on the ordered moment[8,9] were based on a Hartree-Fock theory which contained no quantum spin fluctuations (QSF). The effects, QSF and covalence, are conceptually different. Covalence is a *local* effect: it can occur in one molecule with e.g. a magnetic ion bonded to diamagnetic ions; whereas, the very existence of QSF in an antiferromagnet requires long range order and therefore consideration of *macroscopic* matter. To unify these conceptually disparate effects, we considered the Hubbard model[25] in the strong-coupling regime. The model Hamiltonian is

$$H = \sum_{i,j,\sigma} t_{ij} a_{i\sigma}^{+} a_{j\sigma} + U \sum_{i} n_{i\uparrow} n_{i\downarrow} \; ; \tag{1}$$

where $a_{i\sigma}^{+}$ creates an electron in the (one-particle) state $w_{i\sigma} = w_i(\vec{r})\alpha_\sigma$ and $n_{i\sigma} = a_{i\sigma}^{+} a_{i\sigma}$, $w_i(\vec{r}) = w(\vec{r} - \vec{R}_i)$ is the (normalized) Wannier function at site $\vec{R}_i$ and α_σ is the spin state ($\sigma = \pm 1$ or $\uparrow,\downarrow$). The w_i have the site symmetry of the lattice and are orthonormal. We assume the usual conditions, with the hopping parameter $t_{ij} = t = t^*$ for i and j nearest neighbors, zero otherwise, and the on-site Coulomb interaction is repulsive: $U > 0$. The number of electrons is one per site (half-filled band). We take the external magnetic field to be zero, except insofar as an infinitesimal staggered field is understood for purposes of spontaneous symmetry breaking.[26] Strong coupling means small $|t/U|$.

We calculated the expectation value of the spin density in Ψ, the ground state of (1). The Fourier transform per site of the spin density is

$$s(\vec{q}) = \frac{1}{2N}\sum_{ij\sigma} f_{ij}(\vec{q})\sigma\, a_{i\sigma}{}^{+} a_{j\sigma}\,, \tag{2}$$

where $f_{ij}(\vec{q}) = \int \exp(i\vec{q}\cdot\vec{r})w_i(\vec{r})w_j(\vec{r})d^3r$ and N is the number of sites in the crystal.

This approach differs slightly from (and is simpler than) that of Ref. 1, where we calculated the correlation function $< s(\vec{q})s(-\vec{q}) >$. For the broken symmetry state, one expects that $< s(\vec{q})s(-\vec{q}) > = |< s(\vec{q}) >|^2$ in the thermodynamic limit, for $\vec{q}$ corresponding to a Bragg peak, leading to the same result for the Bragg scattering (see Ref. 26 for discussion).

The treatment of (1) when the hopping parameter is small compared to U and there is a 1/2-filled band, leading to the Heisenberg Hamiltonian, is well known.[27] Essentially the same approach with some formal difference is the following. Writing (1) as

$$H = H_o + V \tag{3}$$

with the unperturbed Hamiltonian H_o chosen as the interaction term proportional to U in (1), we see that the lowest unperturbed level is 2^N - fold degenerate with eigenvalue zero. The ground state of H can be written

$$\Psi = P\Psi + Q\Psi \tag{4}$$

where $P = 1 - Q$ is the projection operator which projects any state on to the ground state manifold of H_o. Then the *effective Hamiltonian* is

$$H_{eff} = PHP - PHQ\frac{1}{Q(H-E)Q}QHP, \tag{5}$$

where E is the eigenvalue of Ψ, can be seen directly to have $P\Psi$ as an eigenstate with eigenvalue E. (Here $Q(H\text{-}E)Q$ for E the lowest eigenvalue of H, is assumed to have an inverse, Löwdin's *inverse of the corner*.[28]) Since $PHQ = PVQ$ and $PHP = 0$, this becomes for small V

$$H_{eff} = -PVQ\frac{1}{H_0}QVP + O(V^3). \tag{6}$$

One readily sees that for (1) this gives

$$H_{eff} = \sum J_{ij}\vec{S}_i\cdot\vec{S}_j + O(t^4/U^3), \tag{7}$$

where $J_{ij} = 2t_{ij}^2/U$ is the intersite exchange interaction and $\vec{S}_j$ is the spin operator for site j. (Here $s_j^z = \frac{1}{2}(n_{j\uparrow} - n_{j\downarrow})$, and $s_j^x + is_j^y = a_{j\uparrow}{}^{+}a_{j\downarrow}$.) The leading term in (7) is the well-known Heisenberg Hamiltonian. The next term contains four-spin terms and has been discussed in detail.[27] Concerning the wave function,[1,2] $P\Psi$ is the lowest eigenfunction of H_{eff}, taken as normalized,

$$P\Psi = \mathcal{A}\Pi_j w_j(\vec{r}_j)\Phi(s_1, s_2, \ldots) \tag{8}$$

and $\vec{r}_j, s_j$ are the space and spin coordinates of the jth electron, $\mathcal{A}$ is the antisymmetrizer, and Φ is the ground state of the spin Hamiltonian (7). To leading order

$$Q\Psi = -\frac{1}{U}QVP\Psi. \tag{9}$$

Using Eqs. (8) and (9) we calculated $< a_{i\sigma}{}^{+} a_{j\sigma} >$, and thus the average of the spin density (2). We assume that the spin ordering is antiferromagnetic. To leading order

$$< s(\vec{q}) >=< S_z >_{Heis} f(\vec{q}) F(\vec{q}), \tag{10}$$

where $< S_z >_{Heis}$ is the expectation value of the spin at a site (an up site) in the symmetry broken ground state of the Heisenberg model in (7). The form factor $f(\vec{q})$ is the Fourier transform of $w(\vec{r})^2$, and $F(\vec{q}) = N^{-1}\sum \exp[i(\vec{q}-\vec{q}_1)\cdot\vec{R}_j]$ is the geometric factor which gives Bragg peaks at the antiferromagnetic wave vectors $\vec{q}_a$, any one of which is $\vec{q}_1$. Note that the normalization of the Wannier functions implies $f(0) = 1$. We also calculated the first correction to (10) [which is $O(t^2/U^2)$], and requires consideration of the corrections in (7)) and showed that with parameters t, U similar to those expected for La_2CuO_4, this correction is negligible.[2] We expect this to be valid for all the materials discussed here. At the Bragg peaks (10) simplifies to

$$< s(\vec{q}_a) >=< S_z >_{Heis} f(\vec{q}_a) \; . \tag{11}$$

Here $< S_z >_{Heis}$ contains the quantum spin fluctuations, and the covalence effects appear in $f(\vec{q})$. The way the covalence effects enter is that the Wannier function underlying the Hubbard model, when applied to a material like La_2CuO_4 or NiO, is chosen as a linear combination of a Cu d-state and the near neighbor O p-states. Then the d-p hybridization, i.e. the covalent mixing,[8] appears in the form factor.

Equation (11) is the result needed to understand the Bragg scattering experiments, which measure essentially

$$m(\vec{q}_a) = g< s(\vec{q}_a)> \; = m_{Heis} \; f(\vec{q}_a) \qquad (\text{in } \mu_B), \tag{12}$$

where g is the g-factor and $m_{Heis} = g < S_z >_{Heis}$. In the approximation which neglects small orbital contributions, the experiments measure precisely $g< s(\vec{q}_a) >$.[9] The straightforward way of applying (12) to measurements is to calculate $f(\vec{q})$ and determine the mean Heisenberg moment as the scale factor needed to agree with experiment, or more simply to divide the measured $m(\vec{q}_a)$ by $f(\vec{q}_a)$ for some reflection. Of course, to do this, one's calculated $f(\vec{q})$ must agree in *shape* with experiment.

A formula like (12) has appeared in the neutron diffraction literature for many years prior to our calculation,[2] often written $m(q) = \mu f(q)$. But neither of the symbols, μ, $f(q)$ were precisely defined; and this lack of definition led to errors. For example, the value of μ obtained by dividing the experimental value of $m(\vec{q}_a)$ at a particular reflection by the corresponding value of $f(\vec{q}_a)$ was thought to contain the covalence effect;[3] this is clearly incorrect in view of (12). (This error is in addition to the use of an inappropriate value of $f(\vec{q}_a)$, already mentioned.)

To generalize the above treatment with one electron per unit cell, i.e. per *cluster*, to cases of interest where there are many electrons per unit cell, we proceed as follows. We first note that the spin density as a function of $\vec{r}$ represented by (10) is

$$S(\vec{r}) =< S_z >_{Heis} \sum_j w(\vec{r}-\vec{R}_j)^2 \exp(i\vec{q}_1\cdot\vec{R}_j) \, . \tag{13}$$

If we had considered the mean field approximation to the Heisenberg Hamiltonian, $< S_z >_{Heis}$ would have been replaced by 1/2 with a corresponding spin density

$$S_{mf}(\vec{r}) = \sum s_{cl}(\vec{r} - \vec{R}_j)\exp(i\vec{q}_1 \cdot \vec{R}_j), \tag{14}$$

where $s_{cl}(\vec{r}) = S_z w(\vec{r})^2$ and S_z is 1/2 in this case. This suggests the following *ansatz*. Given a cluster spin density associated with one magnetic ion,

$$s_{cl}(\vec{r}) = S_z u(\vec{r}) \tag{15}$$

where S_z is the z-component of spin, which is a good quantum number for our cluster models (except for Sec. VI), so $\int u(\vec{r})d^3r = 1$, then the spin density for the antiferromagnetic crystal is

$$S(\vec{r}) =< S_z >_{Heis} \sum u(\vec{r} - \vec{R}_j)\exp(i\vec{q}_1 \cdot \vec{R}_j). \tag{16}$$

where $\vec{R}_j$ locates the jth magnetic ion. Then Eq. (10) holds with $f(\vec{q})$ now given by

$$f(\vec{q}) = \int \exp(i\vec{q} \cdot \vec{r})u(\vec{r})d^3r. \tag{17}$$

Note that normalization of $u(\vec{r})$ implies $f(0) = 1$. This procedure can be seen to be justified in the case of a lattice of magnetic ions only, where Eq. (1) is generalized to more than one electron per ion with a general localized basis set, including intra-ionic electron-electron interactions. In the case where diamagnetic ions are present, we invoke the successful mapping of the 3-band Hubbard model for the CuO_2 planes in La_2CuO_4 on to the 1-band model with appropriate *t*, *U* and $w(\vec{r})$,[29] to give at least a plausibility to our *ansatz*. Comparison with experiment will be, of course, an essential test.

III. FORM FACTORS FOR $KNiF_3$, NiO AND La_2NiO_4 IN THE CLUSTER HARTREE-FOCK APPROXIMATION

In this section we review our Hartree-Fock cluster calculations and comparisons with experiment for these three nickel compounds.[22] The clusters involve a Ni ion surrounded by six ligands. In the cubic materials, $KNiF_3$ and NiO, the ligands are arranged octahedrally, whereas in (tetragonal) La_2NiO_4 the distance from the Ni to the two apical oxygens (which lie along the c-axis containing the Ni) is slightly larger than that to the four oxygens in the ab plane. The cluster formulae are $(NiF_6)^{-4}$ for the fluoride, and $(NiO_6)^{-10}$ for the oxides. We consider the high-spin state (S=1).

We have carried out restricted and unrestricted Hartree-Fock calculations, RHF and UHF respectively, on the clusters. RHF and UHF self-consistent field (SCF) calculations were performed with the COLUMBUS code[30] and the GAUSSIAN92 code,[31] respectively, using contracted Gaussian basis sets. Huzinaga[32] basis sets (9s6p) with an additional diffuse p function are used for F and O. The basis set for Ni is Wachters' (13s9p5d).[33] The basis set (14s11p6d) for Ni with additional diffuse functions suggested by Hay[34] was also used to see how sensitive the calculations are to adding the diffuse functions. These two basis sets, (14s11p6d) and (13s9p5d) give the same form factor and charge density (to within less than 1.5%); so we present the form factor only with the basis (13s9p5d) of Ni (see below for further discussion of this point). All the electrons of the cluster (numbering 86) are explicitly included in these *ab initio* calculations, and the rest of the crystal is treated in the point charge model to give a Madelung potential to the cluster. Formal ionic charges (e.g., 2, -1, and 1 electronic charges for Ni^{+2}, F^- and K^+ respectively) were assigned to the atomic positions for the rest of the system and fractional charge values on the boundary were taken to make the whole system charge neutral according to Evjen's method.[35] We took 482 point charges to obtain the Madelung potential for NiF_6^{-4} in $KNiF_3$, and 722 point charges for NiO_6^{-10} in NiO, after establishing a reasonable convergence in the potential at the center of the cluster (variation < 0.2%). This number of point charges in Evjen's procedure is known to give an almost constant difference from the exact Madelung sum even for the region far from the center of the cluster for the perovskite and fcc structures corresponding to $KNiF_3$ and NiO, respectively.[36]

For a more realistic environment in $KNiF_3$, the point charges which had originally represented the eight nearest K^+'s were replaced by effective core potentials (ECP's).[37] This incorporated Pauli repulsion particularly between electrons in the F^-'s of the cluster and the electrons in the neighboring K^+'s. The effect of this replacement on the form factor of $KNiF_3$ was found to be negligible, and so we did not include ECP's for NiO. Also, we argue in Sec. V. that correlation corrections to the spin density are very small.

The spin density for the cluster,

$$s_{cl}(\vec{r}) = S_z \frac{1}{n} \sum_j^{occ} \sigma_j |\psi_j(\vec{r})|^2 \tag{18}$$

is determined from the results of the HF cluster calculations. In (18), $\psi_j(\vec{r})$ is the jth occupied molecular orbital whose spin is σ_j (= +1 or -1 for up or down spin), $n = n_\uparrow - n_\downarrow$ [$n_\uparrow (n_\downarrow)$ is the number of spin-up (down) electrons]. S_z is the cluster value of the total z-component of spin, namely 1 for the Ni clusters (we found the high-spin state to be of lowest energy). As discussed in Sec. II, when we generate the antiferromagnetic crystal spin density by propagating our cluster result according to (14), we must replace S_z by $< S_z >_{Heis}$ in order to go beyond the mean field approximation to the Heisenberg Hamiltonian to include QSF. Thus we are led to (12) with (15) and (17) to compare with the analogous quantity obtained experimentally.

In our SCF theory, each molecular orbital is a linear combination of atomic centered basis functions,

$$\psi(\vec{r}) = \sum_k \chi_k{}^{Ni}(\vec{r}) + \sum_{j,l} \chi_l{}^{L_j}(\vec{r}) \tag{19}$$

where $\chi_k{}^{Ni}$ is a basis function centered at the nickel, and $\chi_l{}^{L_j}$ is a basis function centered at the jth ligand. For simplicity we omit the label on the molecular orbital, and include the MO coefficient in each basis function. Then

$$|\psi|^2 = \sum_{k,k'} \chi_k{}^{Ni} \chi_{k'}{}^{Ni} + 2\sum_{k,l,j} \chi_k{}^{Ni} \chi_l{}^{L_j} + \sum_{j,l} \sum_{j',l' \neq j,l} \chi_l{}^{L_j} \chi_{l'}{}^{L_{j'}} + \sum_{j,l} (\chi_l{}^{L_j})^2 . \tag{20}$$

When the jth ligand is shared by two symmetrically placed Ni's with antiparallel spins, the contributions to $f(\vec{q})$ from the last term in (20) vanish at $\vec{q} = \vec{q}_a$. This leads to the covalent reduction in the form factor.[8,9] Also the Ni-ligand cross terms lead to a flattening of the form factor in the small-q region.[8,9]

We now present our results and compare with experiment. An interesting failure of Mulliken charge population analysis is discussed in the Appendix.

$KNiF_3$

In Figs. 2a and 2b we compare the RHF and UHF results with the experimental values of Hutchings and Guggenheim.[15] The experimental values of $<s(\vec{q}_a)>=<S_z>_{Heis}\, f(\vec{q}_a)$ are determined from the values of $g<S_z>_{Heis}\, f(\vec{q}_a)/F_{200}$, which include no unknowns and are extracted from the ratios of magnetic and nuclear Bragg scattering intensities at 4.2 K in Ref. 15. Here we took the g-factor as g=2.29 (±0.02) and the nuclear scattering length as F_{200} = 1.218 (±0.020) from the same reference.

The calculated form factor values are multiplied by the factor $<S_z>_{Heis}$ to fit the experimental data. The UHF results, which incorporate core polarization (including the nominally paired t_{2g} electrons), in Fig. 1(b) differ slightly from the RHF results in Fig. 1(a), but this small difference leads to near perfect agreement between the UHF results and the experiment. The best fit to the experiment in Fig. 1(b) gives $<S_z>_{Heis}$ = 0.90 which can be directly compared with the result of spin wave theory for the simple cubic lattice $<S_z>_{Heis}\{\text{spinwave}\}= 0.92$.[7]

We conclude that the theoretical results for the magnetic moment density Eq. (12) in terms of covalence and quantum spin fluctuations, agree well with the experimental data in $KNiF_3$.

NiO

The theoretical AF form factor values from UHF are compared with Alperin's experimental values[16] in Fig. 3. The absolute values of $<s(\vec{q}_a)>$ are not available for NiO[38] so we do not obtain a value of $<S_z>_{Heis}$, the interest here being in the *shape* of $m(\vec{q}_a)$. We scaled the experimental form factor values (by 0.93) to give the best fit to our UHF results, particularly for the small *q* region. The Bragg scattering data in NiO extend to a larger region of *q* than those in $KNiF_3$, allowing a test of the calculated form factor shape. In Fig. 3a we compare our UHF results with the scaled experimental values. Apparently, the UHF results agree very well with the experiment for the first three Bragg peaks and are consistently lower than the experiment for larger *q* values. However, the bumpiness of the data, which results from asphericity of the spin density around each Ni, is traced rather well by our theoretical calculations. The overall agreement between UHF results and experiment in Fig. 3a is reasonable.

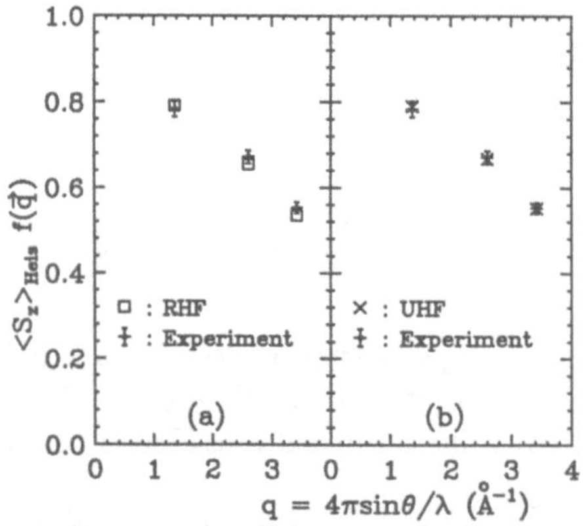

Figure 2. Fourier transform of spin density, $\langle s(\vec{q}_a)\rangle = \langle S_z\rangle_{Heis} f(\vec{q}_a)$, for $KNiF_3$.

A correction to the form factor important in the larger-q region comes from the orbital contribution. We took this contribution for NiO from the work of Khan *et al.* who gave a spherically averaged estimate based on the ionic Ni^{+2} wave function.[39] This orbital contribution is negligible in the small-q region, so we do not expect it to be important in our comparison with $KNiF_3$ experiments, where the data are available only for small q. For larger q, the orbital contribution in NiO helps to give a better fit with experiment, as shown in Fig. 3b. The small discrepancy seen between the calculated and the experimental values might come from the error involved in the spherical averaged estimation of the orbital contribution. With the inclusion of the orbital contribution to the form factor, we conclude that our calculated results agree very well with experiment.

La_2NiO_4

We carried out similar cluster calculations for this material. In this case the ECP for the nearest La^{+3} ions[40] produced a not quite negligible effect and so is included in the results presented. We used 544 point charges for La_2NiO_4; this is the same as PC333 in Table 1 of Ref. 41, and the deviation from the exact Madelung sum is listed in the same table. We noted that this number of point charges for La_2NiO_4 does not give a constant difference from the Madelung sum, however such variation did not appreciably affect the form factor. Increasing the number of point charges, from PC333 to PC553 in the notation of Martin's paper,[41] hardly changed the form factor.[42]

The theoretical AF form factor results from RHF and UHF are compared with the experimental values of Wang *et al.*[12] in Fig.4. Both the RHF and UHF results are seen to disagree seriously with the experiment. Especially, the experimentally observed plateau at small q values is not reproduced in either calculation. The spin fluctuations provide only a scale factor, so this can not help reproduce the shape of the form factor. The plateau at small q in the measured form factor was seen not only in La_2NiO_4, but also in La_2CuO_4.[43] This plus the success of the first attempt by Wang *et al.*[12] to fit the data with a HM type of theory (see Sec. I), among other arguments,[2] seemed to indicate that the flattening in both cases was due to covalence. However, our cluster calculations do not show such an effect in the nickelate. In view of the great success of the same type of calculations for $KNiF_3$ and

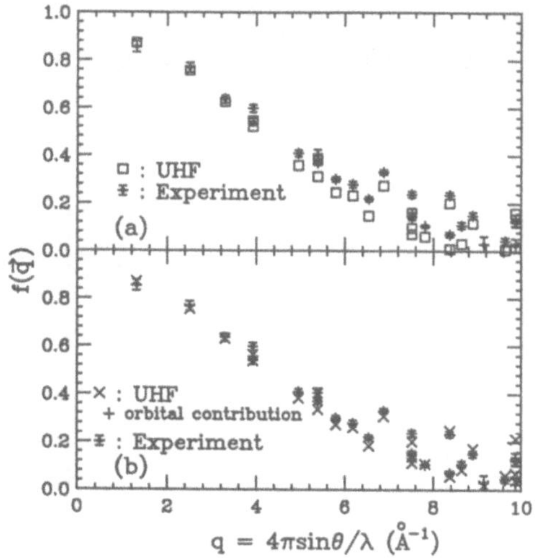

Figure 3. Form factor for NiO (a) UHF,(b) UHF with the corrections from orbital contributions.[39] The experimental values are scaled to give the best fit to the theoretical form factor.

NiO, we conclude that covalence is probably *not* the cause of the dramatic departure of the form factor from free-ion behavior in the case of La_2NiO_4. That is, our cluster calculations for these three Ni compounds suggest that different physics is occurring in the La nickelate as compared to the Ni fluoride and NiO. That such physics might be expected, due essentially to the tetragonal nature of the Ni surroundings in the La nickelate as compared to cubic surroundings in the other two compounds, is discussed in Sec. VI.

IV. FORM FACTOR OF La_2CuO_4 AND $YBa_2Cu_3O_6$ IN THE CLUSTER HARTREE-FOCK APPROXIMATION

Here we discuss the results of HF cluster calculations and comparisons with experiment for the two Cu systems La_2CuO_4 and $YBa_2Cu_3O_6$. The former is isostructural to La_2NiO_4, so the cluster calculations are very similar. For La_2CuO_4, we took a $(CuO_6)^{10-}$ cluster with 544 point charges to calculate the Madelung potential and included the ECP associated with the nearest 8 La^{3+}'s. The structure of $YBa_2Cu_3O_6$, on the other hand is quite different from that of La_2CuO_4 and needs special consideration. It has two CuO_2 planes separated by a Y^{3+} layer. Thus the local oxygen coordination of a Cu^{2+} ion is 5 rather than

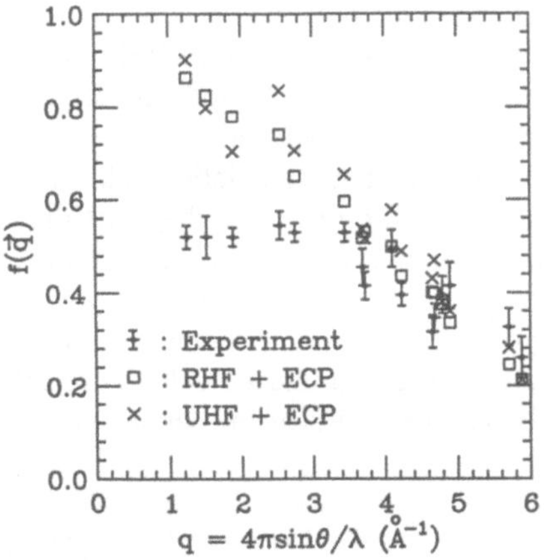

Figure 4. Form factor for La_2NiO_4.

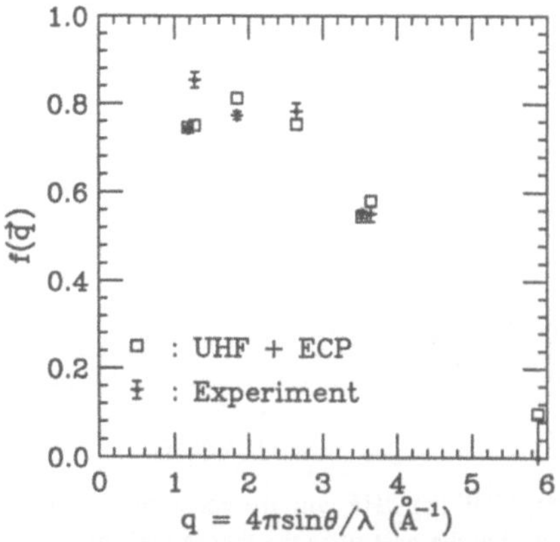

Figure 5. Form factor for La_2CuO_4. The experimental values are scaled to give the best fit to the theoretical values.

6, leading us to consider a $(CuO_5)^{8-}$ cluster. The effect of the crystalline environment has been included through the Madelung potential of 508 point charges and the ECP of 4 nearby Y^{3+}'s and 4 Ba^{2+}'s.[40] Because of the lack of inversion symmetry about the Cu ion, the form factor of the $(CuO_5)^{8-}$ cluster is complex. However, after including the structure factor associated with the two CuO_2 planes per unit cell, the final form factor turns out to be real, consistent with the presence of a center of inversion symmetry.

La_2CuO_4

In Fig.5 we present the UHF results for $f(\vec{q}_a)$ and the experimental values obtained by Freltoft *et al.*[43] The experimental values have been scaled by a factor 0.9 to give the best fit to the theory. Since the orbital contributions were found to be small in the case of Ni compounds, we have omitted them in the present calculation.

We note that the *q*-dependence of the form factor is in much better agreement with experiment in La_2CuO_4 than in La_2NiO_4. This gives some support to our suggestion that a novel mechanism might be operative in the latter case, because the mechanism for the novelty described in Sec. VI holds for two holes in the d-shell, but not one.

If we divide the measured absolute value of $m(\vec{q}_a)$ by our calculated form factor, at (100), we obtain $g\langle S_z\rangle_{Heis}$ = 0.32. This should be compared with 0.66 (g=2.2 for the Cu^{2+} ion) obtained from the spin-wave theory.[7] Possible sources of this large discrepancy will be discussed at the end of this section.

$YBa_2Cu_3O_6$

Fig.6 compares the theoretical UHF form factor $f(\vec{q}_a)$ and the experiment of Shamoto *et al.*;[44] the latter has been scaled to fit the overall form factor. We see that the shape of the calculated form factor agrees with the experiment except at a few points which have large error bars. From the above scale factor we obtain $g\langle S_z\rangle_{Heis}$ = 0.53. As in La_2CuO_4, this Heisenberg moment is appreciably smaller than that obtained[7] in the spin wave approximation to the 2-dimensional Heisenberg Hamiltonian.

We believe the discrepancies in these copper compounds are most likely due to problems in the samples measured in the neutron experiments. To explain, consider first the **theory**. The spin wave approximation is most likely a lower limit to the true Heisenberg ordered spin ($\langle S_z\rangle_{Heis}$): Monte Carlo calculations on the 2-dimensional Heisenberg model[23,24] found an ordered spin very close to the spin wave result, on the assumption[26] that the ratio r of the ordered spin in the symmetry-broken ground state to that in the symmetry-unbroken ground state is $\sqrt{3}$. It has recently been shown[45] that $r \geq \sqrt{3}$ for this model. Further it is believed that small interplanar interactions will increase this Heisenberg ordered moment,[46] although only very slightly for La_2CuO_4. The **experimental** situation is less clear. To discuss this we need to note that what is variously referred to[3-6] as the ordered moment, the average ordered moment, the sublattice magnetization per site or the magnetic moment of a Cu site, is well-defined, namely it is the ratio of the measured $m(\vec{q}_a)$ at the smallest-*q* reflection to the corresponding value of the form factor of K_2CuF_4 [see discussion around Eq. (12)]. As such it is a useful number for *comparison* of the various experimental results, despite its being a physically fictitious quantity (see Sec. I). We will refer to this number as the *moment*.

For La_2CuO_4 (214) the *moment* and Nèel temperature quoted, $0.3\mu_B$ and 185K, are both appreciably lower than the maximum values, $0.6\mu_B$ and 300K, found by Yamada *et al.*[5] in one of their samples, which varied in O concentration. Further it was found that the *moment* and T_N increased with each other.[5] Finally, Keimer *et al.* recently reported a Nèel temperature of 325K in a sample of 214 believed to be very close to stoichiometric.[47]

Extrapolation to 325K of the results of Yamada *et al.*[5] suggest an appreciably higher *moment*. The critical number needed is the value of the *moment* in this recent experiment (it was not reported), or alternatively the value of $m(\bar{q}_a)$; in fact, the full form factor measured on this apparently excellent sample[47] should be compared to our calculated results.[48]

Although the *moment* and T_N were found by both groups [3] and [4] to be essentially constant for x<0.2 in $YBa_2Cu_3O_{6+x}$, and therefore one could justify using the x=0.15 sample as representative of the stoichiometric material, there remains a serious question of consistency. Namely the values of the *moment* found by the two groups differ appreciably: 0.64±0.03 μ_B for the single crystals of [4] with x=0 and 0.15, as compared to 0.46±0.05 μ_B and 0.50±0.05 for the x= -0.01 and 0.15 samples of [3]. Also the more recent x=0.15 sample has 0.55 for the *moment*.[44]

When this discrepancy is cleared up and the form factor for a good stoichiometric sample of La_2CuO_4 is published, a direct determination of $<S_z>_{Heis}$ will be possible. We emphasize the importance of this. At present the only reliable theory of ordering in the Heisenberg ground state is the Monte Carlo analysis,[23,24] but it required the assumption that $r=\sqrt{3}$ to obtain $<S_z>_{Heis}$. Although this value is intuitively reasonable, the only proved result is $r\geq\sqrt{3}$, cited above. This question about r is fundamental, going beyond the specific models considered here. Further, such experimental information is needed to address the puzzle discussed in Sec I.[49]

V. INTRACLUSTER CORRELATION EFFECTS

We have performed multiconfiguration SCF (MCSCF) calculations using the COLUMBUS code[30] to investigate correlation effects on the form factor for $KNiF_3$ and NiO.[22] Here, the correlations are achieved by adding configuration state functions[30] found from the SCF reference determinant (open shell RHF) via electron-hole pair excitations. We allowed the d-electrons to correlate among themselves and with p electrons on the ligands. The results of these calculations, which included 126 configurations, indicated that mainly d-d intra-atomic correlation is important. They also showed that correlation corrections to the natural orbitals corresponding to the unpaired electrons (the e_g Ni states) are negligible. The contribution from the nominally paired electrons (e.g. the t_{2g} electrons) is unfortunately not directly accessible from the code used (COLUMBUS)—only the average of the up and down orbitals is available. Efforts to obtain this information are under way.

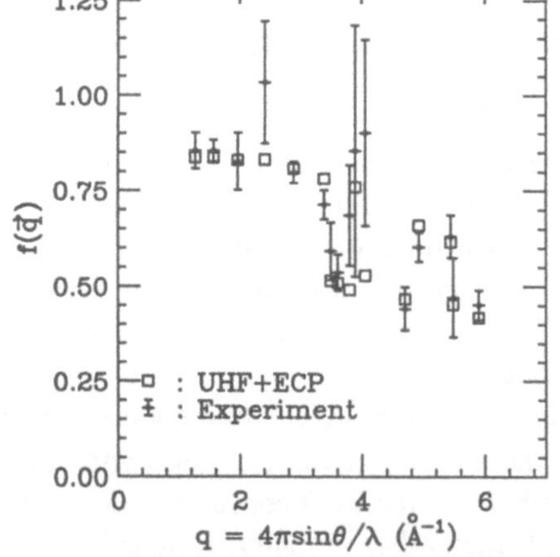

Figure 6. Form factor for $YBa_2Cu_3O_6$. The experimental values are scaled to give the best fit to the theoretical values.

We have also performed correlation calculations where electron-hole excitations from the UHF wave function are carried out, using the GAUSSIAN92 code. That is, corrections to the UHF state are calculated directly. We were able to do this only for a single Ni or Cu ion (for technical reasons). We found that the corrections to the UHF spin density (considered *true correlation effects* by some) are negligible. Since our MCSCF results on the cluster indicated that the most important correlations are intraionic (d-d within the magnetic ion), these single-magnetic-ion results are highly indicative that correlation effects on the spin density are negligible in the full cluster.

VI. NOVEL SPIN ORDERING AND POSSIBLE RELEVANCE TO La_2NiO_4

Given the agreement of our theoretical results with the experimental form factor in the cases $KNiF_3$ and NiO, and the dramatic disagreement for La_2NiO_4 despite the fact that the identical theoretical approach was used in all these cases, suggests that different physics is important in the La nickelate. To our question as to why such a difference might occur, Jan Zaanen gave a very stimulating reply.[50] He noted that because of the tetragonal symmetry of the La compound, the low spin state might be very close in energy to the high spin state. Furthermore, there could be *mixing* between these site-triplet and singlet states, as discussed in a recent article[51] where such mixing is briefly mentioned and attributed to kinetic exchange. We have since been exploring these ideas, and although our considerations are not yet complete, they appear sufficiently promising that we decided to present them here. In addition, the physical consequences of this mechanism are surprisingly novel, namely a new kind of spin ordering.

We first consider the physics of such a state, via a simple model. Let $w_\nu(\vec{r})$, $\nu = 1,2$, be the two e_g orbitals, and consider a two-electron system. Then the triplet state must have both of these occupied, the $S_\zeta = 1$ state having both spins up, for example. In tetragonal symmetry, the one-electron energies are split. With, say $\varepsilon_1 < \varepsilon_2$, the lowest singlet would have $w_1(\vec{r})$ doubly occupied. The reason why this might have energy similar to the triplet is that while it loses the intra-atomic exchange energy, or *Hund's rule energy*, it gains the crystal field splitting $\varepsilon_2 - \varepsilon_1$. A mixture of these states is

$$\Psi = C a_{1\uparrow}^{\ +} a_{2\uparrow}^{\ +} |0> + D a_{1\uparrow}^{\ +} a_{1\downarrow}^{\ +} |0> \tag{21}$$

where |0> is the vacuum, $a_{\nu\sigma}^{\ +}$ creates an electron in $w_\nu(\vec{r})$ with spin σ, and C and D are chosen as real for specificity. Noting that Ψ is a single determinant with $w_{1\uparrow}$ and $Cw_{2\uparrow} + Dw_{1\downarrow}$ occupied, it follows readily that the spin density in Ψ is

$$\vec{s}(\vec{r}) = \hat{\zeta}\frac{C^2}{2}[w_1(\vec{r})^2 + w_2(\vec{r})^2] + \hat{\xi}CDw_1(\vec{r})w_2(\vec{r}), \tag{22}$$

where $\hat{\zeta}, \hat{\xi}$ are orthonormal and $C^2 + D^2 = 1$.

If C=1 we would have the usual picture of the spin density for the high-spin state with spin up. The new, $\hat{\xi}$-component, has a Fourier transform that vanishes at $\vec{q} = 0$ because of the orthogonality of the two e_g states. This gives the possibility that the contribution of the new term to the neutron scattering intensity increases initially with q. Referring to Fig. 5, one sees that that is what is needed to help bring theory into accord with experiment. Covalence effects have a similar qualitative behavior, but as we have seen, they are much too small. In order to see if this promise is born out we need to calculate the neutron intensity for the various Bragg reflections. But to do this one must understand the domain structure, i.e. the degeneracies of the spin ordering.

On the other hand, the spin ordering described by (22), assuming it propagates in a simple antiferromagnetic pattern from site to site (with local spin density alternating in sign), is *highly unusual.* It differs from all previous known magnetically ordered states, in that the spin *direction* varies appreciably *within each magnetic ion.* Even in spiral or helical ordering,[52-54] where the spin density can vary appreciably in direction from ion to ion, the variation within a given ion is negligible. This requires a reanalysis of the neutron data, which traditionally assumes essentially one spin direction for a given ion.

To gain this understanding of the domain structure, plus some insight as to a microscopic mechanism, we have considered a simple, intuitive model, namely an extension of the usual one-band Hubbard model (1) to two bands, including interatomic as well as intraatomic exchange, as follows. As in (3), $H = H_o + V$, where now

$$H_0 = \sum_i \left[\sum_{\nu=1}^{2} \varepsilon_\nu n_{i\nu} + \frac{U}{2} n_i (n_i - 1) - 2j \left(\vec{S}_{i1} \cdot \vec{S}_{i2} + \frac{1}{4} n_{i1} n_{i2} \right) \right] \tag{23}$$

and

$$V = \sum_{i,j}{}' \sum_{\nu,\mu} \left[\sum_\sigma t_{i\nu,j\mu} a_{i\nu\sigma}{}^{+} a_{j\mu\sigma} - 2x_{i\nu,j\mu} \left(\vec{S}_{i\nu} \cdot \vec{S}_{j\mu} + \frac{1}{4} n_{i\nu} n_{j\mu} \right) \right], \tag{24}$$

where $n_{i\nu} = \sum n_{i\nu\sigma}$, and $n_i = \sum n_{i\nu}$, the number of electrons at site i. We assume the one-electron basis functions $w_i(\vec{r})\alpha_\sigma$ are real. The first term of H_0, involving the ε_ν, would be a constant (proportional to the total number of electrons) if the ε_ν were equal; the difference $\varepsilon_2 - \varepsilon_1$, the crystal field splitting, represents the difference in ionization energies of the two electrons in the triplet state. The second U-term is the extension of the usual on-site Coulomb repulsion term: its form, as well as that of the exchange terms (proportional to j and $x_{i\nu,j\mu}$), comes about as follows. Consider the 2-band representation of the electron-electron Coulomb interactions

$$\frac{1}{2} \sum_{i\nu,j\mu} \sum_{l\lambda,m\kappa} \sum_{\sigma,\sigma'} v_{i\nu,j\mu;l\lambda,m\kappa} a_{i\nu\sigma}{}^{+} a_{j\mu\sigma'}{}^{+} a_{m\kappa\sigma'} a_{l\lambda\sigma}\,, \tag{25}$$

where

$$v_{abcd} = \int d^3 r_1 d^3 r_2 w_a(\vec{r}_1) w_b(\vec{r}_2) \frac{e^2}{r_{12}} w_c(\vec{r}_1) w_d(\vec{r}_2).$$

The U terms are those involving $v_{i\nu,i\mu;i\nu,i\mu}$, and assumes these three integrals are equal. The exchange terms involve $v_{i\nu,j\mu;j\mu,i\nu}$ where $j = v_{i1,i2;i2,i1}$, and $x_{i\nu,j\mu} = v_{i\nu,j\mu;j\mu,i\nu}$. The form of the exchange terms comes from the relation

$$\sum_{\sigma,\sigma'} a_{p\sigma}{}^{+} a_{p\sigma'} a_{q\sigma'}{}^{+} a_{q\sigma} = \frac{1}{2} n_p n_q + 2\vec{S}_p \cdot \vec{S}_q \tag{26}$$

where p or q each refer to the double index $i\upsilon$. In the usual triplet case the $n_p n_q$ term is constant and therefore is dropped; in the present case this term must be kept. The hopping terms, involving t_{pq} , are of the usual Hubbard form; we imagine the t_{pq} to be of a similar order of magnitude as that in the one-band model of the La cuprate, modified by proportionality to the corresponding overlap integrals, and that these hopping parameters are small compared to the intra-atomic energies (those in H_0). As motivated above we assume degeneracy of the site-triplet and singlet, which requires

$$\varepsilon_2 - \varepsilon_1 = j. \tag{27}$$

Finally we assume the total number of electrons $\sum n_i = 2N$ (each band is half filled).

To calculate properties of the microscopic model defined by (23), (24) and (27), we first calculate the effective Hamiltonian (5) treating V as a perturbation. The degeneracy of the unperturbed ground state is 4^N (as compared to 3^N when each site is high spin). Writing the hopping and the exchange terms in (24) as V_1 and V_2, respectively, (5) becomes to 2nd order,

$$H_{eff} = PV_2P - PV_1Q\frac{1}{Q(H_0 - E_0)Q}QV_1P, \tag{28}$$

on dropping the constant $E_0 = PH_0P$. This is of course a 4^N by 4^N matrix. Insertion of V_1 leads to considerable complexity, particularly because the intermediate energies depend on the particular initial state being considered (unlike the triplet-only case). Because we are only interested in qualitative results and because the various intermediate energies are of the form $U+\delta$ where δ involves $\varepsilon_2 - \varepsilon_1$ and j, which are expected to be rather smaller than U, we consider the large-U approximation, $Q(H_o-E_o)Q \cong U$. We then obtain

$$H_{eff} = -P\sum_{\langle i,j\rangle}\left\{\sum_{\nu,\mu}\frac{t_{\nu\mu}^2}{U}\left(n_{i\nu}+n_{j\mu}\right)+2\sum_{\nu,\mu}J_{\nu\mu}\left(\frac{n_{i\nu}n_{j\mu}}{4}+\vec{S}_{i\nu}\cdot\vec{S}_{j\mu}\right)\right.$$

$$\left.+\frac{2}{U}t_{12}^2\sum_{\sigma,\sigma'}\left(a_{i1\sigma}{}^{+}a_{i2\sigma'}\,a_{j1\sigma'}{}^{+}a_{j2\sigma}+1\leftrightarrow 2\right)\right\}P \tag{29}$$

We have assumed nearest-neighbor hopping only, as indicated by the sum <i,j> which means to sum over nearest-neighbor pairs, and that $t_{i1,j2}=t_{i2,j1}=t_{12}$, appropriate to the Ni's lying on a square lattice (mimicking a NiO_2 plane in La_2NiO_4, which is what we have in mind). We have put $t_{i\nu,j\mu}=t_{\nu\mu}$ and $x_{i\nu,j\mu}=x_{\nu\mu}$. Also

$$J_{\nu\mu} = (2t_{\nu\mu}^2/U) - x_{\nu\mu} \tag{30}$$

For simplicity we have omitted a set of terms which in general would enter into (29), but which will not contribute in the variational calculation described below.

For the triplet-only case, all the $n_{i\upsilon} = 1$, and the last term in (29) vanishes. Also in general all matrix elements of $\vec{S}_{i\nu}$ are independent of υ. Hence, dropping an additive constant, (29) becomes the familiar nearest-neighbor Heisenberg Hamiltonian

$$H_{Heis} = 2J\sum_{<i,j>}\vec{S}_i\cdot\vec{S}_j, \tag{31}$$

where

$$J = \frac{1}{4}\sum_{\nu,\mu}\left(\frac{2t_{\nu\mu}^2}{U} - x_{\nu\mu}\right), \tag{32}$$

and

$$\vec{S}_i = \sum_{\nu}\vec{S}_{i\nu} \tag{33}$$

is the spin operator for site i. We used the fact that all matrix elements of $\vec{S}_{i\nu}$ in the *P-space* are independent of υ. The interatomic exchange terms $x_{i\upsilon,\,j\mu}$ are positive, i.e. ferromagnetic. The ferro- and antiferromagnetic terms have been called potential and kinetic exchange respectively.[55] The last term in (29) connects a state where a pair of nearest-neighbor sites are in antiparallel triplets to one where both are in singlets.

We now follow a standard first study of a new type of ordering, namely the mean field approximation to H_{eff}. We look for the best wave function of the form

$$\Psi = \prod_{i \subset A} \psi_i^{\uparrow} \prod_{i \subset B} \psi_i^{\downarrow} |0> . \tag{34}$$

We are considering bipartite lattices (e.g. linear chain, square lattice) where *A, B* refer to the two sublattices. Our aim is to show that within this mean field approximation, it is possible to get a mixed site-triplet-singlet ordered state. Thus we make the simple variational choice

$$\psi_i^{\sigma} = Ca_{i1\sigma}^{+} a_{i2\sigma}^{+} + Da_{i1\uparrow}^{+} a_{i1\downarrow}^{+} \tag{35}$$

for $\sigma = \uparrow, \downarrow$. We also considered the replacement of D by σD, but this led to a higher energy. More general would be to include at each site some of the other components of the triplet. A limited investigation along those lines lead to higher energy.

We find for the energy

$$(\Psi, H_{eff} \Psi) = (NZ/2)\varepsilon , \tag{36}$$

where Z is the number of nearest neighbors of a site, and

$$\varepsilon = -\frac{1}{U} \sum_{\nu,\mu} t_{\nu\mu}^{2} + 2(\frac{t_{11}^{2} + t_{22}^{2}}{U} - x_{11} - x_{12})D^2 + 2x_{12}D^4 . \tag{37}$$

Clearly the minimum ε would be at $D = 0$ without potential exchange in H. The important question is, are there values of the parameters such that $J > 0$, as required for the observed antiferromagnetism, **and** that $0<D^2<1$ minimizes ε? The answer can be seen to be yes; the reasons are that the parameters enter into the coefficient of D^2 and J in essentially different ways, and that the potential- and kinetic-exchange-type of terms, $x_{\nu\mu}$ and $t_{\nu\mu}^2 / U$, are expected to be of the same order of magnitude.[56]

Equation (37) shows that the energy is invariant under $D \to -D$; i.e., there will be domains with both signs of the ξ-component for a given ζ-component of the spin density. This will be used in calculations, in progress, of the neutron cross section.

VII. DISCUSSION

We first consider a theoretical matter. The theoretical values of $f(\vec{q}_a)$ at the four smallest-q reflections in Fig.5 (for 214) lie appreciably below 1; this is an indication of covalence. A rough extrapolation (horizontal) of these points to $q = 0$ would suggest a covalent reduction of the ordered moment of ~ 20%. For more comparative precision we will use the precise definition of the ordered moment given earlier,[8,9,2] namely the value of $m(\vec{q}_a)$ at $\vec{q}_a \equiv 0$ after subtracting the last term of (20). This yields a 25% covalent reduction. This disagrees with our earlier result[2] of a 40% reduction, based on the parameters used in the three-band Hubbard model of a CuO_2 plane and the mapping of that model on to a one-band model.[57] A possible source of this discrepancy is that we needed one parameter in addition to those of the 3-band model, namely the overlap S between ionic Cu-3d and oxygen 2p orbitals; the 40% moment reduction came from the estimate S = 0.1. We have checked that, taking the d and p orbitals in two ways, both giving S ≅ 0.08. These choices are free-ion Cu^{+2} and O^{-} orbitals, and projection from the unpaired MO in our cluster calculation. This change in S only reduces the covalent reduction by 10%. Thus there *is* a disagreement between the 3-band-model parameters derived from LDA calculations[57,58] and our cluster calculations, vis à vis the ground state spin density. We have not considered the effect on our ground state calculations of polarizability of ions outside our cluster; this effect, which might increase the covalence, is under investigation. Form factor measurements on stoichiometric La_2CuO_4 could shed light on this question (in addition to the other question raised in Sec. IV). A sensible scaling factor and perhaps even closer shape agreement would increase confidence in our results.

We conclude with the overall picture of covalence obtained by our cluster calculations. The covalent reduction of the ordered moment we find for $KNiF_3$, NiO, La_2CuO_4 and $YBa_2Cu_3O_6$ is 5%, 9%, 25% and 22%, respectively. These numbers are obtained from theory alone; however we listed them only for those materials where a reasonable fit to the experimental form factor shape was obtained. Thus, La_2NiO_4, for which a plausible alternate physics was described, was omitted.

ACKNOWLEDGMENTS

This work was supported by NSF Grant DMR-9213824 and by the Center for Fundamental Materials Research, Michigan State University.

APPENDIX. FAILURE OF THE MULLIKEN CHARGE POPULATION ANALYSIS

In the UHF calculations of NiO and $KNiF_3$, we found the Mulliken charge values obtained with diffuse basis functions in Ni (14s11p6d) to be quite different from the nominal charge values for the ionic material (see Table 1 of [22]). Particularly for NiO, the discrepancy is very large. These values (-0.21 for Ni and -1.63 for O) for NiO seem to contradict the assumption that NiO is highly ionic, which allows us to assign the nominal point charge values, +2 for Ni and -2 for O, when generating the Madelung potential. Therefore we performed the same cluster UHF calculations without diffuse basis functions in Ni (13s9p5d) to see how sensitive the Mulliken charge population was to the choice of basis functions. The Mulliken charge values without diffuse basis functions (13s9p5d) were found to be +1.75 for Ni and -1.95 for O, much closer to the nominal point charge values. Similar values were obtained by Sahoo, Sulaiman *et al.*[59] in their cluster calculations of $YBa_2Cu_3O_6$ and La_2CuO_4.

Even though the results of the two calculations, with different basis sets, look so different in Mulliken charge analysis, we found that physically meaningful quantities, such as charge and spin density, are essentially the same. Our understanding of how this is possible is the following. The diffuse functions in Ni (14s11p6d) are so diffuse that they spread over the neighboring oxygen sites and can mimic the diffuse function on the oxygens. We estimate that the approximately 1/3 of an electron per O assigned to the diffuse Ni orbitals are physically associated with the oxygen ions.

The Mulliken population values in $KNiF_3$ are rather stable with respect to the choice of basis sets and the calculated values are close to their ionic charges (+2 for Ni^{+2} and -1 for F^{-}) even when we use the diffuse basis functions in Ni (see Table 1 of [22]). This is due to the fact that the F^{-} wave function is much more compact compared with the O^{-2} wave function, so that the diffuse Ni functions are simply not appreciably occupied.

In summary, these results show that the Mulliken charges assigned to different ions (such as Ni, F, O etc.) depend not only on the choice of basis set but also on the type of ions in the cluster. (A similar problem was noted by Noell[60] and Bauschlicher and Bagus[61] for the transition metal complexes.) Therefore it sometimes may be misleading to use these charge assignments in describing physical quantities such as the electrostatic potential. As we just noted, our NiO results give an extreme example: clearly the assignment to Ni of electrons in orbitals centered on Ni but so diffuse that most of their weight is at the Ni-O distance, is not sensible. The Mulliken assignment is much more sensible when the orbitals are not so diffuse. The charge density was in fact found to be essentially the same with or without these diffuse Ni functions, so it is clearly reasonable to prefer the Mulliken charges calculated without diffuse functions.

REFERENCES

1. T. A. Kaplan and S. D. Mahanti, J. Appl. Phys. **69** (8), 5382 (1991).
2. T. A. Kaplan, S. D. Mahanti and Hyunju Chang, Phys. Rev. B **45**, 2565 (1992).
3. J. M. Tranquada, A. H. Moudden, A. I. Goldman, P. Zolliker, D. E. Cox, G. Shirane, S. K. Sinha,, D. Vaknin, D. C. Johnston, M. S. Alvarez, A. J. Jacobson, J. T. Lewandowski and J. M. Newsman, Phys. Rev. B **38**, 2477 (1988).
4. P. Burlet, C. Vettier, M. J. G. M. Jurgens, J. Y. Henry, J. Rossat-Mignod, H. Noel, M. Potel, P. Gougeon and J. C. Levet, Physica C 153-155, 1115 (1988); M. J. Jurgens, P. Burlet, C. Vetteir, L. P. Regnault, J. Y. Henry, J. Rossat-Mignod, H. Noel, M. Potel, P. Gougeon and J. C. Levet, Physica B **156 & 157**, 846 (1989).
5. K. Yamada, E. Kudo, Y. Endoh, Y. Hidaka, M. Oda, M. Suzuki and T. Murakami, Solid State Comm. **64**, 753 (1987).
6. J. M. Tranquada and G. Shirane, Physica C 162-164, 849 (1989).
7. P. W. Anderson, Phys. Rev.**86**, 694 (1952).
8. J. Hubbard and W. Marshall, Proc. Phys. Soc. **86**, 561 (1965).
9. W. Marshall and S. W. Lovesey, *Theory of Thermal Neutron Scattering* (Oxford Univ. Press, London, 1971).
10. S. D. Mahanti, T. A. Kaplan, Hyunju Chang and J. F. Harrison, J. Appl. Phys. **73** (10), 6105 (1993).
11. K. Hirakawa and H. Ikeda, Phys. Rev. Lett. **33**, 374 (1974).
12. Xun-Li Wang, C. Stassis, D. C. Johnston, T. C. Leung, J. Ye, B. N. Harmon, G. H. Lander, A. J. Schultz, C.-K. Loong and J. M. Honig, Phys. Rev. B **45**, 5645 (1992).
13. H. Chang, J. F. Harrison, S. D. Mahanti and T. A. Kaplan, unpublished work.
14. W. E. Pickett, Rev. Mod. Phys. **61**, 433 (1989).
15. M. T. Hutchings and H. J. Guggenheim, J. Phys. C **3**, 1303 (1970).
16. H. Alperin, Phys. Rev. Lett. **2**, 55 (1961).
17. D. E. Ellis, A. J. Freeman and P. Ros, Phys. Rev. **176**, 688 (1968).
18. A. J. Wachters and W. C. Nieuwpoort, Phys. Rev. B **5**, 4291 (1972).
19. P. S. Bagus and U. Wahlgren, Molecular Phys. **33**, 641 (1977).
20. G. J. M. Janssen and W. C. Nieuwpoort, Phys. Rev. B **38**, 3449 (1988).
21. J. K. Sharma and D. C. Khan, Phys. Rev. B **14**, 4184 (1976).
22. Hyunju Chang, J. F. Harrison, T. A. Kaplan and S. D. Mahanti, Phys. Rev. B **49**, 15753 (1994).
23. J. D. Reger and A. P. Young, Phys. Rev. B **37**, 5978 (1989).
24. M. Gross, E. Sanchez-Velasco and E. Siggia, Phys. Rev. B **39**, 2484 (1989).
25. J. Hubbard, Proc. Roy. Soc. (London) Ser. A **276**, 238 (1963).
26. T. A. Kaplan, P. Horsch and W. von der Linden, J. Phys. Soc. Jpn. **58**, 3894 (1989).
27. M. Takahashi, J. Phys. C **10**, 1289 (1977).
28. P. O. Löwdin, J. Math. Phys. **3**, 969 (1962).
29. M. S. Hybertson, E. B. Stechel, M. Schluter, D. R. Jennison, Phys. Rev. B **41**, 11068 (1990).
30. R. Shepard, I. Shavit, R. M. Pitzer, D. C. Comeau, M. Pepper, P. G. Szalay, R. Alrichs, F. B. Brown and J. G. Zhao, Computer code COLUMBUS Int. J. Quantum Chem. Suppl. **22**, 149 (1988).
31. M.J. Frisch, M. Head-Gordon, G. W. Trucks, J. B. Foresman, H. B. Schlegel, K. Rhagavachari, M. A. Robb, J. S. Binkley, C. Gonzalez, D. J. Defrees, D. J. Fox, R. A. Whiteside, R. Seeger, C. F. Melius, J. Baker, L. R. Khan, J. J. P. Stewart, S. Topiol and J. Pople, Computer code GAUSSIAN92, Gaussian, Inc., Pittsburgh, PA, 1990.

32. S. Huzinaga, *Approximate Atomic Functions, Research Report*, Division of Theoretical Chemistry, the University of Alberta (1971).
33. A. J. H. Wachters, J. Chem. Phys. **52**, 1033 (1970).
34. P. J. Hay, J. Chem. Phys. **66**, 4377 (1977).
35. J. C. Slater, *Insulators, Semiconductors and Metals* (McGraw-Hill, New York, 1967), pp. 215-220.
36. C. Sousa, J. Casanovas, J. Rubio and F. Illas, J. Comput. Chem. **14**, 680 (1993).
37. W. R. Wadt, and P. J. Hay, J. Chem. Phys.**82**, 284 (1985); **82,** 270 (1985).
38. We are considering only the single crystal values in Ref. 16.
39. D. C. Khan, S. M. Curtane and K. Sharma, Phys. Rev. B **23**,2697 (1981).
40. W. R. Wadt and P. J. Hay, J. Chem. Phys. **82**, 270 (1985); *ibid.* p.284
41. R.L. Martin, *Cluster Models for Surface and Bulk Phenomena*, G. Pacchioni and P.S. Bagus eds. (NATO ASI Series,1991).
42. H. Chang, unpublished.
43. T. Freltoft, G. Shirane, S. Mitsuda, J. P. Remeika, A. S. Cooper and D. Harshman, Phys. Rev. B **37**, 137 (1988).
44. S. Shamamoto, M. Sato, J. M. Tranquada, B. J. Sternlieb and G. Shirane, Phys. Rev. B **48**, 13817 (1994).
45. T. Koma and H. Tasaki, Phys. Rev. Lett.**70**, 93 (1993).
46. T. Matsuda and K. Hida, J. Phys. Soc. Japan **59**, 2223 (1990); K. Kubo and T.A Kaplan, unpublished work.
47. B. Keimer, A. Aharony, A. Auerbach, R. J. Birgeneau, A. Cassanho, Y. Endoh, R. W. Erwin, M. A. Kastner, G. Shirane, Phys. Rev. B **45**, 7430 (1992).
48. Although the approximate spin-wave-based theory presented in [47] gave agreement with the temperature-dependence of the observed scattering intensity at low temperatures, suggesting that the T=0 Heisenberg moment is very close to the spin wave result [7], that is not a direct experimental determination of m_{Heis}. A comparison of the absolute intensities with the calculated form factor would provide a significant check on this important quantity.
49. For the sample of [5] with the maximum 'moment' and T_N, we find m_{Heis} = 0.68±0.06, so this would solve the puzzle. If one assumes the 'moment' vs. T_N trend found in [5] continues to T_N=325K, then the value of m_{Heis} would be >0.8, definitely *above* the spin wave value.
50. Jan Zaanen, private communication, March 1994 APS meeting.
51. J. Zaanen and A. Oles, Phys. Rev. B **48,** 7197 (1993).
52. A. Yoshimori, J. Phys. Soc. Japan **14**, 807 (1959).
53. T. A. Kaplan, Phys. Rev. *116*, 888 (1959).
54. J. Villain, J. Phys. Chem. Solids *11*, 303 (1959).
55. P. W. Anderson, Solid State Physics **14**, 99 (1963).
56. N. Fuchikami, J. Phys. Soc. Japan **28**, 871 (1970).
57. M. S. Hybertsen, M. Schlüter and N. F. Christensen, Phys. Rev. B **39**, 9028 (1989); M. S. Hybertsen, E. B. Stechel, M. Schluter and D. R. Jennison, *ibid.* **41**, 11068 (1990).
58. A. K. McMahan, R. M. Martin and S. Satpathy, Phys. Rev.**38**, 6650 (1989).
59. N. Sahoo, Sigrid Markert, T. P. Das and K. Nagamine, Phys. Rev. B **41**, 220 (1990); S. B. Sulaiman, N. Sahoo, T. P. Das, O. Donzelli, E. Torikai and K. Nagamine, Phys. rev. B **44**, 7028 (1991).
60. J. O. Noell, Inorg. Chem. **21**, 11 (1982).
61. C. W. Bauschlicher, Jr. and P. S. Bagus, J. Chem. Phys. **81**, 5889 (1984).

THE GROUND AND EXCITED STATES OF OXIDES

Paul S. Bagus,[1] F. Illas,[2] C. Sousa,[2] and G. Pacchioni[3]

[1]Institut für Angewandte Physikalische Chemie, Universität Heidelberg
Im Neuenheimer Feld 253, 69120 Heidelberg, Germany
[2]Departament de Quimica Fisica, Universitat de Barcelona
C/Marti i Franques 1, 08028 Barcelona, Spain
[3]Dipartimento di Chimica Inorganica, Metallorganica e Analitica
Universitá di Milano, 20133 Milano, Italy

I. INTRODUCTION

This paper reviews several topics related to the electronic structure of certain bulk oxides; in large part, the results described are based on cluster model studies. The topics considered concern two subjects about oxides. First, the characterization of the ionicity of the oxides; the dominant concern here is for the ground states of the transition metal, TM, oxides MnO and NiO. Second, X-ray photoemission spectroscopy, XPS, measurements are analyzed by using theoretical models for the highly excited states formed when a core level electron is ionized from one of the atoms in an alkaline-earth oxide. However before these topics are presented, the following section of the paper gives a brief review of technical details about the cluster models and cluster wavefunctions.

The section on the Ionic Character of Oxides begins with a discussion of the Mulliken population analysis; this is the most common technique used with Molecular orbital, MO, wavefunctions to measure and quantify ionicity and charge separation.[1] While this method may be satisfactory for simple organics, it has serious limitations for systems that contain metals; in particular, for cluster models of bulk oxides. The nature and the possible reasons for these limitations will be considered. An approach based on orbital projection is presented and shown to give a meaningful characterization of the ionic character of MnO_6 and NiO_6 clusters which model the bulk oxides. This approach, in strong contrast to the Mulliken population analysis, is not strongly basis set dependent. Further, it contains cross-checks which determine the accuracy of the estimates of charge assignments obtained with orbital projection. This approach shows that MnO and NiO are dominantly ionic materials. There is, however, a small amount of covalent bonding between the metal 3d and the O(2p) orbitals.

The interpretation of binding energies, BE's, of both cation and anion core levels is examined in the following section of the paper. Contradictory interpretations have been made of the significance of the trends of the BE shifts for the metal, cation, and oxygen, anion, core level BE's for the series of alkaline-earth oxides from MgO to BaO. The trend of the metal BE's has been interpreted to show that BaO is largely covalent rather than

Edited by T.A. Kaplan and S.D. Mahanti, Plenum Press, New York, 1995

ionic.[2] On the other hand, the trend of the O(1s) BE's has been interpreted to show that BaO is super-ionic.[3] Both of these interpretations assume that shifts in core level BE's can be directly related to changes in the initial charge, Q, of the core ionized atom. While Q is one factor which may shift BE's, another factor, which must not be neglected for ionic crystals, is the Madelung potential. When this factor is taken into account, the apparent contradiction described above is resolved; detailed analyses show that electrostatic effects for purely ionic models of the alkaline-earth oxides dominate to determine the trends of the cation and anion BE's. In particular, the trend of the O(1s) BE's are fully consistent with cluster studies[4] which indicate that the degree of covalent bonding in BaO, although small, is larger than for the other alkaline-earth oxides.

II. CLUSTER MODELS AND WAVEFUNCTIONS

The cluster models and wavefunctions that will be used in this paper are described very briefly below; further details are given in the references cited. The model which is used for the cubic oxides is a cluster with a central metal atom and its six nearest neighbor, NN, oxygens, MO_6. In principle, one can also consider oxygen centered clusters and this has been done in related cluster model studies[5] of the O(1s) BE's. The charge on an MO_6 cluster was chosen to be 10, or $[MO_6]^{10-}$, consistent with six O^{2-} anions and one M^{2+} cation. The clusters are embedded in a set of point charges, PC's, whose values have been optimized to reproduce, in the region of the seven atoms explicitly included in the cluster, the Madelung potential of a perfectly ionic crystal with +2 cations and −2 anions. From 50-300 PC's placed at lattice sites of the cubic crystals were used and, within a given shell from the center, the charges were set equal so that the total MO_6 cluster including the PC's has O_h point group symmetry. The effect of the Madelung potential on the ionicity of the cluster wavefunctions has been studied for the alkaline earth oxides[6] and for Al_2O_3.[4] This was done by reducing the values of the embedding PC's to 75% and to 50% of the values optimized for an ideal ±2 ionic crystal. In other words, the reduced PC's correspond to crystals with ±1.5 and ±1 charges, respectively. The ionic character of the cluster wavefunctions were not changed significantly when the reduced PC's, even reduced by 50%, were used.[4,6] This shows that the applied Madelung potential does not bias the results by forcing the cluster wavefunctions to be ionic; rather the ionic character of the clusters and of the crystals that they model results from chemical forces.

The MO_6 clusters considered in this paper are for the cubic alkaline-earth oxides, MgO, CaO, SrO, and BaO, and for the TM oxides, MnO and NiO. The alkaline-earth oxides are closed shell systems while MnO and NiO have open d shells which are coupled high spin on each metal cation; further, the d shells are crystal field split into t_{2g} and e_g. For the alkaline-earth oxides, purely ionic cluster models[7] of the crystals are also used. These models involve replacing the six NN O anions in MO_6 by PC's with a -2 charge and placing an M^{2+} cation, with all its electrons, at the center; this model is denoted $[M^{2+}/Mad]$. For these clusters, SCF wavefunctions obtained with fairly extended contracted Gaussian basis sets are used in this paper. More information about details of the calculations are given elsewhere; for the alkaline-earth oxides, see Refs. 4 and 7 and for MnO and NiO, see Refs. 8 and 9.

III. IONIC CHARACTER OF OXIDES

The most common approach to determine bonding character[1] is to use the gross charges and the decomposition of molecular orbitals into atomic components given by a Mulliken population analysis of electronic wavefunctions or by a closely related variant of the Mulliken approach. While it is known that these approaches may not give precise values for properties such as bond ionicity, it is believed that they give reliable values, especially for trends, except for pathological cases such as might arise, for example, with a one-center basis set.[1] In fact, Mulliken population analyses often give a seriously incorrect description of the bond character and several methods have been developed which do not suffer from the artifacts of population analyses.[10-12] In particular, the projection operator method[12] can be used to study the covalent and the ionic character of the bonding in metal oxides. However before describing the results obtained with projection operators, it is useful to review briefly a few examples where the results of Mulliken population analyses are incorrect and seriously misleading.

Limitations of Mulliken Population Analyses

The first example, which shows that populations can give very misleading results about the extent of charge transfer, concerns the ionicity of NiO. It is taken from the work of Chang et al.[13] who described their population analysis results as a *failure of the Mulliken population analysis*. They considered Mulliken population analyses for self-consistent field, SCF, wavefunctions for an NiO_6 cluster model of NiO where two different basis sets were used to represent the SCF orbitals for the cluster. One basis set was of modest size. With this basis set, the Mulliken analysis indicates that Ni is an anion! It is not simply that Ni has a positive gross charge which is less than +2.0; the gross charge of Ni is, in fact, -0.2 electrons. When the most diffuse basis functions centered on Ni and with atomic angular momentum character of s, p, and d symmetry are removed from the basis set, the Mulliken analysis indicates that Ni is a cation but with a charge, Q, still less than 2 electrons, Q=+1.75. It will be shown below that the projection of Ni and Oxygen orbitals on the NiO_6 cluster wavefunction indicate that the charge of the Ni cation in NiO is even closer to +2 than the result of the population analysis with this smaller basis. This example indicates that using basis functions which allow the orbitals of the Ni in NiO to have the possibility of being different from those of the isolated atom or cation, in particular to form covalent bonds with the O ligands, leads to a Mulliken population which is absurd. In order to obtain a population which, although not necessarily correct, is at least physically reasonable, one must severely reduce the flexibility of the basis set used to obtain the Ni06 cluster wavefunction. In general, there is a danger that this limitation of the basis set may bias the calculation and lead to non-physical results. Change et al.[13] were careful to guard against this danger by directly comparing physical properties of the SCF wavefunctions obtained both with and without the diffuse Ni basis functions. They found that the charge and spin densities for these two SCF wavefunctions are essentially indistinguishable.[13] Thus, the two wavefunctions yield the same physical properties. However, it is not clear that the populations for either basis set correctly reflect the charge state of Ni in the cluster. As will be shown later, projection of the orbital character does provide meaningful assignments of charge, even giving an estimate of the uncertainties of the assignments.

A somewhat similar problem was found for the population analysis of the $Ni(CO)_4$ molecule. Spangler et al.[14] reported that a Mulliken population analysis give about a full electron of Ni 4s and 4p character in the wavefunction and concluded, based on these populations, that the role of the 4sp in the bonding of Ni to CO is significant. Further, the CO populations showed that the σ donation from CO to Ni and the π back-donation from Ni to CO were both large; they were ~0.2 electrons for each of the four CO groups in $Ni(CO)_4$. Bauschlicher and Bagus[15] obtained quite similar results for the populations in $Ni(CO)_4$; however, they also made a direct test of the participation of the Ni 4sp orbitals in the chemical bonding between Ni and CO by constructing a basis set which could not represent either the 4s or 4p orbitals of Ni. This basis set was constructed by removing from the original basis the functions that could represent the outer maxima of the Ni 4s and 4p atomic orbitals. When these functions are deleted, the basis set can only describe the Ar core and 3d orbitals of Ni; clearly, the Ni 4s and 4p orbitals cannot be described. The two basis sets are denoted by WSP for the original basis containing the functions needed to describe the Ni 4s and 4p orbitals and by DONLY for the basis with these functions removed. It is stressed that the population analysis for the SCF wavefunction computed with the WSP basis set showed large occupations of the Ni 4s and 4p orbitals. Since the wavefunction computed with the DONLY basis set, Ψ(DONLY), cannot contain these Ni orbitals, it will have to have a very different population analysis and, if the populations are at all meaningful, Ψ(DONLY) will give a much poorer description of the chemical bonding between Ni and CO than that obtained with Ψ(WSP). In particular, the energy of Ψ (DONLY) should be very different from that of Ψ(WSP).

The population analyses for the $Ni(CO)_4$ wavefunctions obtained with the two different basis sets are given in Table I. Exactly as expected, they indicate a completely different bonding character for the two wavefunctions. With the DONLY basis set, there is no Ni 4sp character in the SCF wavefunction; this is necessary by construction. Further, with the DONLY basis set the population analysis indicates that there is no σ donation but only π back-donation while the analysis of `Ψ(WSP) indicates that the σ donation and the π back-donation for $Ni(CO)_4$ are of comparable importance. However, the differences between Ψ(WSP) and Ψ(DONLY) are apparent and not real. This is seen from the fact that the energies of the two wavefunctions differ by less than 0.45 eV; as discussed in Ref. 15, even this small value is an upper bound to the importance of the Ni 4s and 4p orbitals for the bonding in $Ni(CO)_4$. Another clear measure of the similarity of the WSP and DONLY wavefunctions is given by the overlap of the two 84 electron wavefunctions; this is essentially 1, $\langle\Psi(WSP)|\Psi(DONLY)\rangle=0.994$. Thus the dramatically different populations shown in Table 1 are obtained for wavefunctions which are essentially identical. Clearly, in this case also, the Mulliken population analysis cannot be used to obtain a reliable view of the character of the chemical bonding.

In the final example of the limited value of population analyses, another case of the metal-carbonyl bond is examined. However in this case, the reason for the errors in the analysis are identified and a test is suggested which may be used to determine if serious uncertainties are present. The linear molecule AlCO in its $^2\Sigma^+$ state was one of the models used as to study the interaction of CO with an Al surface;[16] of course, the principle interest in the present work is to understand the population analysis artifacts for the AlCO molecule. The gross populations for AlCO with reasonable bond lengths,[16] are quite consistent with the conventional Blyholder model for the metal surface to CO interaction arising from σ donation balanced by π back-donation.[15,16] The Al has a gross population of 13.3 electrons and is -0.3 electrons negatively charged; the CO has a gross population of 13.7 with 9.7 σ and 4.0 π electrons. The CO σ donation to Al is significant and no π back-donation is allowed because of the $^2\Sigma^+$ symmetry of the AlCO state. The $^2\Sigma^+$ state of AlCO had been chosen to study the repulsive contributions that the CO(5σ) makes to the metal-carbonyl bond. In fact, the Al-CO potential curve is purely repulsive and at the distances used for the population analysis, the interaction is repulsive by 5.2 eV! The gross

Table 1. Mulliken population analysis for the WSP and DONLY basis set SCF wavefunctions for $Ni(CO)_4$.

	Ni Character			CO Character	
	3d	4s	4p	σ	π
Ψ(WSP)	9.13	0.29	0.77	9.77	4.18
Ψ(DONLY)	9.14	0.00	-0.01	10.03	4.19

population analysis which appears to describe a large σ dative covalent bond does not indicate that the interaction is strongly repulsive. The repulsive character of the bond can be seen from the net and overlap populations. The net charge on Al is 16.0 and the net charge on CO is 16.5 electrons; recall that neutral Al has 13 and neutral CO has 14 electrons. The large overlap population between Al and CO of -5.5 gives the gross populations of 13.3 and 13.7 for Al and CO respectively because it is divided equally, -2.75 electrons, between the two units. There are two important things to note about the overlap population. The first is that the strong anti-bonding character shown by the large **negative** overlap population is fully consistent with an interaction repulsive by over 5 eV. The second is that the arbitrary equal division of the overlap population between Al and CO introduces a large uncertainty and makes the gross populations meaningless.

Large overlap populations, and the consequent uncertainty of the gross populations, are a general difficulty of the Mulliken population analysis. The overlap populations are likely to be large whenever there is significant overlap between the atomic or molecular orbitals of one sub-unit with those of another sub-unit in an interacting system. This is certainly true for all the three examples discussed in this section. The overlap between the O 2p and the Ni 4s and 4p orbitals in NiO is obviously large; this is also true for the overlap of the CO 5σ with the Ni 4s and 4p orbitals in $Ni(CO)_4$ and for CO 5σ with Al 3s and 3p in AlCO. It is important to note that the statement above does not address the use of large basis sets to describe the fine details of the orbitals. It addresses the chemical problem that the outermost orbitals of atoms at equilibrium distance from each other in a molecule or a solid usually have a large overlap. It is quite possible that population analysis artifacts may arise whether either large or small basis sets are used. The basis set size may, however, introduce further artifacts related to limitations of the basis set on one or both of the sub-units, i.e. to the basis set superposition error,[10] this superposition error can also lead to large Mulliken overlap populations. The important conclusion is that large overlap populations are a clear indication that the Mulliken gross populations are unreliable and probably seriously misleading. The overlap populations provide a valuable check for the occurrence of serious artifacts in the Mulliken gross populations; more extensive use should be made of this check.

The fact that Mulliken population analyses may have serious artifacts is generally known.[1] Unfortunately, the possibility that artifacts may be present in a particular analysis is almost always neglected. It is thought that these problems will only arise in pathological cases, for example when all the basis functions are on a single center, and that they will not occur in a normal calculation. The examples discussed above show that large errors for Mulliken population analyses do occur in standard molecular orbital calculations. The large orbital overlap which is probably responsible for many of these errors actually limits the ability to assign electrons in an extended molecular or solid state system in a unique way to the atoms which comprise this system. However, the assignment obtained from the Mulliken population analysis has, as shown above, serious artifacts. In the following, a method which has fewer artifacts will be described and used to examine the ionicity given by cluster models of several oxides.

Projection of Orbital Character

A measure[12] of the extent to which an orbital, φ, is contained in a general many electron wavefunction, Ψ, can be obtained with a projection operator, $P_{op}(\varphi)$;

$$P_{op}(\varphi) = \varphi(x)\varphi^{\dagger}(x) \tag{1}$$

where x can, in principle, denote spin and spatial coordinates. In the work discussed below,

φ is taken as the space component of the spin orbital and x denotes only the spatial coordinate; a summation over the spin coordinate is implicit. In general the atomic or molecular oribital φ is a physically or chemically interesting orbital of one of the interacting sub-units; below, it will be a ligand 2p orbital or a metal 3d orbital. The measure of interest is the expectation value of $P_{op}(\varphi)$ taken with the cluster or molecular wavefunction, Ψ,

$$P(\varphi) = \langle\Psi|\varphi\varphi^{\dagger}|\Psi\rangle. \quad (2)$$

Neglecting orbital degeneracy, $P(\varphi)\leq 2$ and three cases can be distinguished. For case 1, $P(\varphi)\approx 2$ and φ is fully contained in Ψ or φ is fully occupied in Ψ. For case 2, $P(\varphi)\approx 0$ and φ is not contained or occupied in Ψ at all. For the third case, $0<P(\varphi)<2$ and φ may be involved in covalent bonds. The projection operator method has been applied, in particular to characterize the ionicity, for the analysis of a large number of cluster models of extended systems. In particular, it has been used to study adsorbates on metal[17,18] and semi-conductor[19] surfaces and to characterize the ionicity of alkaline-earth oxides[4] and of CuO.[20] In the following, new results are presented to show that MnO and NiO are dominantly ionic systems but that they involve a small amount of covalent bonding between the metal 3d and the O 2p orbitals.

Table 2. Projection of the Mn^{2+} 3d orbitals, P(3d), on the SCF wavefunction for $[MnO_6]^{10}$; individual contributions to the total P(3d) are given. The values of P(3d) correspond to numbers of Mn 3d electrons. For the higher lying t_{2g} and e_g orbitals, the %d composition, as given by projection, and the Mn-O bonding or anti-bonding character are also given.

Contribution	P(3d)	%d	Orbital Character
t_{2g}^3 - open shell	2.96	98.5	Anti-bonding
t_{2g}^6 - dominantly O(2p)	0.09	1.4	Bonding
Total t_{2g}	3.04	---	---
e_g^2 - open shell	1.92	95.8	Anti-bonding
e_g^4 - dominantly O(2p)	0.14	3.5	Bonding
e_g^4 - other	0.03	---	---
Total e_g	2.08	---	---
Total d	5.12	---	---

Table 3. Projection of the Ni^{2+} 3d orbitals, P(3d), on the SCF wavefunction for $[NiO_6]^{10-}$; individual contributions to the total P(3d) are given. The values of P(3d) correspond to numbers of Ni 3d electrons. For the higher lying e_g orbitals, the %d composition, as given by projection, and the Ni-O bonding or anti-bonding character are also given.

Contribution	P(3d)	%d	Orbital
Total t_{2g}	6.00	---	---
e_g^2 - open shell	1.87	93.7	Anti-bonding
e_g^4 - dominantly O(2p)	0.21	5.3	Bonding
e_g^4 - other	0.04	---	---
Total e_g	2.12	---	---
Total d	8.12	---	---

First, the projections of the 3d orbitals of the Mn^{2+} and Ni^{2+} ions on the wavefunctions for the $[MnO_6]^{10-}$ and $[NiO_6]^{10-}$ clusters, respectively, are considered. The atomic ions are embedded in the point charge, PC, field which represents the Madelung potential of the ionic crystal. This takes into account the small modifications of the 3d orbitals to form t_{2g} and e_g levels in the cubic field of the crystal. The projection operator $P_{op}(3d)$, see Eq. (1), is formed by taking the sum of the five d components. It is recalled that the configurations of the cluster wavefunctions are

$$[MnO_6]^{10-}(^6A_{1g}): \quad (cores)t_{2g}{}^3(^4A_{2g})e_g{}^2(^3A_{2g}) \quad \text{and} \tag{3}$$
$$[NiO_6]^{10-}(^3A_{2g}): \quad (cores)t_{2g}{}^6e_g{}^2(^3A_{2g}),$$

where the cores are closed shell. The orbitals specifically shown are dominantly metal 3d with a small anti-bonding mixture of O 2p character. The highest core levels of t_{2g} and e_g symmetries are dominantly O 2p with a small bonding mixture of metal 3d. The projection of the metal 3d orbitals provides quantitative information about the extent of these bonding and anti-bonding combinations; these projections are given in Tables 2 and 3.

The contributions to P(3d), see Eq. (2), are divided into several parts to show the origin of the deviation of P(3d) from the values of 5 and 8 that indicate perfect ionicity for MnO and NiO, respectively. The contributions of the dominantly metal 3d and dominantly O 2p orbitals of t_{2g} and e_g symmetry are shown. For e_g symmetry, there are also other closed shell orbitals and their contributions are summed under the entry $e_g{}^4$-other. For each of these orbitals or groups of orbitals, the contribution to P(3d) is shown in Tables 2 and 3; for the dominantly metal 3d and O 2p orbitals, the percent of d character in the orbital, as given by the projection, and the metal-O bonding or anti-bonding character is also shown. Finally, sums over all orbitals of t_{2g} and e_g symmetry as well as the total d projection are shown. The contribution from individual closed shell orbitals is not unique since the SCF wavefunction is invariant to arbitrary rotations of the closed shell orbitals with each other and the sum over the closed shells is the only unique quantity.[1] This is most important for the e_g orbitals where, for both MnO and NiO there are 3 closed shell orbitals of dominantly O 1s, 2s, and 2p character. Although given the energy differences of these shells, the 3d mixing with the orbitals of dominant O 1s and O 2s character is quite small but not entirely negligible; see Tables 2 and 3. On the other hand, the contributions from the open shell orbitals are unique.[1]

The projections of $Mn^{2+}(3d)$ on the $[MnO_6]^{10-}$ cluster are given in Table 2. The total P(3d) and the sub-totals for t_{2g} and e_g symmetries are extremely close to their nominal values of 5, 3, and 2, respectively, for an ideal ionic crystal. However, they are slightly larger than these nominal values by ≤ 0.1 electrons indicating a small amount of covalent bonding between the metal 3d and O 2p. The open shell t_{2g} and e_g orbitals are >95% d and have anti-bonding character between Mn 3d and O 2p. On the other hand, the dominantly O 2p closed shell orbitals of either t_{2g} and e_g symmetry have small amounts of Mn 3d character, ~0.1 electrons for both the $t_{2g}{}^6$ and the $e_g{}^4$ shells (see Table 2), and are bonding. Figs. 1 and 2 give contour plots for one of the degenerate components, the xy component, of the bonding and anti-bonding e_g orbitals of $[MnO_6]^{10-}$. The plots are in the xy plane and contain 4 O atoms near the corners and the Mn atom in the center. Solid lines indicate contours where the orbital, φ, is positive and dashed lines contours where $\varphi<0$; dotted lines indicate $\varphi=0$. The nodes along the x and y axes show that both orbitals have xy symmetry. The closed shell orbital, Fig. 1, is dominantly O 2p and the open shell orbital, Fig. 2, is dominantly Mn 3d. The contour increment is $\Delta\varphi=0.025$ a.u.. The first non-zero contour corresponds to a density of 0.0006 electrons/bohr^3 and the second contour to a density of 0.0025; both are very small densities. For both orbitals, the small extent of the mixing of O 2p and Mn 3d is dramatically shown by the small number of contours around the atom with

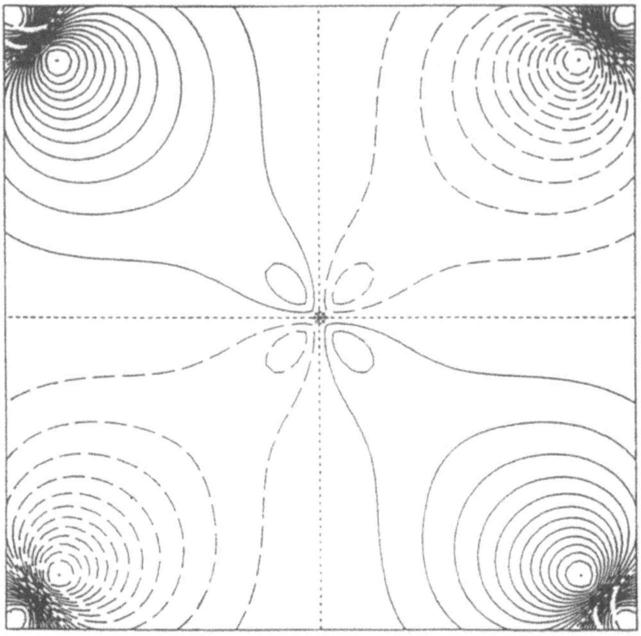

Figure 1. Contour plot of the closed shell, bonding, dominantly O(2p), e_g orbital of the $[MnO_6]^{10-}$cluster SCF wavefunction. The Mn atom is at the center of the plot and the plane used, denoted the xy plane, includes 4 O atoms near the corners of the plot. Solid lines indicate contours where $\varphi>0$ and dashed lines contours where $\varphi<0$; the increment for the contours is $\Delta\varphi=0.025$ a.u.

the minority density. The bonding character of the dominantly O 2p orbital, Fig. 1, is shown by the fact that the orbital is additive between Mn and O while the anti-bonding of the open shell, dominantly Mn 3d, orbital, Fig. 2, is shown by the node between Mn and O. Both the projections and, in a graphical fashion, the orbital contour plots show that the d character, P(3d)=5.1, is close to the value of 5 for a perfectly ionic MnO crystal and that the Mn(3d) with O(2p) covalent bonding, while present, is rather small.

Similar results are found for NiO from the projections given in Table 3. The t_{2g} contribution to the projection is 6.00. This shows that the t_{2g} electrons do not participate in covalent bonding with O 2p and, thus, the character of the individual cluster orbitals of t_{2g} symmetry is not given. This absence of bonding character is entirely expected since both the Ni and O_6 t_{2g} shells are completely filled with 6 electrons. There is some bonding character in e_g, perhaps slightly larger than for MnO (see Table 2), but the total d projection is 8.1 only slightly larger than 8 for a perfectly ionic NiO crystal.

The projections of the O orbitals on the MO_6, M=Mn or Ni, cluster wavefunctions provide complementary information to that which is obtained from the projection of the metal d orbitals. In particular, it allows one to study symmetries where the d orbitals are not present. An O_6^{12-} cluster is used to describe the O orbitals for ideal ionic O anions and the SCF orbitals for this cluster are projected on the MO_6 cluster wavefunctions. For projection on the MnO_6 wavefunction, the O_6 positions are the same as for MnO_6 and the O_6 cluster is surrounded by the same external set of PC's. Parallel choices are made for the O positions and PC's for the O_6 cluster used for the projection on the NiO_6 wavefunction. In previous work,[4] two different choices have been made to include the effects of the metal cation which in the metal oxide is at the center of the O_6^{12-} cluster. The first choice involved replacing the cation by a point charge and the second choice involved replacing an Al^{3+} cation by a pseudo potential. In the present work, either a frozen Mn^{2+} or Ni^{2+} cation, as appropriate, is placed at the center of the O_6 cluster and the O_6^{12-} SCF orbitals are

Figure 2. Contour plot of the open shell, anti-bonding, dominantly Mn(3d), e_g orbital of the $[MnO_6]^{10-}$ cluster SCF wavefunction; see caption to Fig. 1.

computed in a variational space which is orthogonalized to the cation orbitals. This choice has several advantages: (1) The use of a central cation which has a 2+ charge is realistic as shown by the metal 3d projections described above. (2) Full account of the finite size of the metal cation is taken into account. In particular, potentially unrealistic large polarizations of the anion charge distributions toward the cation[21] are avoided because a frozen metal dication rather than a PC is used. (3) Any possible difficulties related to the use of a pseudo potential[22] are avoided.

The projections of the O_6^{12-} cluster orbitals from the MnO_6 and NiO_6 wavefunctions are given in Table 4. They are given as ΔP, the deviation of $P(O_6^{12-})$ from the value of 60 for an ideal ionic system, $\Delta P(O_6^{12-})=P(O_6^{12-})-60$. The increase in the ΔP(3d), shown in Tables 2 and 3, should be reflected as a decrease in $\Delta P(O_6^{12-})$. In addition, the projections to the total $P(O_6^{12-})$ from the a_{1g} and t_{1u} O_6^{12-} cluster orbitals are given separately in Table 4; the ΔP for these symmetries are taken with respect to the ideal, O_6^{12-}, values of 6 for a_{1g} and 24 for t_{1u}. The metal 4s orbital has a_{1g} and the metal 4p orbital has t_{1u} symmetry. Reductions of the a_{1g} and t_{1u} ΔP's from zero may indicate that there is participation of the metal 4s and 4p in covalent bonding with O.

Table 4. Projection of O_6^{12-} cluster orbitals on the SCF wavefunctions for $[MnO_6]^{10-}$ and $[NiO_6]^{10-}$. The total projection and the contributions from only O_6^{12-} orbitals of a_1 or of t_{1u} symmetry are given. The values are given as $\Delta P(O_6^{12-})$, the deviation of the projection from the value for an ideal ionic system; the $\Delta P(O_6^{12-})$ are compared to ΔP(3d). Negative values of ΔP indicate fewer electrons than in an ideal ionic system while positive values indicate more electrons.

	ΔP(3d) Total	$\Delta P(O_6^{12-})$ Total	$\Delta P(O_6^{12-})$ a_1 - only	$\Delta P(O_6^{12-})$ t_{1u} - only
$[MnO_6]^{10-}$	+0.12	-0.18	-0.01	-0.04
$[NiO_6]^{10-}$	+0.12	-0.15	-0.01	-0.01

For both MnO and NiO, the results for ΔP in Table 4 show that the decrease of $P(O_6^{12-})$ from 60 is comparable to the increase of P(3d) from its ideal ionic value. This indicates that the primary contribution to the covalent bonding in these crystals is from the metal 3d with O(2p) interaction. The a_{1g} symmetry contribution to $\Delta P(O_6^{12-})$ is negligible and shows that the metal 4s contribution to the covalent bonding is either non-existent or very small. The same is true for the Ni 4p contribution to the covalent bonding in NiO. For MnO, the t_{1u} contribution to the decrease of $\Delta P(O_6^{12-})$ is slightly larger and there may be a small Mn 4p involvement in the covalent bonding with O(2p). However, this may be an over interpretation of the projections, especially because 0.04 electrons is a small change. It is best to focus on the large changes. These show that there is a small metal d covalent bonding with O 2p for both MnO and NiO but that the crystals are dominantly ionic; the number of d electrons is within ~0.1 of the ideal ionic value. It is possible that even this small covalent character in the MO_6 clusters is an upper bound to the covalent bonding in the MO crystals because while the metal to oxygen stochiometry is 1:1 in the crystal, it is 1:6 in the cluster.

IV. CORE LEVEL BINDING ENERGIES IN ALKALINE-EARTH OXIDES

As may be seen from the discussion in the preceding section, the ionicity and the degree of covalent bonding for metal oxides is an important and challenging theoretical problem. It is a problem which has important consequences for the electronic structure and the materials properties of these crystals.[20,23] X-ray Photoelectron Spectroscopy, XPS, data for the metal and O core level binding energies, BE's, especially the trends of these BE's along a series of related oxides, should provide information about the ionicity. A commonly used approach for the analysis of XPS core level BE shifts[24] is to relate the change in the BE, ΔBE, to the change in the charge, ΔQ, of the atom which has been ionized by

$$\Delta Q \propto \Delta BE, \tag{4}$$

where, of course, ΔQ refers to the charge on the atom before it is core ionized. Thus an increased BE means that the charge on the ionized atom is more positive and a decreased BE means that the charge is less positive or more negative.

Unfortunately analyses based on the relationship of Eq. (4), have led to contradictory conclusions about the ionicity along the series of alkaline-earth oxides: MgO, CaO, SrO, and BaO. Wertheim[2] examined the trend of the metal core level BE's along this series; in particular, he based his analysis on the shift between the core level BE's in the bulk metals and in the oxides, ΔBE(MO-M). For Mg, ΔBE is +2.2eV; i.e., it is ~2eV harder to remove a core electron from an Mg atom in MgO than from an Mg atom in the bulk metal. The direction of the shift is certainly consistent with Mg being a cation in MgO and neutral in Mg metal; the relatively small size of the shift, which is much smaller than one would expect between a neutral Mg and Mg^{2+}, will be discussed further below. However, most important is the trend of the ΔBE(MO-M) as the alkaline-earth metal becomes heavier. The value decreases monotonically along the series and becomes negative for Ba. If the relationship of Eq. (4) is used, then $\Delta BE<0$ found for Ba is not consistent with Ba being close to Ba^{2+} in BaO. This led Wertheim[2] to argue that there was a strong covalent bond between O(2p) and the normally empty Ba(5d) levels such that Ba in BaO is essentially neutral. On the other hand, Barr and his collaborators[3] studied the O(1s) BE's along the series of the alkaline-earth oxides. They noted that the O(1s) BE decreased monotonically from ~530 eV for MgO to ~527.5 eV for BaO. It was accepted that MgO had substantial ionic character containing Mg cations and O anions. Since the O(1s) BE decreases along the series, O must become more anionic and hence metal more cationic for heavier alkaline-earth metal oxides; in fact, BaO was described as *super-ionic*. A detailed theoretical analysis of cluster models for this series of oxides[4] gave strong evidence that the bonding is largely ionic in all alkaline-earth oxides with the metal being essential M^{2+} and O being essentially O^{2-} for all the oxides in this series. However, there is a small but increasing covalent character as the metal becomes heavier; MgO is purely ionic while BaO has the largest, although still small, covalent character. These results disagree with the arguments both of Wertheim[2] and of Barr.[3] However, a proper analysis of the core level BE shifts for both the metal core levels[7] and the O (1s) level[5] shows that the observed trends are fully consistent with the theoretical results.[4] An important feature of the analyses of the BE shifts[5,7] is that they take into account the BE changes due to the Madelung potential of the oxides.

It is possible to divide the contributions to the BE's into initial state and final state terms.[25] The initial state terms arise from the chemical environment of the ionized atom. In order to obtain the BE from the initial state effects only, it is necessary to freeze all the orbitals as they are when the core level is completely filled and compute the Koopmans' theorem, or KT, BE with these frozen orbitals. While the KT BE's do fully reflect the consequences of the chemical bonding and the chemical environment of the ionized atom, they completely neglect the response of the electrons in the system to the new potential because a core electron is removed. This response, or relaxation, is a final state effect which exists only for the core ionic state. It is taken into account by determining a separate SCF wavefunction for the final core hole state as well as an SCF wavefunction for the initial, unionized state. The ΔSCF BE is computed as the difference of the SCF energies for these two states and includes both initial and final state contributions to the BE. The difference between the KT BE and the ΔSCF BE is called the relaxation energy, E_R, and it is a measure of the importance of final state contributions to the BE. These contributions are quite important for the absolute value of the BE; the relaxation energy is large and, depending on the nuclear charge, Z, of the ionized atom, on the size of the system, and on the core level ionized, ranges from ~10 eV to more than 50 eV. However, there is strong evidence (Ref. 25 and references therein) that the dominant contributions to BE shifts arise from initial state effects. In particular for XPS spectra with a dominant main peak and a minor satellite structure, E_R while large is nearly constant for all atoms with the same Z in a given system. The near constant values for E_R is very important because it proves that the observed BE shifts are indeed *chemical* shifts[24] and that they do, indeed, probe and provide information about the initial chemical state and the initial chemical environment of the core ionized atom. In the following, the focus will be on the KT BE's ; however, some ΔSCF results will be presented to show that the shifts are indeed dominated by initial state chemistry rather than final state relaxation.

Metal Core Level BE's

The analysis of the metal core level BE shifts[7] was based, in large part, on two simple cluster models of ideal ionic crystals. The first model was an isolated metal di-cation, denoted, M^{2+}, and the second model was M^{2+} surrounded by PC's chosen to reproduce the Madelung potential and is denoted [M^{2+}/Mad]. The BE shifts obtained with these models are taken with respect to the BE's of the free atom; thus the shifts used are $\Delta BE(M^{2+}-M_{at})$ and $\Delta BE([M^{2+}/Mad]-M_{at})$. This choice is made to more easily identify and study the individual mechanisms which are responsible for the BE shifts. Because KT BE's are used and because the theoretical shifts are taken with respect to the isolated atom and not the bulk metal, it is not possible to directly compare the absolute values of the theoretical and experimental shifts. However, it is expected that the trends of the ΔBE from Mg to Ba obtained with the theoretical analysis should be rather similar to the observed shifts; this was indeed found to be the case.[7a] Finally, the theoretical results are for ionizations of a metal 2p electron while ionization of higher lying core levels are measured for Sr and Ba. However, there is experimental[2] and theoretical[7b] evidence that the BE shifts, ΔBE, depend only rather weakly on the particular core level which is ionized.

Consider first the KT BE for the ionization of the metal dication compared to the neutral atom; these ΔBE's are given in Table 5. It is harder by ~10-20 eV to core ionize the M^{2+} dication than the neutral M atom. These differences can be related to the electrostatic potential due to the two outermost s electrons, ns^2, which are present in the neutral atom but not in the di-cation The potential which a core electron sees due to the ns^2 shell is, to a very good approximation, the potential of the ns electrons at the atomic nucleus,

$$V = 2e^2\langle 1/r\rangle_{ns} . \qquad (5)$$

The size of the ns orbital is smallest for Mg 3s and largest for Ba 6s; this size is shown by the $\langle r\rangle_{ns}$ given in Table 5 for the atomic SCF orbitals of the neutral atoms. Also given in Table 5 are the values of the potential at the nucleus due to the ns electrons, Eq. (5); these are within 10% of the $\Delta BE(M^{2+}-M_{at})$ for all the alkaline-earths. This establishes that the trend of ΔBE to smaller values, by almost a factor of 2, from Mg to Ba is due to the increasing size of the outermost, ns, orbital of the neutral atom. However, the ΔBE ($M^{2+}-M_{at}$) are an order of magnitude larger than the experimentally observed BE shifts,[2] ΔBE(MO-M). This is a strong indication that there is another important factor which, in the oxides, reduces the value of the metal core level BE's.

This factor is the influence of the Madelung potential, V_{Mad}; its effect on core level BE's is, to a good approximation, the value of the potential at the nucleus of the ionized atom. It acts to reduce the cation BE's while it raises the anion BE's. The sign of the effect on the BE's clear if one recalls that V_{Mad} at the cation nucleus is dominated by the nearest neighbor, NN, anions while V_{Mad} at the anion nucleus is dominated by the NN cations. In the following sub-section, the influence of V_{Mad} on the anion BE's will be considered; here its influence on lowering the cation BE's is examined. In these cubic oxides, the Madelung potential at the atomic nuclei depends inversely on the metal to oxygen distance, R(M-O),

$$V_{Mad} \propto 1/R(M\text{-}O). \qquad (6)$$

Table 5. Koopmans' Theorem BE shifts of the cation 2p level for various ionic models of the alkaline-earth oxides. Data for the size and electrostatic potential of the outermost ns electrons and for the lattice constant and the Madelung potential are also given.

	Mg	Ca	Sr	Ba
$\Delta BE(M^{2+}-M_{at})$/ev	+19.9	+15.0	+13.4	+11.7
$\langle r\rangle_{ns}$/bohrs	3.3	4.2	4.6	5.3
$V=2e^2\langle 1/r\rangle_{ns}$/eV	+21.6	+16.3	+14.6	+12.8
$\Delta BE([M^{2+}/Mad]-M^{2+}$/eV	-24.0	-20.9	-19.6	-18.2
R(M-O)/bohrs	4.0	4.5	4.9	5.2
V_{Mad}/eV	-24.0	-20.9	-19.6	-18.2
$\Delta BE([M^{2+}/Mad]-M_{at})$/eV	-4.1	-5.9	-6.2	-6.5

Thus as the lattice constant increases, the effect of V_{Mad} on the BE's decreases. The influence of the Madelung potential on the cation BE's is simply the difference of the metal BE for the [M^{2+}/Mad] cluster and the isolated M^{2+} cation; values of this difference, ΔBE([M^{2+}/Mad]-M^{2+}), are given in Table 5. Also given are R(M-O) and V_{Mad} at the M^{2+} nucleus. As expected from the small size of the 2p orbital, the approximation of using the value of the Madelung potential at the metal nucleus is very good and $\Delta BE \approx V_{Mad}$. However because R(M-O) grows monotonically along the series of alkaline-earth oxides, the size of V_{Mad} and its effect to lower the cation BE is reduced monotonically. However the reduction in magnitude of V_{Mad} is only ~25% from Mg to Ba while the reduction in $<1/r>_{ns}$ is ~50%; this distinction is quite important for understanding the trend of the total BE shifts of the cation BE's.

The two contributions to the BE shift arising, one, from the atomic term in forming M^{2+} and, two, from the Madelung potential cancel,

$$\Delta BE(\text{Atomic}) = \Delta BE(M^{2+}\text{-}M_{at}) > 0 \text{ and}$$
$$\Delta BE(\text{Madelung}) = \Delta BE([M^{2+}/\text{Mad}]\text{-}M^{2+}) < 0.$$

The total shift, ΔBE([M^{2+}/Mad]-M_{at}), is simply the sum of these two terms. Both these terms decrease in magnitude as the alkaline-earth becomes heavier but they change at different rates. The ΔBE(Atomic) term goes as $<1/r>_{ns}$ and decreases rapidly. The ΔBE(Madelung) term goes as $1/[R(M^{2+})+R(O^{2-})]$, where the R are effective radii for the metal cation and oxygen anion, and this term decreases less rapidly, see Table 5. Thus, ΔBE(Madelung) becomes more important as the atom becomes heavier and the trend in the total ΔBE([M^{2+}/Mad]-M_{at}) is to smaller values for heavier metal atoms; it is largest, -4.1 eV, for MgO and smallest, -6.5 eV, for BaO. For the reasons discussed above, the values of these ΔBE are smaller by ~6 eV than the experimentally measured XPS BE shifts, ΔBE(MO-M). However, the trends of the theoretical and experimental ΔBE are very similar.[7] From Mg to Ba, the theoretical values change by -2.4 eV to lower ΔBE and the change in the experimental values is -2.7 eV.[2] This agreement gives strong confidence that the important factors responsible for the shifts of the cation BE's in these oxides have been identified.

The models described so far are purely ionic models of the oxides; no chemical interaction or covalent bonding is possible. The trend of the ΔBE to smaller values arises entirely from the canceling electrostatic effects of forming a dication from a neutral atom, Δ BE(Atomic), and of the Madelung potential, ΔBE(Madelung), due to the ±2 ions in the crystal. If it is assumed that the oversimplified relation of Eq. (4) which directly relates BE shifts to the charge state of the ionized atom holds, then the trend of the ΔBE to smaller values could be incorrectly ascribed to greater covalency in the oxides of the heavier alkaline-earth metals.[2]

Two other, more extended, methods to determine the ΔBE have been used. First, an MO_6 cluster was used to describe the oxides in addition to the purely ionic models described above; this cluster allows covalent bonds to form between the metal and oxygen. Second, the BE shifts for the [M^{2+}/Mad] and the MO_6 clusters have been determined with ΔSCF BE's in addition to the KT BE's used above; ΔSCF BE shifts indicate whether final state relaxation effects significantly alter the ΔBE from their KT values. The change in the ΔBE from Mg to Ba are listed in Table 6 for the various clusters and for both KT and ΔSCF BE's. Clearly, the differences among the various models are small compared to the average value of the change in ΔBE; they are all in acceptable agreement with the change in the experimental ΔBE(MO-M) between Mg and Ba. The results shown in Table 6 for the

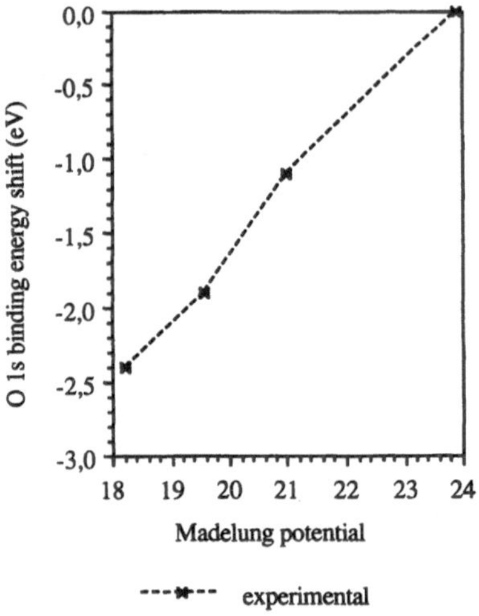

Figure 3. XPS measurements (see Ref. 26) of the O(1s) BE's for the alkaline-earth oxides as a function of V_{Mad} at the oxygen nucleus. Values of V_{Mad} for the various oxides are: MgO, +23.9 eV; CaO, +21.0 eV; SrO, +19.6 eV; and BaO, +18.2 eV. The O(1s) BE's are shown as shifts from the BE for MgO which is taken at zero.

MO_6 cluster were obtained with a basis set for the metal atoms which did not allow the possibility of participation of the empty nd orbitals of the metal cation in the covalent metal-oxygen bonding.[7a] The metal atom basis sets used for the MO_6 clusters have also been extended to include the possibility of nd involvement in the cluster wavefunctions;[4,7b] it is found that they do contribute a small amount to the metal-oxygen covalent bond. This small bonding contribution leads to an additional increase in the change of the ΔBE from Mg to Ba.[7b] Thus, chemistry does make a small contribution to the trend of the ΔBE; however, it is not major as proposed[2] based on the oversimplified relationship, Eq. (4), that the ΔBE directly reflect the change in charge, ΔQ, on the ionized atom. Electrostatic effects are the dominant factors which determine the trend of ΔBE among the series of alkaline-earth oxides. However, it is essential to include the important influence of the Madelung potential which is comparable in magnitude and acts to reduce the shifts in the BE for the ionized cation.

Oxygen 1s BE's

Detailed cluster model studies of the O(1s) BE's in the alkaline-earth oxides have been made using several types of O centered clusters ranging from [O^{2-}/Mad] to M_6O_{13} clusters which include the nearest shell of metal neighbors and the next nearest shell of O

Table 6. Trends of the ΔBE from MgO to BaO for [M^{2+}/Mad] and MO_6 cluster models and for KT and ΔSCF BE's. The change of ΔBE, ΔBE(Ba)-ΔBE(Mg), is given in eV.

	Change of ΔBE
ΔBE([M2+/Mad]-M_{at})-KT	-2.4
ΔBE([M2+/Mad]-M_{at})-ΔSCF	-3.1
ΔBE(MO6-M_{at})-KT	-3.1
ΔBE(MOb-M_{at})-ΔSCF	-2.7
ΔBE(MO-M)-Experiment[a]	-2.7

[a]See Ref. 2

neighbors of the central O anion.[5] However, rather than review the cluster results, it will be shown that the importance of the Madelung potential, demonstrated in the previous sub-section, allows one to correctly interpret the change in the O(1s) BE's along the series of the oxides. For a pure ionic model, there were two competing electrostatic effects for the cation core level BE's; these are the atomic effect of going from a neutral atom to a di-cation and the effect of the Madelung potential of the ionic crystal. For the anions, the concern is for the changes of the O(1s) BE among the different oxides. For a pure ionic model, only one electrostatic effect contributes to these changes of the O(1s) BE; this effect arises from the Madelung potential at the O nucleus. This potential is positive and it acts to increase the O(1s) core level BE.

In Fig. 3, the experimental XPS O(1s) BE's for the alkaline earth oxides[26] are plotted. The O(1s) BE's are shown relative to the BE in MgO. The O(1s) BE decreases monotonically from MgO to BaO; it is smaller by ~2.5 eV for BaO than for MgO. The Madelung potential, which makes the O(1s) BE larger, also decreases monotonically from MgO to BaO. However, it decreases by 5.7 eV, more than twice as much as the decrease of the O(1s) BE. In other words, something else is acting to make the O(1s) BE in BaO larger by ~3 eV than it would be in the pure ionic model. Covalent bonding involving charge donation from the anion to the unoccupied levels of the cation acts in the correct direction to explain this limitation of the pure ionic model since when O is less than O^{2-}, its 1s BE will become larger. Such an explanation is certainly consistent with the analysis of the bonding in these oxides[4] which shows that the covalency in these dominantly ionic oxides grows from MgO to BaO. Unfortunately, a precise study of the chemical effects on the O(1s) BE using the cluster model[5] is complicated because of cluster artifacts which appear to give uncertainties of ~1-2 eV for the O(1s) BE's. It appears that larger clusters than used in Ref. 5 are required to reduce this uncertainty in the O(1s) BE. However, there are critical points clearly demonstrated by a proper analysis, explicitly including the Madelung potential, of the experimental XPS data[3,26] for the O(1s) BE's. The increase in the O(1s) BE's from MgO to BaO does not indicate an increase in the ionicity of the oxides. In fact, the increase of the O(1s) BE is not as large as would be expected from the change in the Madelung potential from MgO to BaO. This is consistent with the cluster model results[4] which show that BaO has more covalent character bonding than the essentially purely ionic MgO.

All of the analyses presented show that the trends of the BE's for both the metal and oxygen core levels of the alkaline-earth oxides are dominated by electrostatic effects. While chemistry and covalent bonding do contribute to the trends of the BE's, attempts to relate the BE shifts to deviations from ideal ionicity in the crystals will fail unless the crystal Madelung potential is taken into account. It remains a challenge to use clusters to determine the effects of the covalent bonding on the BE's. This is because the largest clusters used up to the present contain artifacts; in particular, related to the fact that atoms at the edge of the cluster are not in the proper cubic environment of the crystal.[5,27]

VI. CONCLUSIONS

Several important matters related to the use of clusters to determine materials properties of bulk oxides which depend on their electronic structure have been discussed in this paper. These matters are in three categories. The first category involves the methodology needed to properly interpret the properties of the cluster wavefunctions. The second category concerns analyses of the ionicity of MnO and NiO. The third category concerns the significance of the trends of BE shifts in alkaline-earth oxides.

For the methodology, it has been shown that Mulliken populations can give very misleading descriptions of the charge separation and, hence, the chemical bonding, especially when metal atoms are involved. A large orbital overlap population is a clear indication that the Mulliken gross populations have limited meaning. These overlap populations, or at least those for the orbitals with the largest overlap populations, should be presented in all cases, in particular in all publications, where Mulliken gross populations are used to characterize chemical bonding. It has been shown that orbital projections are able to give reliable information, avoiding the artifacts of the Mulliken population analysis. This is due, in large part, to two facts. First, the orbital projection depends only weakly on the basis sets used to expand the orbitals while the Mulliken populations are very sensitive to the choice of basis sets. Second, the orbital projection approach contains tests which can be used to estimate the uncertainty in the assignment of charges which may result from the overlap of the orbitals of the interacting units. Finally, ideal bonding situations can be simulated. In particular for ionic materials, it is possible to construct clusters which model ideal ionic bonding and to determine the consequences of this ionicity; then it is possible to add covalent or chemical effects. Such separation of individual terms is only possible through the use of theoretical models.

It was shown that MnO and NiO are dominantly ionic materials. There is a small amount of covalent bonding between the metal 3d and the O(2p) which leads to an increase in the number of metal 3d electrons by +0.1 over the ideal ionic limits of 5 3d electrons for Mn^{2+} and 8 for Ni^{2+}. Further, it was established that the participation of the metal 4s and 4p in forming covalent bonds with O was extremely small, if present at all. A detailed analysis of the BE shifts for the alkaline-earth oxides shows that the trends of these shifts are dominated by electrostatic effects for these largely ionic systems. However, the proper interpretation of the chemical significance of these trends must take into account the influence of the Madelung potential on the core level BE's. Once this is done, the BE trends can be used to obtain information about the ionicity of, for example, BaO compared to MgO.

This paper has discussed one of the important challenges for cluster model theory. This is to obtain proper interpretations of the cluster wavefunctions and of the consequences of various electrostatic and chemical bonding interactions for materials properties. One of the advantages of theory for this purpose is that, very often, specific terms can be separated and their importance can be individually determined; for the oxides, this has involved constructing models of purely ionic systems. The *ab initio* molecular orbital cluster model has another powerful advantage that makes it especially suited for the study of condensed matter. *Ab initio* wavefunctions are determined without the use of any parameters which have been adjusted to fit experiment. Of course, the properties of these wavefunctions may not correctly reproduce observed properties; if essential physics has been neglected when the cluster model was constructed, then the results will not be correct. However, *ab initio* cluster wavefunctions do provide an unbiased test of the effects that are included in the model. The advantage of *ab initio* models is that one can often directly relate features of the model to results which are obtained with the cluster wavefunctions.

ACKNOWLEDGMENT

One of us, P.S.B., wishes to express his gratitude to H. Kuhlenbeck for very stimulating and productive discussions. In particular, Dr. Kuhlenbeck pointed out that, for the alkaline-earth oxides, the slope of the O(1s) BE's plotted against the oxide Madelung potentials is not +1 and that another effect, besides the Madelung potential, must be present to explain the BE shifts.

REFERENCES

1. H. F. Schaefer, *The Electronic Structure of Atoms and Molecules* (Addison-Wesley, Reading, 1972) and *Methods of Electronic Structure Theory* edited by H. F. Schaefer (Plenum, New York, 1979).
2. G. K. Wertheim, *J. Elec. Spec. and Related Phen.*, **34**, 303 (1984).
3. T. L. Barr and Y. L. Liu *J. Phys. Chem. Solid*, **7** 657 (1989) and T. L. Barr and C. R. Brundle, *Phys. Rev. B*, **46**, 9199 (1992).
4. G. Pacchioni, C. Sousa, F. Illas, F. Parmigiani, and P. S. Bagus, *Phys. Rev. B*, **48**, 11573 (1993).
5. G. Pacchioni and P. S. Bagus, *Phys. Rev. B*, **50**, 2576 (1994).
6. F. Illas, A. Lorda, J. Rubio, J. B. Torrance, and P. S. Bagus, *J. Chem. Phys.* **99**, 389 (1993).
7. (a) P. S. Bagus, G. Pacchioni, C. Sousa, T. Minerva, and F. Parmigiani, *Chem. Phys. Lett.* **196**, 641 (1992) and (b) C. Sousa, T. Minerva, G. Pacchioni, P. S. Bagus, and F. Parmigiani, *J. Elec. Spec. and Related Phen.*, **63**, 189 (1993).
8. P. S. Bagus, T. Minerva, G. Pacchioni, And F. Parmigiani, to be published.
9. P. S. Bagus, G. Pacchioni, and F. Parmigiani, *Chem. Phys. Lett.* **207**, 569 (1993).
10. P. S. Bagus, K. Hermann, and C. W. Bauschlicher, *J. Chem. Phys.* **80**, 4378 (1984).
11. P. S. Bagus and F. Illas, *J. Chem. Phys.* **96**, 8962 (1992).
12. C. J. Nelin, P. S. Bagus, and M. R. Philpott, *J. Chem. Phys.* **87**, 2170 (1987).
13. H. Chang, J. F. Harrison, T. A. Kaplan, and S. D. Mahanti, *Phys. Rev. B*, **49**, 15753 (1994).
14. D. Spangler, J. J. Wendoloski, M. Dupuis, M. M. L. Chen, and H. F. Schaefer, *J. Am. Chem. Soc.* **103**, 3985 (1981).
15. C. W. Bauschlicher and P. S. Bagus, *J. Chem. Phys.* **81**, 5889 (1984).
16. P. S. Bagus, C. J. Nelin, and C. W. Bauschlicher, *Phys. Rev. B*, **28**, 5423 (1983).
17. P. S. Bagus and F. Illas, *Phys. Rev. B*, **42**, 10852 (1990).
18. P. S. Bagus, G. Pacchioni, and M. R. Philpott, *J. Chem. Phys.* **90**, 4287 (1989).
19. I. P. Batra and P. S. Bagus, *J. Vac. Sci. Technol. A*, **6**, 600 (1988).
20. F. Parmigiani, G. Pacchioni, F. Illas and P. S. Bagus, *J. Elec. Spec. and Related Phen.*, **59**, 255 (1992).
21. G. Pacchioni and P. S. Bagus, *Surf. Sci.*, **269/270**, 669 (1992) and P. S. Bagus and G. Pacchioni, *Phys. Rev. B*, **48**, 11573 (1993).
22. P. S. Bagus, C. W. Bauschlicher, C. J. Nelin, B. C. Laskowski, and M. Seel, *J. Chem. Phys.* **81**, 3594 (1984).
23. J. Zaanen and G. A. Sawatzky, *J. Solid State Chem.* **88**, 8 (1990).
24. K. Siegbahn, C. Nordling, A. Fahlman, R. Nordberg, K. Hamrin, J. Hedman, G. Johansson, T. Bergmark, S. E. Karlsson, I. Lindgren, and B. Lindberg, in *ESCA-Atomic, Molecular, and Solid State Structure Studied by Means of Electron Spectroscopy* (Almqvist and Wiksells, Uppsala, Sweden, 1967).
25. P. S. Bagus, C. R. Brundle, G. Pacchioni, and F. Parmigiani, *Surface Science Reports*, **19**, 265 (1993).
26. M. I. Sosulnikov and Y. A. Teterin, *J. Elec. Spec. and Related Phen.*, **59**, 111 (1992).
27. J. Q. Broughton and P. S. Bagus, *Phys. Rev. B*, **36**, 2813 (1987).

COVALENT CARBON COMPOUNDS: FROM DIAMOND CRYSTALLITES TO FULLERENE-ASSEMBLED POLYMERS

Mark R. Pederson

Complex Systems Theory Branch
Naval Research Laboratory
Washington DC 20375-5345

I. INTRODUCTION

As an illustration of how density-functional-based calculations on clusters are being used to help with the understanding, design and characterization of solid state properties a variety of recent calculations on hydrocarbon, carbon and other clusters are discussed. The density-functional formalism[1] and our implementation of a local-orbital-based approach to the quantum-mechanical solution of Schrödinger's equation is presented in the first section.[2] In addition to reviewing the solution to the Kohn-Sham equations within a Gaussian orbital method we review the recent numerical advances and discuss the calculation of cohesive energies,[3] Hellmann-Feynman forces, and vibrational modes.[4,5] The following sections summarize recent applications of this formalism within the present context of solid-state applications of cluster calculations. From the standpoint of enhancing our understanding of how a crystal grows from a collection of clusters, I review our work on hydrocarbon interactions at the diamond vapor interface and show that the results of these calculations adequately account for several of the atomistic interactions that impact growth rates in diamond-chemical vapor deposition.[3] The importance of a predictive theory for surface reaction barriers is underscored and recent work toward describing such phenomena within the density-functional framework is highlighted.[5] Another way in which cluster calculations can impact solid-state physics is to aid in the determination of cluster assemblies with interesting optical and electrical properties. Examples of such phenomena which are strongly influenced by the properties of the constituent clusters include linear and nonlinear susceptibilities of crystals and may also include ferroelectric materials. Within this context we review our work on the calculation of the static linear and hyperpolarizabilities of hydrocarbon clusters and fullerene molecules and show how these properties can also affect the Hubbard U parameters.[6,7] A final application, in Sec. IV, will be a discussion on how calculated cluster properties can be used to help characterize a material's make up. In this regard I discuss recent calculations on the geometries, electronic structure and vibrational modes of fullerene dimers and compare the predicted results to those of experiment.[8]

Edited by T.A. Kaplan and S.D. Mahanti, Plenum Press, New York, 1995

II. THEORETICAL AND COMPUTATIONAL METHODS

The calculations presented in this work are based on the Hohenberg-Kohn density-functional theory[1] which relates the total energy of a system of electrons to the electronic *density* rather than the electronic *wavefunction.* In principle this leads to a tremendous simplification to the many electron problem since the electronic density is significantly less complex than the many-electron wavefunction. However, in contrast to wave-function based formalisms of quantum mechanics, the relationship between the ground state energy and the electronic density is not obvious and the development of the computational formalism was dependent on first obtaining good approximations to the energy functional. This has been the subject of a great deal of investigation for many years. In this work, we discuss results that have been obtained within two parametrizations of the density-functional theory. For obtaining vibrational modes, equilibrium geometries, polarizabilities and electronic density of states, we have primarily used the local Perdew-Zunger parametrization of the Ceperley-Alder energy functional[9] which is referred to as the local-density approximation (LDA). However in cases where very accurate energetics are required a higher level parametrization which is due to Perdew and coworkers,[10,11] referred to as the generalized gradient approximation (GGA), is used. The generalized gradient approximation depends on the electronic densities and gradients of the electronic densities in contrast to the local density approximation which depends only on the densities.

Regardless of which parametrization is used, the total energy for a cluster consisting of N electrons and M nuclei is represented by a sum of the kinetic energy of the Kohn-Sham orbitals, the coulomb interactions between the combined electronic and nuclear charge distributions and the parametrized exchange-correlation energy which is due to the electronic densities. Once this ansatz for the total energy is accepted, the variational principle tells us that the ground state of the system is found by self-consistently determining a solution in which the Kohn-Sham orbitals satisfy a Schrödinger equation and the Hellmann-Feynman-Pulay forces on each nucleus vanish.

In this work, the solution to these equations is found by expanding the Kohn-Sham orbitals in terms of a linear combination of Gaussian-type orbitals. Such representations of single-electron orbitals are common in traditional quantum-chemistry formalisms and were first applied to solid-state applications by Lafon and Lin.[12] Since the existing approximations of the density functional depend nonlinearly on the electronic density, it is not possible to reduce the requisite integrals to simple analytic forms and some form of numerical integration is necessary regardless of the form of the basis functions. To perform the numerical work that is necessary for this work we developed a variational integration mesh[2a] which allows one to efficiently and accurately calculate the secular matrix, total energies and Hellmann-Feynman forces.

As discussed in Ref. [2b], this scheme starts by placing the entire cluster system in a box that is large enough to enclose the electronic wavefunctions. The box is then subdivided into three different type of regions which are referred to as atomic spheres, interstitial parallelepipeds and excluded cubic regions. Integrations within the atomic spheres are accomplished by using angular integration meshes of Stroud[13] and a radial quadrature mesh that is specially adapted to functions with multiple length scales.[2a] The excluded cubic region refers to the region of space that is outside a sphere but inside of an inscribing cube. Integrations in this region are accomplished by first breaking the region up into forty-eight subregions that are equivalent to an elementary region defined by $r>L>x>y>z>0$, writing the integral over this region in spherical coordinates, and then using three iterated Gaussian quadratures to obtain integration weights. Integrations within the remaining interstitial regions are performed by three iterated Gaussian quadrature schemes that are variationally adapted to the environment of the neighboring atoms. The numerical integration scheme is discussed in detail in Ref. [2a].

With the variational integration scheme discussed above, we find the forces on the nuclei are easily calculated from the Hellmann-Feynman-Pulay theorem.[2b] The Hellmann-Feynman-Pulay theorem states that, once a self-consistent solution to the Kohn-Sham equation is found, each nucleus feels a force that is represented by a sum of two terms. The first term, referred to as the Hellmann-Feynman term, corresponds exactly to the force that a nucleus would feel from a classical charge density constructed from the Kohn-Sham density and the remaining nuclei. The second term, referred to as the Pulay correction, results due to the constraint that the basis functions in a Gaussian formulation are constrained to follow the nuclei as they evolve through space. A discussion of these points within the framework of Gaussian-orbital based electronic structure calculations appears in Ref. [2b]. Given the ability to calculate forces, geometrical optimizations of complex clusters are significantly simplified. A variety of algorithms exist for the optimization of cluster geometries from Hellmann-Feynman forces. We use the conjugate gradient method discussed in Ref. [14] which allows one to optimize a geometry of an N-atom cluster with O(3N) total energy and force calculations. The calculation of vibrational modes depends on the determination of second order derivatives of the energy surface with respect to the nuclear degrees of freedom. We have recently developed the *method of mutually orthogonal displacements* for this problem. The approach combines a forward and backward finite-differencing algorithm with group theory and allows for the first-principles determination of the vibrational modes with a minimal number of calculations. Details of this method are discussed in Ref. [15].

With the numerical methods discussed above, it is possible to determine from first principles a variety of cluster properties that are of interest to solid state applications. We now turn to some examples of this method and begin by discussing the results of Table 1 and Table 2.

III. HYDROCARBONS AND DIAMOND CRYSTALLITES

Approximately five years ago many researchers in the fields of chemistry, materials science and physics began searching for new ways of growing diamond. The resurgence of interest in diamond emerged because of all the extreme properties that it possesses. Here we discuss some of the recent cluster-based simulations on diamond crystallites that have been performed to help identify some of the important interactions for growth. We start by discussing calculations on a variety of hydrocarbons which were performed to determine the accuracy of density-functional methods for hydrocarbon chemistry.

Table 1. Selected bondlengths as calculated within the local density approximation and as measured experimentally. The above results show that the local-density approximation accurately describes all trends observed in hydrocarbon molecules. These results and a more complete discussion of other geometries are discussed in greater detail in Ref. [3].

Molecule	Bond	Re (LDA)	Re (Expt)
CH_4	C-H	2.08	2.07
C_2H_6	C-H	2.09	2.10
C_2H_2	C-H	2.03	2.00
C_6H_6	C-H	2.09	2.05
C_2H_2	C C	2.30	2.28
C_2H_4	C=C	2.56	2.53
C_2H_6	C-C	2.91	2.91
C_2 (S=0)	C C	2.37	2.35
C_2 (S=1)	C=C	2.54	-----
H_2 (S=1)	H-H	1.46	1.41
C_6H_6	C:C	2.65	2.64

Presented in Table 1 are a compilation of LDA-based data on hydrocarbon molecules. As shown in this table, the LDA reproduces the experimental geometries rather accurately. In accord with earlier applications of the density-functional theory to bulk crystals, we observe that hydrocarbon bondlengths are reproduced to an accuracy of 1-2 percent. While the geometries are calculated quite accurately within the local-density-approximation, cohesive energies are overestimated within this approximation. To illustrate this, we present hydrocarbon cohesive energies in Table 2 as calculated within the local-density approximation, the generalized gradient approximation and as measured experimentally. For clarity we note that this corresponds to the classical energy required to break each molecule into separated spin polarized atoms. Again, in accord with past applications of the density-functional theory to periodic systems, the cluster cohesive energies overestimate experiment by approximately 0.7 eV per atom. Recently, these energies have been recalculated within the generalized gradient approximation of Perdew *et al.* and the results of these calculations are also included in Table 2. The agreement with experiment is significantly improved. In contrast to the overbound results obtained within the local density approximation, the generalized gradient approximation reproduced experiment to approximately 0.1 eV per atom. With these points in mind, we now turn to several different calculations of interest to the problem of diamond chemical vapor deposition.

Density-Functional-Based Energetics of Surface Adsorbates

Pictured in Fig. 1 is the model we have used for the substrate of a diamond surface. The cluster is composed of nine four-fold coordinated carbon atoms that surround a three-fold coordinated carbon atom and has been developed to reproduce a dangling bond environment that is observed on the otherwise hydrogenated diamond <111> surface. This model is a reasonable reproduction of the reactive parts of the growth surface. One property of interest is the relative stability of adsorbates or groups of adsorbates on the diamond surface. We have calculated the adsorption energies for methyl radicals (CH_3), acetylenic radicals (C_2H), acetylene molecules (C_2H_2) and for gaseous hydrogen atoms. The quantitative results are discussed in detail in Ref. .[3] Here the main qualitative findings are presented. All of the adsorbates studied here are found to bind with reasonably large binding energies to the diamond surface. Within LDA the acetylenic radical (C_2H) is the most strongly bound adsorbate with an adsorption strength that is 2.1 eV larger than the methyl radical, 1.7 eV larger than the hydrogen atom and 4.7 eV larger than an acetylene molecule. This point suggests that any acetylenic radicals would readily displace other

Table 2. Cohesive energies for several different hydrocarbon molecules as calculated within the local-density approximation and the generalized-gradient approximation. These results were obtained using 18 single s-gaussians, 9 single p-gaussians and 4 single d-gaussians. Accounting for the degeneracies of the p and d sets, this leads to sixty-five gaussians on each hydrogen and carbon atom. This basis set leads to converged results. See Refs. [3] and [11] for a more complete discussion of these results.

Molecule	LDA (eV)	GGA (eV)	Expt (eV)
H_2	4.89	4.55	4.75
C_2	7.51	6.55	6.36
CH_4	20.09	18.33	18.40
C_2H_2	20.02	18.09	17.69
C_2H_4	27.51	24.92	24.65
C_2H_6	34.48	31.24	31.22
C_6H_6	68.42	61.25	59.67

adsorbates on the diamond surface. The linear character of the acetylenic radical would allow for rather simple growth models and their presence in the vapor phase would most likely be beneficial since the addition of a C_2H adsorbate would introduce two strongly bound carbon atoms to the diamond surface. However, our calculations also predict that production of the C_2H radical costs approximately 1.0 eV more than CH_3 so it is not expected to be very abundant in the growth gas.

Of the hydrocarbon adsorbates we have studied, the next most strongly bound adsorbate is the methyl radical which is bound by a typical C-C bond energy (4.5 eV within LDA and ~3.8 eV experimentally). With respect to a single adsorption event, methyl radicals lead to the ideal placement of the carbon atom upon adsorption. The new C-C bond is found to be 2.92 Bohr in excellent agreement with the experimental C-C separation in diamond. Also the adsorbed carbon atom is tetrahedrally coordinated as it should be in the diamond lattice. While the methyl radical is an ideal single adsorbate, further calculations show that the umbrella-like structure of the adsorbate can be a severe limitation to methyl-mediated diamond chemical vapor deposition. That is, the repulsion between hydrogen atoms on neighboring adsorbates leads to strong steric repulsions which can significantly hinder growth. To illustrate this problem we constrained islands of N=1, 2 and 3 methyl radicals to be situated at ideal diamond lattice sites and allowed the 3N backbonded hydrogen atoms to relax under the action of their Hellmann-Feynman forces. Analysis of the results shows that the steric repulsions between neighboring adsorbates significantly decreases the removal energy of a single methyl radical. For two neighboring adsorbates, the energy required to remove the first methyl radical is reduced to 3.9 eV (LDA) due to the steric repulsions. However for an island of three neighboring methyl radicals the energy required to remove the first methyl radical decreases dramatically to 1.5 eV. For comparison we note that an acetylene molecule (C_2H_2) is bound to the diamond <111> surface by approximately 1.7 eV. Since the acetylene molecule is significantly more abundant than the methyl radical in the growth vapor, these energetic results suggest that the enhanced reactivity of gaseous methyl radicals and the good stability of isolated methyl adsorbates do not necessarily lead to good growth rates. The steric repulsions between neighboring adsorbates significantly alter the substrate's affinity for a high density of such adsorbates and lead to the conclusion that adsorbed acetylene molecules are as stable as islands of methyl adsorbates.

While the reactivity of a CH_3 molecule is not as favorable from the standpoint of carbon addition, the presence of vaporous CH_3 radicals and hydrogen atoms is important from the standpoint of maintaining the substrate. As pictured in our surface model (Fig. 1), the density of surface dangling bonds is rather small in comparison to the density of hydrogenated surface carbons. It is generally accepted that the surface-dangling-bond density must be reasonably small so that ordered growth of a diamond surface can occur. As the density of surface bonds increases, prospects for surface reconstruction and amorphous growth are expected to increase also. The emergence of either of these surface attributes would then have to be reversed to grow ideal diamond. If the surface dangling bond density is reasonably low, an isolated dangling bond will quickly bond to a vaporous entity such as methyl radicals, acetylene molecules or hydrogen atoms. Since typical growth conditions have a very high density of hydrogen in the growth gas, and since the LDA hydrogen adsorption energy (4.9 eV) is roughly the same as the methyl adsorption energy, it is reasonable to expect that hydrogen adsorption will occur much more often than hydrocarbon adsorption. Therefore, to obtain reasonable growth rates, a mechanism for hydrogen abstraction from the surface must be present. It has been suggested that the hydrogen surface abstraction is the rate limiting step for diamond growth.[16]

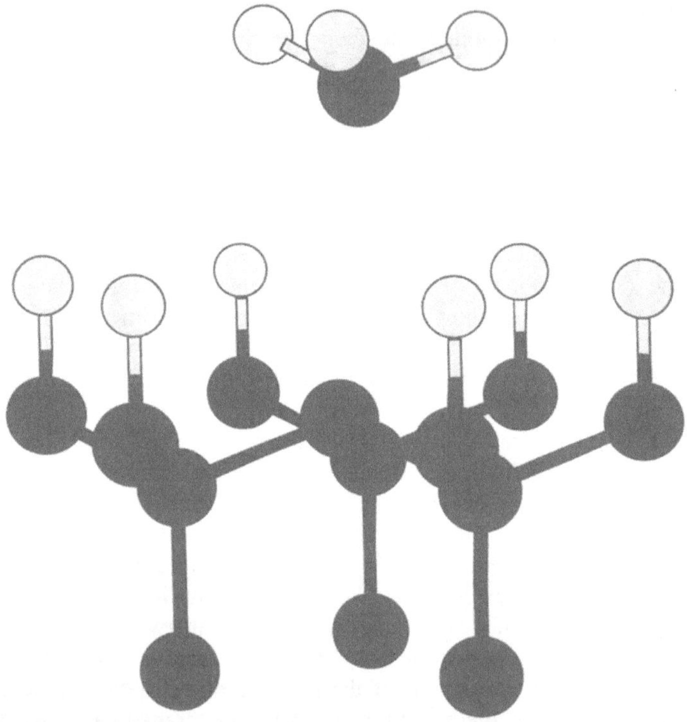

Figure 1. Schematic picture of a model used for studying the interaction between a gaseous adsorbate and an incompletely passivated diamond surface. As discussed in text, the presence of a dangling bond (H vacancy) on the diamond <111> surface allows a methyl radical, acetylenic radical and an acetylene molecule to bind to the surface. In addition to an emphasis on single site adsorption, it is necessary to consider repulsions between neighboring adsorbates.

Prospects for Abstraction Barriers within Density-Functional Theory

The rate at which surface and other reactions occur is largely governed by the reaction barrier rather than a reaction energy. Such barriers are generally on the order of 0.5-1.0 eV which is on the same order as the average error in LDA bond energies shown in Table 1. Since the overbinding that is present in the LDA is approximately equal to a reaction barrier it is reasonable to expect that transition states might also be overbound within the LDA. While such an overbinding is not expected to be very important for the understanding of higher energy phenomena, it is clear that qualitatively incorrect reaction barriers could be obtained from the local-density-approximation. Since reaction barriers can be very important to the understanding of fabrication, it is worth studying this problem in greater detail.

Pictured in Fig. 2 is a schematic diagram on how a simple hydrogen abstraction reaction might occur. In this case a CH_3 radical is being used to abstract a hydrogen atom from a CH_4 molecule. The object is to remove the black hydrogen atom from the methane molecule (Configuration A) and allow it to rebond with the methyl radical. This leads to the energetically equivalent mirror image of the original configuration (Configuration C). If the methyl and methane carbon atoms are well separated it is necessary to completely break a C-H bond before reforming it which leads to an upper bound for this reaction energy of a C-H methane bond (approximately 3.8 eV). Nature follows a different path during the abstraction of a hydrogen atom. As illustrated in the figure, the two carbon atoms come

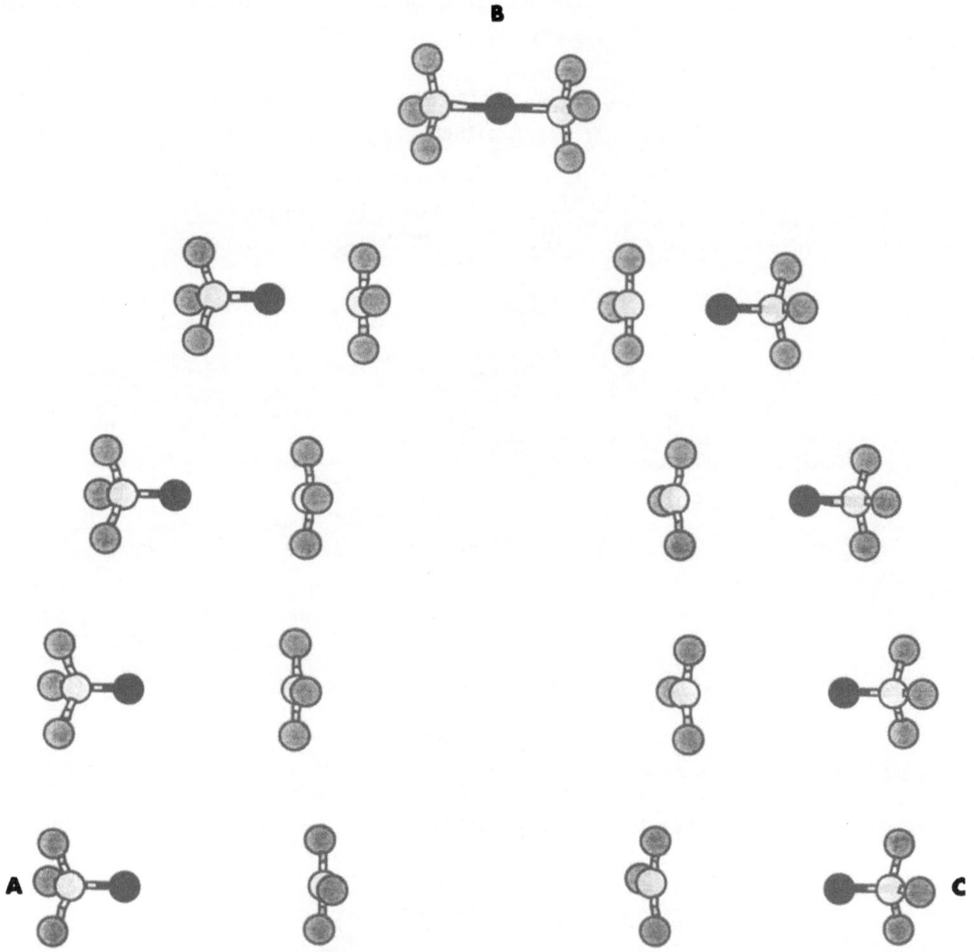

Figure 2. One of the growth limiting steps in diamond is believed to be the hydrogen abstraction reaction. One of the low energy hydrogen abstraction reactions is illustrated above. The lowest energy pathway between configuration A and configuration C is determined by allowing the two carbon atoms to be separated by 2.677 Å from one another at the transition state (configuration B). The remaining six hydrogen atoms relax to accommodate the environment mandated by the C-C separation and the intrabond H location. At the transition state configuration (B), there is one imaginary vibrational mode and the classical equilibrium geometry corresponds to a broken symmetry with the intrabond H atom relaxing off center. The experimental barrier is known to be approximately 14 kcal/mole.

close together, the six passive hydrogen atoms relax and the active hydrogen slowly pulls away from its original parent until it is able to overcome a very small barrier at the transition state (Configuration B). Once the hydrogen travels over the barrier the complex relaxes downhill until it reaches Configuration C.

To obtain the abstraction barrier we have used the C-C separation (X) and the position of the intrabond hydrogen atom (Y) as reaction coordinates. For many different (X,Y) configurations, we have minimized the total energy by allowing all the other nuclear degrees of freedom to relax under the action of their Hellmann-Feynman forces. The LDA and GGA energy surfaces as a function of configuration are presented in Fig. 3. Both of these surfaces illustrate the primary effects reasonably well. As noted above, for large C-C separations (X) the energy required to abstract the hydrogen is large since one of the C-H

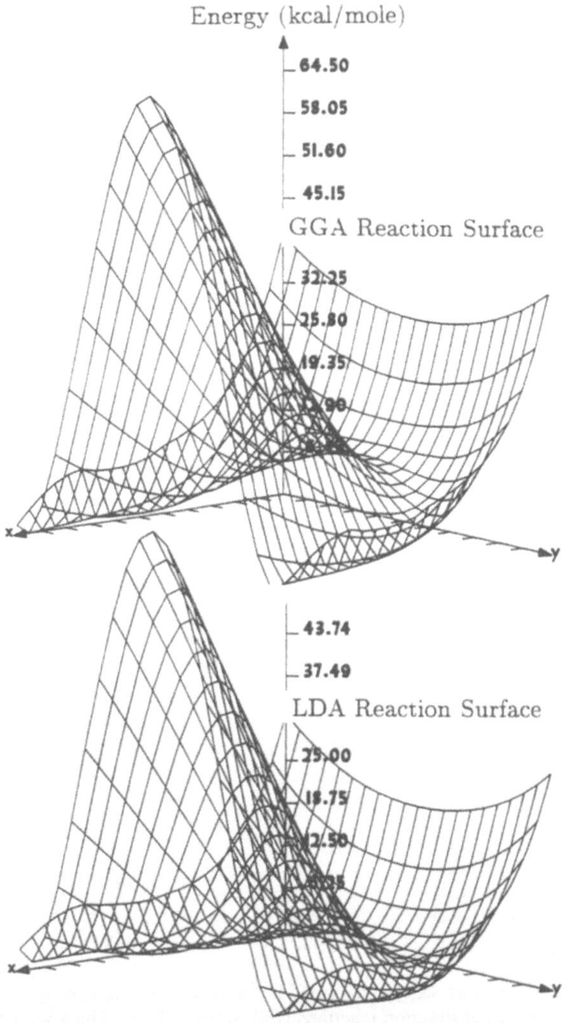

Figure 3. The reaction surfaces as generated within the local-density approximation and the generalized gradient approximation are compared. The x-axis corresponds to the separation between the two carbon atoms and the y-axis corresponds to the placement of the intrabond hydrogen atom. All other nuclear and electronic degrees of freedom have been relaxed. The local density approximation predicts a negligibly small (0.7 kcal/mole) abstraction barrier which is a factor of twenty smaller than the experimentally determined reaction barrier. In contrast the generalized gradient approximation predicts a barrier of approximately 9.4 kcal/mole.

bonds is completely broken before the other is reformed. For small C-C separations the kinetic energy repulsions between the carbons and the hydrogen atom dominate and the hydrogen prefers to sit at the center of the bond. In this regime, transfer of the hydrogen requires no energy but a large amount of energy is required to force the two carbon atoms close to one another. Within the LDA, the transition state occurs for a C-C separation of 2.677 Å. However, as expected from the above qualitative discussion, the LDA overbinding does indeed limit the accuracy of the reaction barrier and predicts an overall energy barrier of only 0.7 kcal/mole which is twenty times smaller than experiment. The generalized gradient approximation leads to a significantly improved reaction barrier with an overall barrier of 9.3 kcal/mole. The difference between these two energies (~0.3 eV) is on the order of a typical LDA bond energy.

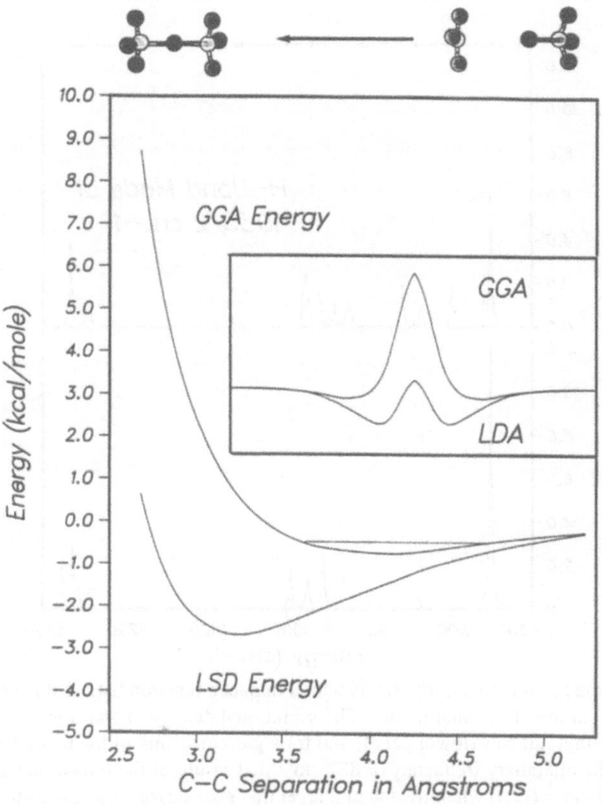

Figure 4. The LDA and GGA reaction pathways for the methyl-methane hydrogen abstraction barrier are compared above. The curve corresponds to the minimum energy pathway through the valleys of Fig. 3. The geometries are schematically presented above the top of the curve. In addition to dramatically underestimating the reaction barrier, in comparison to GGA, the LDA leads to a significantly more compact and overbound reactive complex. Experimental data on this complex is difficult to obtain. However we note that the GGA equilibrium separation of 4.1Å and cohesive energy of 0.4 kcal/mole is very close to the experimentally observed geometries and energies for the methane-methane interaction.

For further comparisons of the GGA and LDA we turn to Fig. 4 which shows the energy along the classical reaction path. Comparison of the two curves shows that both LDA and GGA lead to weakly bound reactive complexes with binding energies on the order of Van der Waals binding. The local density approximation leads to a more compact complex with a C-C separation of 3.27 Å and a binding energy of 2.8 kcal/mole. The generalized gradient approximation leads to a less compact complex with a C-C separation of 4.1Å and a binding energy of 0.7 kcal/mole. Consideration of zero point energies decrease the binding energies to 2.4 and 0.4 kcal/mole respectively. It is difficult to determine from experiment whether GGA or LDA is more accurately describing the reactive complex. To partially address this point we note that because the two systems are well separated it is expected that in this regime the CH_3-CH_4 interactions would be similar to CH_4-CH_4 interactions which are well characterized. The van der Waals parameters for two CH_4 molecules would predict a C-C separation of 4.2 Å and a binding energy of 0.3 kcal/mole. This analysis suggests that the GGA improves the description of the reactive complex as well as the transition state.

To determine the energy scale of the changes due to the zero point vibrational energy of the nuclei we have calculated the vibrational modes of a CH_3 radical, the CH_4

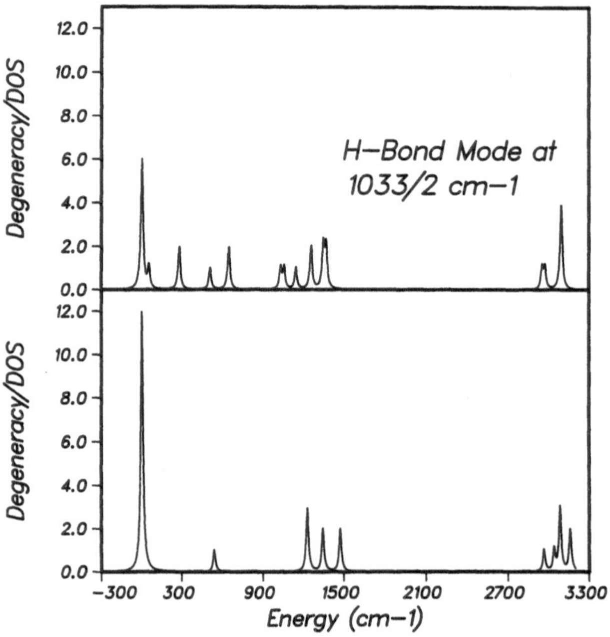

Figure 5. To obtain all the zero-temperature effects which impact a reaction barrier it is necessary to calculate the zero-point motion from the vibrational modes. The vibrational density of states for a gas consisting of isolated methane and methyl radicals (lower panel) and for a gas consisting of the methyl-methane transition state is shown above. The imaginary frequency of 883 cm^{-1} that results at the transition state geometry is determined to be 1033/2 cm^{-1} (real) when treated at a level that goes beyond the quadratic approximation. With respect to the isolated molecules, the transition state acquires an excess zero-point energy of 0.67 kcal/mole. The increase in zero point energy is largely driven by a reduction in the number of zero energy translational and rotational modes that occurs when the two isolated systems congregate.

molecule and the CH_3-CH_4 transition state. The resulting vibrational densities of states are compared in Fig. 5. As expected, within a quadratic approximation, the transition state yields one imaginary vibrational mode that is associated with the intrabond hydrogen atom relaxing off-center. To more accurately determine the energy associated with that mode I have solved the Schrödinger equation numerically for this vibration. The single-particle potential that is observed by the hydrogen and the ground and lowest excited state associated with this potential are presented in Fig. 6. At the higher level of approximation, we predict the intrabond hydrogen mode to have a real frequency of 516 cm^{-1} rather than an imaginary frequency. Including effects due to zero point motion enhances the zero temperature barrier by approximately 0.67 kcal/mole. See Ref. [5] for additional discussion.

Due to the fact that density-functional based calculations are intrinsically less complicated and faster than traditional quantum-chemical methods, the ability to perform reaction calculations within such a framework is very desirable. The example of this section suggests that density-functional based predictions of cluster-cluster reactions are of sufficient accuracy to help with an understanding of reactions that are important for an understanding of crystal growth and catalysis. While this example is indeed promising, additional calculations are required before a full understanding is possible. Possible outcomes are that existing approximations to the density-functional theory will prove to be accurate enough for such studies or that an analysis of the small errors will lead to improved functionals. It is also possible that synergistic approaches which rely on many

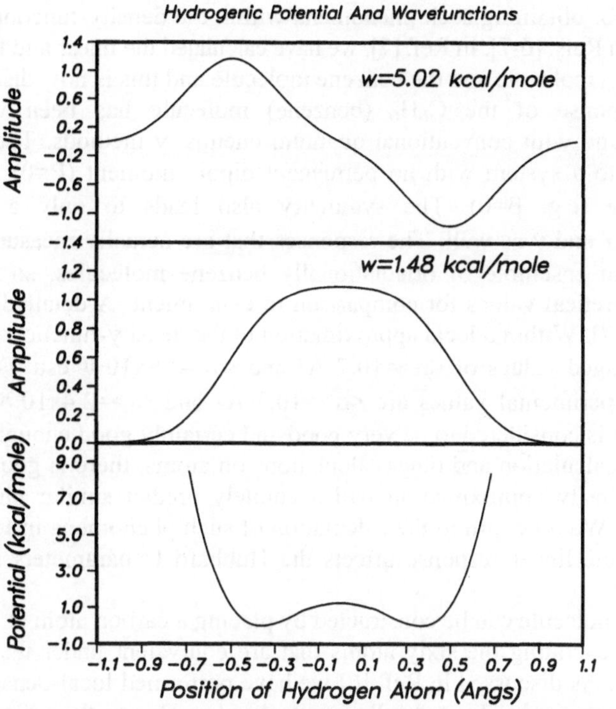

Figure 6. The potential (a) observed by the active hydrogen molecule as it travels along the axis between two carbon atoms. In contrast to the quadratic approximation that leads to an imaginary vibrational mode that is comprised of 73 % intrabond H motion, numerical solution of the one-dimensional Schroedinger equation lead to a gerade occupied bound state at 1.48 kcal/mole (b) and a lowest ungerade excited state at 5.02 kcal/mole (c).

density-functional calculations for obtaining pathways and traditional quantum chemical approaches for a few final configurations will prove to be the best approach.

IV. THE OPTICAL RESPONSE OF CLUSTERS AND CLUSTER ASSEMBLIES

The density-functional theory allows for the determination of the ground state energy for a system of electrons in a static external potential. As a potential it is possible to determine the response of a cluster to a slowly varying external potential. In this section we review the formalism and applications of the density-functional theory to this problem. A more complete discussion of this work may be found in Refs. [6-7].

To obtain static and linear hyperpolarizabilities within the density-functional framework, one appends a field-dependent term to the usual energy functional and then minimizes the total electronic energy [W(**E**)] for different values of the field strength (**E**). Due to the interaction of the electronic charge distribution with the electric field the systems energy changes according to:

$$W(E)=Wo+\Sigma_i P_i E_i+ \Sigma_{ij} E_i\, \alpha_{ij} E_j+ \Sigma_{ij}\, \beta_{ijk} E_i E_j E_k+ \Sigma_{ijkl}\, \gamma_{ijkl} E_i E_j E_k E_l \qquad (1)$$

In the above equation, W_o is the LDA total energy at zero field, **P** is the system's permanent dipole moment and α, β and γ are the first, second and third order polarizability tensors. Different conventions on multiplicative prefactors in front of these tensors exist.

The formalism for obtaining such phenomena within the density-functional framework has been discussed in Refs. [6-7]. In Ref. [7], we have calculated the linear and hyperpolarizability tensors of the C_{60} molecule and the benzene molecule and this is now discussed.

The response of the C_6H_6 (benzene) molecule has been well characterized experimentally and with conventional quantum chemistry methods. The symmetry of the molecule leads to a system with no permanent dipole moment (**P**=0), and no odd-order optical response (e.g. $\beta=0$). The symmetry also leads to only a few independent components of α and γ as well. The responses that are actually measured experimentally correspond to an ensemble of orientationally benzene molecules, so it is necessary to average the theoretical values for comparison to experiment. A detailed discussion of this appears in Ref. [7]. Within a local approximation to the density-functional theory we obtain spherically averaged values of $<\alpha>=10.7$ Å^3 and $<\gamma>=1.9\text{x}10^{-36}$ esu. For comparison we note that the experimental values are $<\alpha>=10.3$ Å^3 and $<\gamma>=2.4\text{x}10^{-36}$ esu respectively. This comparison is considered to be very good and certainly good enough for technological uses. From this calculation and other calculations on atoms, there is good reason to expect that the local-density approximation will accurately predict similar phenomena in other systems as well. We now turn to the calculation of such phenomena in fullerene molecules and show how the linear response affects the Hubbard U parameters in pure and alkali doped fullerenes.

The C_{60} molecule can be constructed by placing a carbon atom at R=(0.0, 6.57,1.32) Bohr and then generating the sixty atoms that are equivalent under the icosahedral point group symmetry. As discussed in Ref. [6] we have performed local-density-calculations on the neutral and negatively charged fullerene molecules. Due to the very high symmetry of the isolated molecule, there is no permanent dipole moment, no odd polarizability tensors and the lowest nonvanishing response terms are isotropic with respect to field orientation. We have performed calculations as a function of field strength to determine the linear- (α) and hyper-polarizabilities (γ) of a fullerene molecule. By performing calculations at six different field strengths ranging between 0.0 and 0.032 Hartree/Bohr and analyzing the dependence of the dipole moment and total energy on field strength we are able to unambiguously determine these polarizabilities to be $\alpha=82.7$ Å^3 and $\gamma=7\text{x}10^{-36}$ esu. The linear polarizability has been measured by to be 83.5-84.9 Å^3.[17,18] Recent work by Ecklund *et al.* has attributed approximately 2 Å^3 of the polarizability to the nuclear relaxations so the agreement between theory and experiment appears to be very good indeed.[19] With respect to the nonlinear polarizability, the agreement would not be characterized as excellent but there is good reason to expect that the agreement is good enough to encourage further joint theoretical and experimental investigations of nonlinear phenomena. The experimental result for the hyperpolarizability is $\gamma=313\text{x}10^{-36}$ esu.[20]

Polarization Contributions to the Hubbard U Parameters in Fulleride Crystals

In order to understand superconductivity in the solid state, researchers generally reduce the full Hamiltonian to a simplified Hubbard or extended Hubbard Hamiltonian that folds a variety of physical phenomena into nearest-neighbor hopping integrals and on-site repulsions between electrons at the same site. In an isolated system the Hubbard U parameter is related to the ionization energy and the electron affinity. Alternatively, by assuming the electrons are being incrementally added to a single band, the Hubbard U parameter is simply related to the second derivative of the total energy with respect to the electronic occupation numbers. In general, the Hubbard U parameter associated with a cluster decreases as it congregates with other clusters. In the limit of zero overlap, the decrease of the Hubbard U parameter is due primarily to the polarization of neighboring clusters and subsequent interaction of their dipole moments. By placing an excess charge on a fullerene molecule an electric field $\mathbf{E}=Q\mathbf{r}/r^3$ results at all other points in space. Neglecting other effects would lead to an induced dipole moment $p=\alpha\mathbf{E}$ at each point in space and a subsequent lowering of energy. However since the resulting lattice of dipole moments interact with one another , the response of each fullerene molecule is modulated. In Ref. [6] a model for a loosely knit assembly of polarizable clusters was introduced and solved as a function of cluster size and the number of charge defects. From the results of this work, the dielectric constant and the screened Hubbard U parameters have been determined. We find that the Hubbard U parameter is reduced from the isolated value of 3.0 eV to 1.27 eV when the fullerene molecules are placed on an FCC lattice.

V. VIBRATIONAL MODES OF CLUSTERS AND ASSEMBLIES

As the research and development of complex cluster-assembled materials continues there will be a greater need for theoretically aided characterization of such materials. Experimental techniques for the materials characterization include the measurement of electronic density of states, the determination of geometries from an analysis of the x-ray data and the measurement of the infrared (IR) and Raman vibrational spectra. For simple materials it is not always necessary to perform all three measurements on a sample to confidently determine the system's structure. However, for complex materials many different techniques may be required to confidently determine the structure of a material. In this section we discuss a recent application of the density-functional theory which aided in characterizing the structure of the monoalkali doped fullerene crystals.

Recent work by Ecklund *et al.* [21] and Stephens *et al.* [22] have suggested that under certain conditions fullerene molecules can polymerize. As pictured in Fig. 7, polymerization may be achieved by allowing the short polar dimer bonds to break. This would lead to two adjacent three-fold coordinated carbon atoms that each have a dangling bond extending radially outward. By allowing another similar fullerene molecule to come into close proximity the extramolecular dimer bonds rehybridize which leads to a four-membered ring of four-fold coordinated carbon atoms. With respect to the stability of the dimer shown in Fig 7, conventional arguments can lead to two different conclusions. For example if the C atoms on a fullerene molecule are viewed as resonating graphitic bonds one would conclude that it would be energetically unfavorable to create the four-membered ring. Alternatively if the fullerene molecule is viewed as a collection of atoms with single and double bonds, it is expected that polymerization via four-fold coordinated rings should be energetically favorable. Since the fullerene molecules exhibit a definite albeit reduced bond alternation, it is clear that the resonating picture is not completely valid and that a stable polymer might be preferred. To address this point we have performed calculations on two four-member analogs of the polymers. The first structure consists of the $[C_2H_4]_2$ dimer and is also pictured in Fig. 7. After fully relaxing this structure we find that the C-C bond distances are stretched from their normal double-bond lengths to a slightly strained single bondlength of 1.548 Å. The hydrogen atoms relax out of the plane of the carbon dimers and lead to a local environment that closely resembles tetrahedrally coordinated carbon atoms. Our calculations show that this system is bound within LDA by 1.71 eV. The lowest energy pathway between these structures has not yet been determined. However we note that if the two molecules follow a simple path that extrapolates between the isolated and bound states a LDA barrier of 0.59 eV is observed. Since the hydrogen atoms naturally relax out of the plane during the formation of the four-membered ring, it is reasonable to expect that this type of dimerization would occur between two bucky balls.

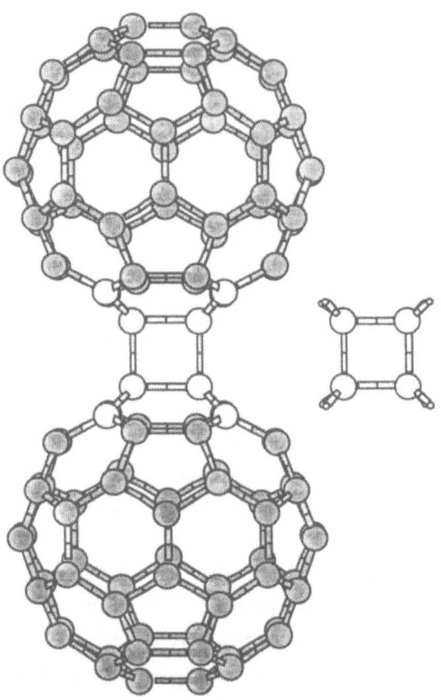

Figure 7. A picture of fullerene molecules that are dimerized by the introduction of a four-membered carbon ring. Also pictured is a smaller $[C_2H_4]_2$ dimer that is a simpler analog of this ring. We find that the C-C bonds associated with four-membered rings to be in the range of 1.55-1.60Å suggesting that local charge states and second neighbor interactions do not significantly influence these bondlengths. Our calculated center-to-center distance of 8.81Å is in good agreement with the x-ray data of Stephens et al. The small deviations between our results and their measurements are very likely due to Rb-C interactions.

To further address the questions of the energetic stability of fullerene polymers we have relaxed the shaded atoms in Fig. 7 to the positions shown in the figure and found that the energy of the resulting 180 atom complex to be more stable than two isolated fullerene molecules by 0.2 eV. The resulting density of states is compared to the isolated fullerene density of states in Fig. 8. An analysis of the Hellmann-Feynman forces suggest that the center of mass of each of the fullerene molecules is in the correct location and that the forces on the atoms that are more than three layers from the four-membered ring observe negligible Hellmann-Feynman forces. This suggests that further relaxations would lead to small albeit rigid downward-displacements of the lower two-thirds of the hemisphere and slight upward and non-rigid displacement of the upper third of the fullerene molecules. To estimate the additional energy gained by fully relaxing the fullerene dimer we have used a simple one-parameter model that depends on an approximate force constant matrix and our calculated Hellmann-Feynman force. Assuming the dynamical matrix is diagonal, the additional relaxation is given by 60 $(Frms)^{+}/2h_d$ where Frms is the root-mean-square of the force and h_d is the value of the diagonal elements of the force constant matrix. A reasonable estimation of the relaxation energy is then determined by using the average of the diagonal components of the 180x180 C_{60} force constant matrix(4) which yields a value of $h_d = 0.45\ au$. Within this approximation the additional relaxation is given by $60|F_{rms}|^2/2h_{avg}$ where F_{rms} is the root-mean-square of the force and h_{avg} is the average of the force constant matrix. Within this approximation an additional 0.8 eV stabilization occurs suggesting that there is a net stabilization of 1.0 eV per four membered ring in the polymer. This number is in reasonable agreement with the results of Ecklund which suggest that it requires 1.25 eV to

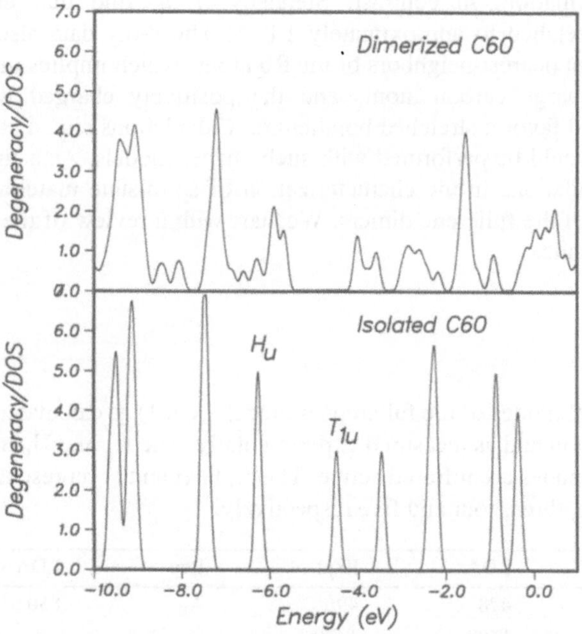

Figure 8. A comparison of the electronic density of states associated with an isolated fullerene molecule and a dimerized fullerene molecule illustrates several universal expectations about the electronic and vibrational energy spectrum. For a chain of fullerene molecules that are polymerized along the z axis, the highly degenerate spectrum observed in the C_{60} molecule splits into eight different 1-fold representations. However, the perturbations along the x and y axes are reasonably mild which causes some of the degeneracies present in C_{60} to reappear as accidental degeneracies in the chain. For example, the three-fold T_{1u} level splits into a one-fold and quasi two-fold level.

remove a fullerene molecule from an oligimer chain. We note that better agreement is not necessarily expected since an activation energy is expected to be larger than a cohesive energy. However since LDA overestimates cohesive energies our theoretical energy is enhanced in the same direction as the experimental energy.

There are two types of oligimers that are assumed to be created. As discussed in Ecklund *et al.*[21] are making a variety of oligimers of different lengths. The recent x-ray data of Stephens *et al.*[22] pertains to chains of infinite length and allows for a determination of some of the important structural parameters. The most striking result of Stephens *et al.* is that the fullerene center-to-center distance is significantly shorter than that found in the usual cubic fullerene crystals. Stephens *et al.* predict a center-to-center distance of 9.12 Å which is only 0.3 Å larger than we find for the fullerene dimer. We believe that the results from our calculations are in accord with the experimental measurement for the following reason. First, the structures studied by Stephens *et al.* contained a concentration of Rb. The Rb atoms are ionized and lead to a hard-sphere repulsion between the Rb atoms and the surfaces of the fullerene molecules. For the Rb_3C_{60} compound, the center-to-center distance is known to be 0.24 Å larger than that of the pure C_{60} crystal. Further, the x-ray analysis suggests that the fullerene atoms would be sandwiched between adjacent six-membered rings (shown in Fig. 7) and would lead to a repulsion between the two bucky balls that was not present in our calculation.

The presence of the Rb atoms might also explain the only other deviation between our theoretical structure and that of Stephens *et al.* Our results for both the four-membered rings lead to a more-or-less normal C-C bond for all of the atoms that participate in four-membered-ring formation. In contrast, Stephens *et al.* find that the polar fullerene bondlengths are stretched to approximately 1.8 Å. The x-ray data also shows that these carbon atoms are not nearest-neighbors of the Rb atoms which implies interactions between these negatively charge carbon atoms and the positively charged Rb ions would be attractive and would favor a stretched bondlength. Calculations aimed at understanding the Rb-C interactions could be performed with such cluster models. As a final example of the use of cluster calculations in the characterization of solid-state materials, we discuss the vibrational modes of the fullerene dimers. We start with a review of the vibrational modes of a fullerene molecule.

Table 3. Vibrational modes of the fullerene molecule (cm^{-1}) as calculated within the local density approximation and as measured experimentally. The A_g and H_g modes are Raman active and the F_{1u} modes are infrared active. The A, F, G and H representations have degeneracies of one, three, four and five respectively.

Rep	LDA	Expt	Rep	LDA	Expt
A_g	478	496	A_u	850	
	1499	1470			
			F_{1u}	547	527
F_{1g}	580			570	577
	788			1176	1183
	1252			1461	1428
F_{2g}	547		F_{2u}	342	
	610			738	
	770			962	
	1316			1185	
				1539	
G_g	486				
	571		G_u	356	
	759			683	
	1087			742	
	1296			957	
	1505			1298	
				1440	
H_g	258	273			
	439	437	H_u	404	
	727	710		539	
	767	774		657	
	1093	1099		737	
	1244	1250		1205	
	1443	1428		1320	
	1576	1575		1565	

The vibrational modes of a fullerene molecule, as calculated by Quong *et al.*,[4] are presented in Table 3. Since the fullerene molecule consists of sixty atoms, classical arguments suggest that there will be a total of 174 non-zero vibrational modes. The six zero-energy vibrational modes are due to the three rotational and translational degrees of freedom. Due to the icosahedral symmetry of the fullerene molecule the vibrational spectrum is split into two nondegenerate representations (A_g and A_u), four triply degenerate representations (F_{1g}, F_{1u}, F_{2g} and F_{2u}), two four-fold degenerate representations (G_g and G_u) and two five-fold degenerate representations (H_g and H_u). There are a total of ten Raman active modes, two of which have A_g symmetry and eight of which have H_g symmetry. The four F_{1u} modes are infrared active. A comparison of our theoretically determined vibrational modes with experiment leads to the following conclusions. Typical discrepancies between experiment and the local-density-approximation appear to be 20-30 cm^{-1} for the fourteen modes IR or Raman active modes. We expect that the optically silent modes exhibited in Table 3 are of equal accuracy. While 20-30 cm^{-1} should be considered as good accuracy, it does place limitations on experimental and theoretical comparison. For example, if the symmetry of a fullerene molecule is broken by the presence of a carbon-13 atom, an adsorbed hydrogen, or by dimerization, the mixing that occurs between optically silent modes and optically active modes leads to an enriched IR spectrum. However, since there is an inherent uncertainty of 20-30 cm^{-1} in theoretical modes, the relative intensities of the peaks in the resulting IR spectrum may be shifted and juxtaposed when compared to experiment. In some parts of the energy spectrum the number of peaks could appear to be incorrect if the experimental line widths are also of this order or if theory and/or experiment have symmetrically distinct modes that are accidentally degenerate.

To obtain a reasonably simple model for the vibrational modes of a chain of fullerene molecules, we have performed a series of local-density-calculations to extract the couplings between atoms on different fullerene molecules and to determine the dimerization induced perturbations of the dynamical matrix. Heuristically, one expects that the primary on-ball perturbations of the dynamical matrix will be to change the diagonal elements of the dynamical matrix near the poles of adjacent C_{60} molecules. The four-fold coordination of the ring carbons is expected to sphericalize the diagonal block of the dynamical matrix. A reasonable model is to replace the strong directional dependence associated with the diagonal block of an isolated matrix by a diagonal matrix with elements $h_{xx}=h_{yy}=h_{zz}=h_{avg}=(991\ cm^{-1})^2$, where h_{avg} is the average value of the dynamical matrix found in Ref. [8]. By transforming the Hessian matrix elements associated with the dimer carbons at the north and south poles of a fullerene molecule, one obtains a dynamical matrix that exhibits the symmetry of a long fullerene n-mer. The vibrational spectrum associated with this one-ball model is exhibited in Fig. 9. In addition to the total density of states, the projection of the density of states onto the Raman-active and IR -active manifolds are also included.

The vibrational spectrum is complicated and there are some energy ranges where this simple treatment is not expected to account for the changes in IR activity. Each energy regime must be analyzed from several perspectives when comparing to experiment. Here we limit the discussion to the part of the IR spectrum in the energy range of the $F_{1u}(4)$ energy. Analysis of the polymerized vibrational states in this energy range reveals that the twin-peaked structure of the projected $F_{1u}(4)$ density of states is due in entirety to a mixing between the optically silent $G_u(6)$ levels with the optically active $F_{1u}(4)$ levels. In Ref. [8], a simple two state analysis was used to show that the effect of dimerization causes a splitting between the two peaks of $\sqrt{(\varepsilon_1-\varepsilon_2)^2+(\Delta/\varepsilon_c)^2}$ where ε_1 and ε_2 are the energies of the unperturbed $F_{1u}(4)$ and $G_u(6)$ states, Δ is the coupling constant and ε_c is the average of the

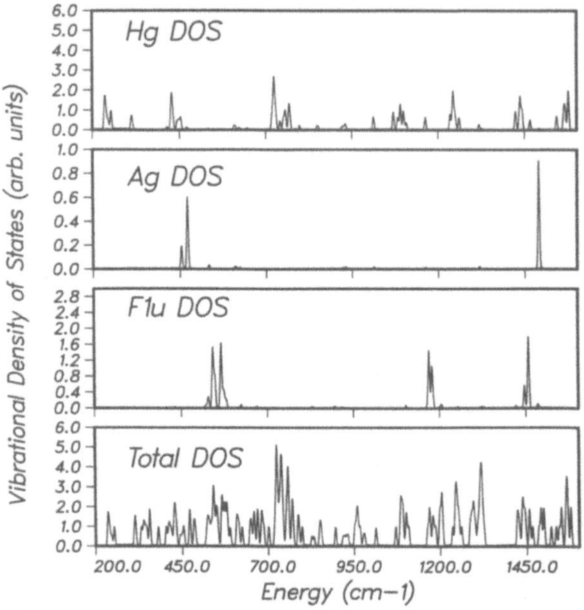

Figure 9. The total and projected vibrational density of states associated with a fullerene dimer. Upon dimerization, the optically active F_{1u} level mixes with optically silent modes and leads to a richer infrared spectrum. Many of the qualitative features observed by Mihaly and Martin[23] are reproduced by the projected F_{1u} vibrational density of states.

unperturbed phonon energies. The coupling constant is calculated to be approximately (Δ/ε_c)=135 cm-1.

The appearance of the double peak for the polymerized model is then explained by noting that the (x,y) rows of the $F_{1u}(4)$ and $G_u(6)$ are essentially unperturbed by the dimerization but the z rows of the $F_{1u}(4)$ and $G_u(6)$ mix strongly, leading to two peaks that are 10 and 105 cm-1 below the original $F_{1u}(4)$ peak. The twin-peaked structure that is experimentally observed corresponds to the observation of the more-or-less pure (x,y) rows of the $F_{1u}(4)$ state and a second peak which is due to the mixed $G_u(6)$-$F_{1u}(4)$ z rows. Experimentally the low-intensity peak is 15 cm-1 above rather than 10 cm-1 below the high intensity peak.[23] We have verified that this observation implies that the energy of the pure $G_u(6)$ level is higher than the pure $F_{1u}(4)$ level. While heuristic arguments have been used in entirety here, we note that they are based on results from density-functional based cluster calculations and the details of this work can be found in Ref. [8].

VI. CONCLUSION

As is clear from the work discussed here and elsewhere in this book there are a variety of ways in which cluster calculations are impacting the understanding and design of solid state properties. Also illustrated in the above discussion is the fact that each theory and numerical algorithm has specific subfields where it is most appropriate and the discussion of this topic necessitates equal emphasis on both the specific theory and the application. With respect to density-functional based calculations several known results have been reiterated here. First, it is possible to obtain very accurate geometries, vibrational modes, electronic densities, polarizabilities and hyper polarizabilities within a local approximation of the theory. Second, the overbinding that is present in existing local approximations to the density-functional theory can be circumvented if the cluster energies are calculated within the generalized gradient approximation. Finally, we presented results that show one example where the generalized gradient approximation significantly improves results. In contrast to the local-density-approximation which leads to qualitatively incorrect results for the methyl-methane hydrogen abstraction energy, we show that the generalized gradient approximation leads to results that are almost quantitatively correct.

With respect to applications we have shown that the density-functional method can be used to learn about growth processes that occur during the fabrication of solid-state materials, calculate barriers associated with formation, understand the optical response of an assembly of clusters, and use theoretical geometries, electronic structures and vibrational modes to aid experimentalists in the characterization of solids composed of clusters. While the examples discussed here touch upon a reasonably wide range of uses for cluster calculations in solid state physics many other uses have been discussed elsewhere. There is little doubt that calculations on such systems will become increasingly important as theorists continue to improve their computational means for exploring such problems and as experimentalists enhance their abilities to fabricate cluster based materials.

ACKNOWLEDGMENTS

The work reviewed in this text is a compilation of several different collaborations between myself and others in the Complex Systems Theory Branch at the United States Naval Research Laboratory. Much of the diamond work was performed in collaboration with Dr. Koblar A. Jackson and Dr. Warren E. Pickett. and was supported by ONR-SDIO-IST. The work on vibrational modes and polarizabilities of C_{60} was performed in collaboration with Dr. Andrew A. Quong.

REFERENCES

1. P. Hohenberg and W. Kohn, Phys. Rev. **136**, B864 (1964); W. Kohn and L. J. Sham, Phys. Rev. , A1133 (1965).
2. (a)M. R. Pederson and K. A. Jackson , Phys. Rev. B 41, 7453 (1990); (b)K. A. Jackson and M. R. Pederson, Phys. Rev. B. 42, 3276 (1990); M. R. Pederson and K. A. Jackson, Phys. Rev. B 43, 7312 (1991).
3. M. R. Pederson, K. A. Jackson and W. E. Pickett, Phys. Rev. B 44, 3891 (1991).
4. A. A. Quong, M.. R. Pederson and J. L. Feldman, Sol. Stat. Comm. 87, 535 (1993).
5. M. R. Pederson, Chem. Phys. Lett. 230, 54 (1994).
6. M. R. Pederson and A. A. Quong, Phys. Rev. B 46, 13584 (1992).
7. A. A. Quong and M. R. Pederson, Phys. Rev B 46, 12906 (1992).
8. M. R. Pederson and A. A. Quong, Phys. Rev. Lett. 74, 2319 (1995).
9. J. P. Perdew and A. Zunger, Phys. Rev. B 23, 5048 (1981).
10. J. P. Perdew, Phys. Rev. Lett. 55, 1665 (1985);J. P. Perdew and Y. Wang, Phys. Rev. B. 33, 8800 (1986); J. P. Perdew, Phys. Rev. B 33, 8822 (1986), Y. Wang and J. P. Perdew, Phys. Rev. B 44, 13298, (1991); J. P. Perdew and Y. Wang, Phys. Rev. B 45, 13244 (1992).
11. J. P. Perdew, J. A. Chevary, S. H. Vosko, K. A. Jackson, M. R. Pederson, D. J. Singh and C. Fiolhais, Phys. Rev. B 46, 6671 (1992).
12. E. E. Lafon and C. C. Lin, Phys. Rev. 152 , 579 (1966).
13. A. H. Stroud, *Approximate Calculations of Multiple Integrals* , (Prentice-Hall, Englewood Cliffs, NJ, 1971).
14. W. H. Press, B. P. Flannery, S. A. Teukolsky, and W. T. Vetterling, *Numerical Recipes: The Art of Scientific Computing* (Cambridge University Press, Cambridge, 1988).
15. M. R. Pederson, A. A. Quong, J. Q. Broughton and J. L. Feldman, Comp. Mat. Sci. 2, 129 (1994).
16. For a compilation of articles on subjects related to diamond chemical vapor deposition see: J. Mater. Res. 11 (1990).
17. P. Ecklund, Bull. Am. Phys. Soc. 37, 191 (1992).

18. A. F. Hebard, R. C. Haddon, R. M. Fleming, and A. R. Korton, Apl. Phys. Lett. 59, 2109 (1992).
19. P. Ecklund *et al.* (to appear).
20. Z. H. Kafafi, J. R. Lindle, R. G. S. Pong, F. J. Bartoli, L. J. Lingg and J. Milliken, Chem. Phys. Lett. 188, 492 (1992); Z. H. Kafafi, F. J. Bartoli, and J. R. Lindle, Phys. Rev. Lett. 68, 2702 (1992).
21. P. Zhou *et al.*, Chem. Phys. Lett. 211, 337 (1993); Y. Wang *et al.*, Chem. Phys. Lett. 217, 413 (1994); A. M. Rao *et al.*, Science 259, 955 (1993).
22. O. Chauvet, G. Oszlanyi, L. Forro, P. W. Stephens, M. Tegze, G. Faigel and A. Janossy, Phys. Rev. Lett. 72, 2721 (1994).
23. M. C. Martin, D. Koller and L. Mihaly, Phys. Rev. B 47, 14607 (1993); M. C. Martin *et al.*, Phys. Rev. B 51, 3210 (1995).

QUANTUM MONTE CARLO FOR ELECTRONIC STRUCTURE OF SOLIDS

Luboš Mitáš

National Center for Supercomputing Applications
University of Illinois at Urbana-Champaign
Urbana, IL 61801

I. INTRODUCTION

Currently, most of the *ab initio* electronic structure calculations of solids are based on the Local Density Functional Theory which in practice employs the Local Density Approximation (LDA).[1] The LDA approach has tremendously increased our ability to predict electronic and atomic structures, cohesive energies and other fundamental quantities for a large variety of condensed and molecular systems. However, there are still open problems with the accuracy of this method like underestimated band gaps and band widths for many solids or difficulties in describing systems with transition metal elements. LDA also produces different systematic errors when describing systems with different electronic structure (like atoms and solids) which often results in significant overbinding. The experience in correcting these discrepancies by using, for example, the Generalized Gradient Approximation is somewhat ambiguous.[2-5] Moreover, there seems to be no general consensus on what might be the most efficient and robust approach to improve the LDA deficiencies.

As an alternative to the Local Density Functional Theory, a high degree of accuracy is offered by quantum chemistry methods like the Configuration Interaction or Coupled Cluster approaches.[6] These methods, while exceedingly accurate for small systems (10 to 15 correlated electrons), are rather difficult to apply to solids or large molecular systems. There are essentially two sources of these difficulties: unfavorable scaling in the number of electrons and the necessity to use a sufficiently rich basis set. The combination of these two factors makes calculations of larger systems very problematic.

In recent years yet another alternative for accurate electronic structure calculations has emerged in the field of quantum Monte Carlo (QMC).[7] This approach is based on stochastic methods which are used to evaluate multi-dimensional integrals and also to solve the many-body Schrödinger equation with a high degree of accuracy. Applications to (relatively) simple systems like the homogeneous electron gas, hydrogen and helium systems, and small atoms and molecules were very successful,[8-12] and several of these calculations have become benchmarks and references for the calibration of other, mostly approximate, methods. However, the development of QMC methods for systems with heavy elements was not straightforward because of the problems with large energy contributions coming from core electrons. Fortunately, once it was understood that this was

Edited by T.A. Kaplan and S.D. Mahanti, Plenum Press, New York, 1995

the key obstacle in expanding this powerful approach to more complicated systems, several research groups developed valence-only approaches to deal with the atomic cores.[13-18] This development has opened entirely new possibilities for studying the electronic structure of large molecules, clusters and solids.

However, most of the QMC calculations so far have been devoted to the evaluation of ground state total energies. Very recently, the first QMC calculation of an electronic excited state in a solid was carried out,[19] opening new possibilities for band structure analysis by this method.

In this paper, we would like to give a brief overview of these developments. The paper is organized as follows. In the next section we will briefly introduce the variational and diffusion Monte Carlo (VMC and DMC) methods. Then we will mention the techniques for elimination of core electrons and a way to deal with the valence-only Hamiltonians in these methods. We will then present the form and optimization of the trial (variational) wave function. Next, we will apply this method to (*i*) nitrogen systems which include two types of solids, and (*ii*) to diamond. In this part we also describe calculations of excited states which give information about the band structure. Finally, we will conclude with a discussion about the scaling of this approach with the number of valence electrons.

II. VARIATIONAL AND DIFFUSION MONTE CARLO METHODS

In the variational Monte Carlo (VMC) method one evaluates the expectation value of a given Hamiltonian H of a system with N electrons,

$$E_V = \frac{\int \Psi^*(R) H \Psi(R) dR}{\int \Psi^*(R) \Psi(R) dR} = \frac{\int |\Psi(R)|^2 [H\Psi(R)/\Psi(R)] dR}{\int |\Psi(R)|^2 dR} \tag{1}$$

where $\Psi(R)$ denotes the trial function, and the integration over $3N$-dimensional space $R = (r_1, r_2, \ldots, r_N)$ with the probability density $|\Psi(R)|^2$ is carried out by the Metropolis Monte Carlo technique. The Monte Carlo method of integration allows us to construct the trial function as a true many-body object which includes the electron-electron and also higher order correlations in an explicit form. In particular, it is very easy to incorporate the electron-electron cusp, a non-analytic behavior of the wave function at the point where two electrons coalesce.[20-22] It is well-known that this non-analytic feature is the main reason for a slow convergence of the wave function expansion in quantum chemistry methods.[20] The VMC method is robust and relatively fast; however, the variational energy is biased by the choice of the trial function, number and optimality of variational parameters, *etc.* Thus, except for few-electron systems, it is rather difficult to construct a trial function for which the variational bias is negligible. The problem of the variational bias must be addressed and is particularly important, for example, when comparing energies of two different systems (like an electron gas in two different phases) because the quality of trial functions in the two cases might be different.

Fortunately, the diffusion Monte Carlo (DMC) or projector method, which solves stochastically the Schrödinger equation with a high degree of accuracy, is capable of removing a large part of the variational bias. It is not very complicated to show that[7]

$$\lim_{t\to\infty} \exp(-tH)\Psi(R) \propto \exp(-tE_0)\Phi_0(R), \tag{2}$$

where $\Phi_0(R)$ is the ground state of the prescribed symmetry, E_0 is the corresponding eigenvalue and we suppose that $\langle\Psi|\Phi_0\rangle \neq 0$. The long time solution of this projection can be found by solving the imaginary time Schrödinger equation. It is presented here in a form which includes the importance sampling by the trial function

$$f(R,t+\tau)=\int G(R,R',\tau)f(R',t)dR' \tag{3}$$

where

$$f(R,t)=\Psi(R)\Phi(R,t) \tag{4}$$

and

$$G(R,R',\tau)=\frac{\Psi(R)}{\Psi(R')}\langle R|\exp(-\tau H)|R'\rangle. \tag{5}$$

while $\Phi(R,t)$ is the solution of the time-dependent Schrödinger equation. In an actual simulation the solution of (3) is found iteratively because the Green's function $G(R,R',\tau)$ is known in an analytical form only for small τ.[7] The fermion problem is treated by using the fixed-node approximation which enforces the nodes of the resulting wave function $\Phi(R,t)$ to be identical to the nodes of $\Psi(R)$ so that the product $\Phi(R,t)\Psi(R)$ is non-negative everywhere (the node is a subset of (R)-space defined by $\Psi(R)=0$).[7,9,11] In most cases the impact of the fixed-node approximation on the calculated quantities is rather small (on the order of several percent of the correlation energy). However, there are cases where the fixed-node error can be significant if an adequate trial function is not used, *e.g.*, when there is a strong near-degeneracy effect so that the trial function has a multi-configurational character.[20,22]

The result of the stochastic solution of (5) is a set of samples of the product $\Phi(R)\Psi(R)$ where $\Phi(R)=\Phi(R,t\to\infty)$, and the energy of the fixed node solution can be evaluated by the so-called mixed estimator

$$E_{DMC}=\frac{\int\Phi(R)H\Psi(R)dR}{\int\Phi(R)\Psi(R)dR}=\frac{\int\Phi(R)\Psi(R)\left[H\Psi(R)/\Psi(R)\right]dR}{\int\Phi(R)\Psi(R)dR} \tag{6}$$

Further details of the simulation techniques of Eq. (5) can be found elsewhere.[7,9,22]

III. TREATMENT OF THE CORE ELECTRONS

The contribution from the core electrons to the total energy scales as $\approx Z^2$ where Z is the atomic number. Thus the total energy increase is very rapid as one goes to heavier elements ($Z>10$), and it has been shown that the total computational demands grow like $\approx Z^6$.[23] To some extent, this unfavorable scaling can be improved by using more sophisticated variants of VMC and DMC,[22] but there is little doubt that the core electrons make the Monte Carlo calculations very inefficient: most of the computational time is spent on sampling large energy fluctuations in the core. However, it is well-known that for most purposes the core electrons are chemically inert and have negligible impact on valence effects like binding, valence excitations, *etc.* Therefore, it has become an important task to either suppress the core energy fluctuations or to eliminate the core electrons from the simulations altogether.

The core elimination techniques that have been used for solids can be divided into two types. The first one, formulated by Bachelet *et al.*,[16] constructs an effective inverse mass tensor of the electron inside the core (pseudo-Hamiltonian) in such a way that it reflects the different action of the core on the valence electrons in different symmetry channels (s,p or d). The actual construction of this operator is rather non-trivial[24-26] because of several conditions which it has to fulfill: the tensor must be positive definite and go smoothly to the ordinary scalar value outside the core, the pseudo-wave functions have to match exactly the original atomic wave functions beyond the core radius and the pseudo-charges must be equal to the valence charges within the core radius. It turns out that this set of conditions is contradictory, and it seems that there is only a limited number of elements for which this operator can be constructed with acceptable compromises on the accuracy. On the other hand, once the pseudo-Hamiltonian is successfully constructed, its use is straightforward both in VMC and DMC since the resulting operator is *local.* Using this approach, the first DMC calculation of a solid has been carried out and an excellent agreement with experiment for the cohesive energy of silicon crystal was found.[27] For recent progress in pseudo-Hamiltonian techniques see Ref. 26.

The second possible way to eliminate the core electrons is based on nonlocal pseudopotentials (or effective core potentials), which are often used even within mean-field calculations.[28] Initially, there were a few attempts to implement nonlocal pseudopotentials in DMC[13] which were not entirely convincing because only two-electron systems were calculated. Then, pioneering VMC calculations of carbon and silicon solids[14] showed that for these systems an excellent agreement with experiment for the cohesive energy can be obtained. Another step forward in the application of nonlocal pseudopotentials in QMC was made by us,[17] in expanding the use of nonlocal pseudopotentials in an approximate way within the DMC method. In general, the use of a nonlocal operator within the diffusion Monte Carlo is very difficult since the Green's function given by (5) is not guaranteed to be nonnegative for a nonlocal Hamiltonian. On the other hand, we have shown that it is possible to evaluate the nonlocal terms with a sufficiently accurate trial function and that excellent estimations of the fixed-node energies can be found by this technique. Strictly speaking, the DMC energy is not guaranteed to be an upper bound to the exact energy but it converges quadratically to the exact fixed-node value as the trial function converges to the exact eigenfunction. We have carried out several test calculations of this technique on transition metal atoms and small *sp* molecules which demonstrated that it is possible to construct sufficiently accurate trial functions such that the resulting DMC energies are very close to the fixed-node values.[19,29]

Finally, we mention a third approach for dealing with the core without eliminating the corresponding electrons but rather by trying to suppress the energy fluctuations.[18] This damped-core method is based on a combination of VMC in the core region and DMC in the valence region. It certainly reduces the impact of core energy fluctuations in DMC simulations; however, it has been successfully applied only to a few heavy atoms and it remains to be seen how accurate and efficient it will be for larger systems.

IV. THE TRIAL FUNCTION

The trial function is one of the most important factors in obtaining accurate results by the VMC and DMC methods in an efficient way. There has been a remarkable development of the functional forms for the trial function which has brought a significant improvement in efficiency and accuracy of QMC calculations in recent years. One of the general state-of-art functional forms is given by[20,21,29,30]

$$\Psi = \sum_n e_n D_n^{\uparrow} D_n^{\downarrow} \exp\left[\sum_I \sum_{i<j} u(r_{iI}, r_{jI}, r_{ij})\right] \tag{7}$$

where i, j, denote electrons, I corresponds to ions, $D_n^{\uparrow}$ and $D_n^{\downarrow}$ are Slater determinants of spin-up and spin-down one-electron orbitals, e_n are expansion coefficients and $u(r_{iI}, r_{jI}, r_{ij})$ is the function which describes electron-electron and, possibly, electron-electron-ion correlations. Orbitals both from LDA and Hartree-Fock (HF) calculations have been used for building the Slater determinants and also several functional forms and parametrizations of the correlating function have been proposed.[20,29,30]

The variational parameters which enter into the function $u(...)$ are found by minimization of the variational energy or by minimization of the variance of the local energy given by

$$\sigma_H^2 = \frac{\int |\Psi(R)|^2 \left[H\Psi(R)/\Psi(R) - E_V\right]^2 dR}{\int |\Psi(R)|^2 dR} \tag{8}$$

or a combination of both. The number of variational parameters used has varied between a few and several tens.[22]

V. QMC SIMULATIONS OF SOLID SYSTEMS

The QMC simulations of solid systems are usually carried out in a periodically repeated supercell which consists of a certain number of primitive cells of the given crystal structure. There are several important features which are special to these types of calculations.

First, it is necessary to calculate the Coulomb potential energy term of the periodically repeated supercells using the Ewald summation.[15,32] Second, because of the finite size of the supercell only a restricted set of **k**-points from the Brillouin zone of the given structure is occupied. Usually, these are the points which fold into the Γ point of the corresponding supercell lattice. Clearly, the smaller the supercell the more limited "sampling" of the Brillouin zone by occupied states. Thus, the energy exhibits finite size effects, and in order to get the infinite size value one has to correct the simulation results. The finite size correction is evaluated within a mean-field method like LDA by calculating the difference between the fully converged energy, and the energy found by using the occupation which is identical to the occupation in the QMC simulation. The corrections are much less pronounced for large gap insulators where the QMC simulation with a few atoms in the supercell can give an excellent estimation of the bulk total energy. For metals, especially those with a complicated shape of the Fermi surface, one might need to go to very large supercells with several hundred atoms so that the trial function reflects the important features of the corresponding Fermi surface. Very recently, Rajagopal et al.[31] proposed to use the non-Γ point occupation of the states in order to get more balanced sampling of the bands and much faster convergence towards the bulk limit.

The last point is the range of the correlation terms in the trial function. In several calculations a long-range pair correlation term, based on the Random Phase Approximation solution for the homogeneous electron gas, was used to describe the electron-electron correlations.[15,31,32] In order to maintain the periodicity, one has to use the Ewald summation to calculate contributions to the correlation factor from the periodically repeated supercells. However, it is possible to restrict the correlation term to a finite range. In our papers,[19,33] we have used short-range correlation terms which were rather efficient to evaluate and produced good quality trial functions.

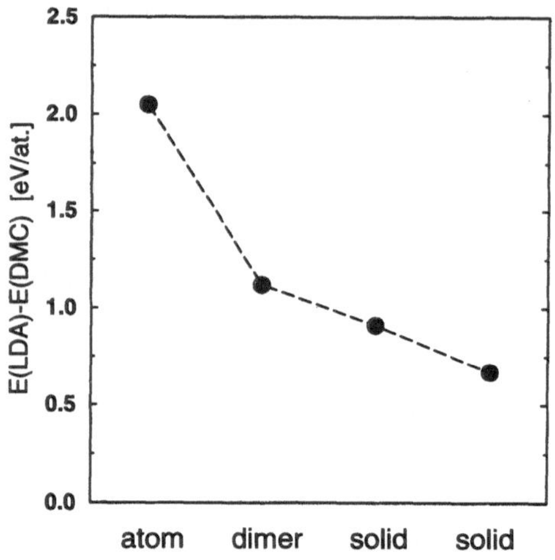

Figure 1. Total energy differences between LDA [34] and DMC results for the nitrogen atom, dimer, molecular and atomic solids (Pa3 and $I2_13$).

VI. CALCULATIONS OF NITROGEN SYSTEMS AND DIAMOND

It is well-known that the strong molecular bond of the nitrogen dimer is the reason that condensed nitrogen will form a molecular crystal at low temperatures. However, several theoretical calculations have appeared which pointed out that at high pressure the molecular structure would become energetically unstable against some type of atomic ordering. Very recently, using the LDA method for an extensive structural search, a new type of atomic covalent structure has been proposed as a candidate for the high-pressure nitrogen phase.[34] Our primary interest in nitrogen systems was motivated by the desire to understand the accuracy of the LDA method for predicting energy differences between systems with different types of bonding (molecular and atomic). Thus, we carried out systematic calculations of the nitrogen atom, dimer, compressed molecular and atomic solids using both the VMC and DMC methods.[19] We used the Hartree-Fock pseudopotentials[35] and the valence one-body orbitals were found by the Hartree-Fock method. The orbitals were expanded in localized basis sets (Gaussian and numerical) for both localized (atom and molecule) and periodic cases. We constructed rather accurate trial functions which in VMC gave ≈ 75-90 % of the correlation energy. Simulations of the nitrogen solids in molecular Pa3 and atomic $I2_13$ structures have been performed in 8-atom supercells which correspond to the one and two **k**-point occupation of the Brillouin zone, respectively. The volumes of the supercells were identical and corresponded to the LDA energy minimum of the $I2_13$ atomic structure.[34] The difference of LDA and DMC

Table 1. The DMC binding energies [eV/at.] for the nitrogen dimer and two compressed solid systems compared with the results of the HF and LDA[38] approaches and with experiment.[19]

	HF	LDA	VMC	DMC	Exper.
dimer	2.61	5.85	4.52(3)	4.87(3)	4.96
Pa3	0.01	4.45	2.0(1)	3.3(1)	...
$I2_13$	0.25	4.75	2.1(1)	3.4(1)	...

total energies are plotted in Fig. 1. It is interesting to observe the dependence of the LDA error on the character of the electronic structure. Evidently, the largest error is found for the atom, which can be considered just the opposite case of the homogeneous electron gas, used as a fundamental model for deriving the LDA approach. However, even if the error for solids is significantly smaller, it is still surprising that there is a small difference between atomic and molecular structures; this is presumably caused by the different character of correlation in molecular and atomic solids.

The binding energies which we have found are compared with HF and LDA results in Table I. We have recovered about 98% of the experimental binding energy for the nitrogen dimer. This dimer is very often used as a testing case for the quantum chemical methods as it has a very large correlation energy due to the triple bond and an extremely small interatomic distance. The accuracy of the DMC result is on the level of the best quantum chemical calculations. It is also clear that because of very high energy of the atom the LDA method exhibits a large overbinding both for the molecule and solids.

Very recently, we started calculations of the diamond crystal. Again we calculated only a small supercell with 8 atoms which resulted in a finite size correction of about 0.5 eV/atom. Other details of the calculation were similar to the nitrogen case and will be published elsewhere.[33] The DMC binding energy is 7.2(1) eV/atom, which compares favorably with the experimental value of 7.36 and is significantly better than the LDA result of 8.6 .

VII. EXCITED STATES IN SOLIDS

It is well-known that many-body correlations have a significant impact on important band structure features such as band gaps, band widths, excitations, *etc.* One of the typical excitations in insulating solids is the promotion of the valence electron to a conduction band creating thus an electron-hole pair: an exciton. If we suppose that the exciton is of the Mott-Wannier type, *i.e.* weakly localized, then its energy is given by the conduction - valence band difference minus the exciton binding energy. (The exciton binding energy in large gap insulators is of order ≈ 0.1 eV or smaller.) The exciton can be easily created in the quantum Monte Carlo scheme by promoting the electron from the valence into conduction state in the Slater determinant. However, it is important to consider the symmetry of the ground and excited states. First, we have the translational symmetry which is labeled by the crystal momentum k. In addition, the orbitals can be classified according the crystal symmetry group. If the symmetries of the excitonic and ground state are different we can use the DMC method directly. On the other hand, if these symmetries are identical (so that the exciton is not the ground state of the given symmetry) use of the DMC method is rather difficult, and one has to use a much more complicated version of this method to reliably calculate the energy difference.[36]

Unless the simulation cell is very large, the exciton is artificially localized. This effect increases the interaction energy of the hole-electron pair which is then dependent on the size of the simulation cell. Using the hydrogen-like description of the Mott-Wannier exciton we can express its interaction energy as

$$E_{exc} = \frac{1}{2\varepsilon r_0} \tag{9}$$

where ε is the static dielectric constant and r_0 is an average distance between the hole and electron. In our simulation r_0 is roughly equal to the edge length of the simulation cell. As

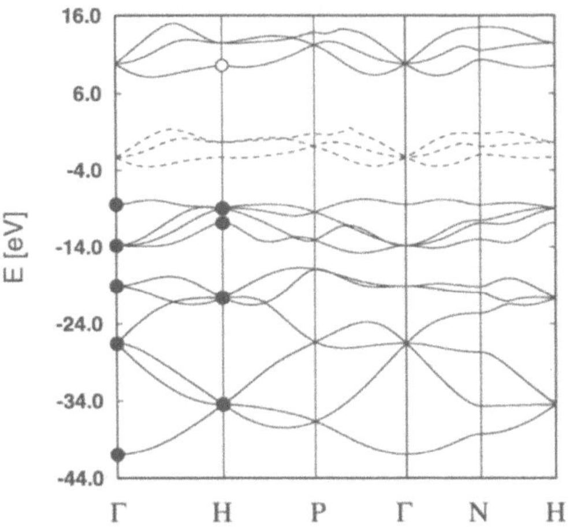

Figure 2. The Hartree-Fock band structure of the atomic nitrogen solid in the $I2_13$ structure. The dashed lines are inserted LDA conduction bands using the highest HF occupied band as a reference. The occupied states are denoted by the filled circles, the excitation by the unfilled circle.

the size of the simulation cell increases, the finite size effects become less pronounced and for sufficiently large cell the exciton size is determined by the minimum of the total energy.

The HF bands of the atomic nitrogen solid in $I2_13$ structure are given in Fig. 2. For the excited state calculation the highest occupied Γ state was exchanged by the lowest unoccupied *H* state. In Table 2 the value of the Γ→H excitation is compared with the results from mean-field approaches. The exciton correction for this case was estimated to be ≈ 0.4 eV. Another example (Fig. 3) of a similar calculation is for diamond for the Γ→X excitation. After correcting the DMC results for the excitonic effect, which for the diamond 8-atom simulation cell is $E_{exc} \approx 0.4$ eV, the agreement with experiment is excellent.

For the sake of completeness, we mention that there is also another possibility of calculating an excited state without creating the exciton. We can add an additional electron to the system together with a charged background to maintain charge neutrality. One complication with this approach is that the background interacts with all electrons and this creates an artificial contribution to the energy. Fortunately, this contribution is a constant which is exactly known and it is therefore possible to estimate the excitation energy.[37] Moreover, by a combination of the exciton and add-an-electron techniques one can study excitonic effects which for many systems (like copper-oxygen superconducting compounds or C_{60}) are of the highest scientific interest.

Table 2. Excitation energies [eV] for the nitrogen atomic solid in the $I2_13$ structure and diamond compared with experiment.[39] The DMC values are corrected for the excitonic effects (see text).

	HF	LDA	DMC	Exper.
nitrogen Γ→H	18.0	6.1	8.5(4)	...
diamond Γ→X	13.2	4.6	5.9(4)	6.1

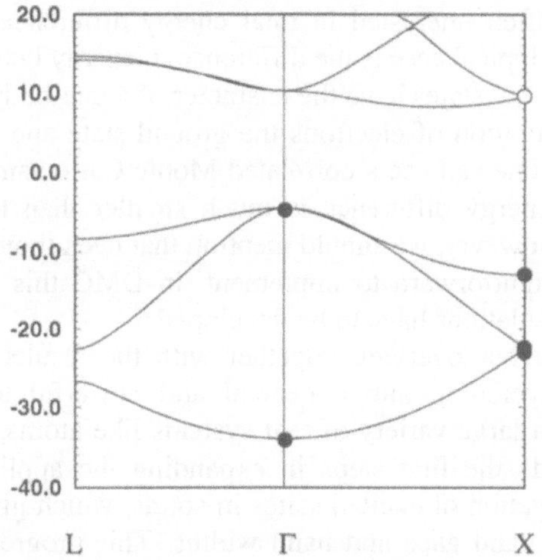

Figure 3. The Hartree-Fock band structure of diamond. Otherwise the notation is the same as on Fig. 2.

VIII. DISCUSSION AND CONCLUSIONS

Having in mind future applications with an increasing number of electrons, it is important to consider the scaling properties of the QMC approach. The basic scaling formula for estimating the computer time $T(N)$ [23] required for evaluating the total energy of the system of N valence electrons with accuracy ε is given by

$$T(N) \propto T_s(N) T_f(N) / \varepsilon^2 \tag{10}$$

where $T_s(N)$ denotes the computational demands to generate an uncorrelated energy sample in the stochastic process *i.e.*, generate a statistically uncorrelated configuration and evaluate its local energy), while $T_f(N)$ measures the increase in local energy fluctuations with increasing N. Calculation of pair quantities (*e.g.*, potential energy, correlation factors) scales quadratically with N. Also, the filling of the Slater matrices scales quadratically in N as long as the basis set which is used is localized. The calculation of the Slater determinant scales as N^3. Thus, we can write approximately[23]

$$T_s(N) = \alpha_2 N^2 + \alpha_3 N^3 \tag{11}$$

where α_2, α_3 are constants and in real calculations we have $\alpha_2 \gg \alpha_3$. Therefore, for small and medium systems (say, up to few hundred electrons) the quadratic term dominates. If we suppose that for a sufficiently large system the contributions to the energy fluctuations by the electrons are approximately independent, we can write for the other term

$$T_f(N) = \beta N \tag{12}$$

where β is a constant. The scaling relation (10) is significantly better than for the high quality quantum chemistry methods ($\approx N^6 - N^8$), and it is even more favorable if we are interested in intensive quantities or in relative energy differences. First, the total energy is usually calculated per small number of electrons (like per atom or per stoichiometric group of atoms). In such cases the computer time scaling (10) improves by a factor of 1/N. Second, we are very often interested in total energy differences rather than in the total energies themselves. A typical case is the difference in energy between the ground state and excited state. Many excited states have the character of a one-body excitation, which means that for a given configuration of electrons the ground state and excited state are strongly correlated. In this case one can use a correlated Monte Carlo sampling technique in which the error bar for the energy difference is much smaller than the error bar for the two separate calculations. However, we should mention that even though in VMC the correlated sampling is rather straightforward to implement, in DMC this is more difficult and the techniques for such calculations have to be developed.[38]

We believe that our overview together with the results presented show that the QMC methods are developing into a general and powerful tool for investigating the electronic structure of a large variety of real systems like atoms, molecules and solids. In addition, we have made the first steps in expanding the applicability of this powerful approach to the investigation of excited states in solids, which gives information about the band structure such as band gaps and band widths. This progress opens completely new possibilities for the future; in particular, we believe that we will be able to carry out high accuracy calculations of systems with transition metal elements, large molecules and clusters with hundreds of valence electrons and thus enter an age of theoretical and computational design of real materials.

ACKNOWLEDGMENTS

The discussions and encouragement by David Ceperley and Richard M. Martin are gratefully acknowledged. I would like to thank Jeffrey C. Grossman and Thomas A. Kaplan for reading the manuscript. This work is supported by the NSF grant ASC 92-11123. The calculations have been done on the NCSA Cray-YMP and HP9000/715 cluster.

REFERENCES

1. P. Hohenberg and W. Kohn, Phys. Rev. **136**, B864 (1964); W. Kohn and L.J. Sham, Phys. Rev. **140**, A1133 (1965).
2. B. Barbiellini, E.G. Moroni and T. Jarlborg, J. Phys.: Condens. Matter **2**, 7597 (1990).
3. M. Korling and J. Häglund, Phys. Rev. B **45**, 13293 (1992).
4. A. Garcia, E. Elsässer, J. Zhu, S.G. Louie and M.L. Cohen, Phys. Rev. **46**, 9829 (1992).
5. C.J. Umrigar and X. Gonze, in *Proceedings of the Conference on Concurrent Computing in the Physical Sciences* 1993.
6. A. Szabo and N.S. Ostlund, *Modern Quantum Chemistry* (McGraw-Hill, New York, 1989) and references therein.
7. D.M. Ceperley and M.H. Kalos, in *Monte Carlo Methods in Statistical Physics*, edited by K. Binder (Springer, Berlin, 1979); K.E. Schmidt and M.H. Kalos, in *Monte Carlo Methods in Statistical Physics II*, edited by K. Binder (Springer, Berlin, 1984).
8. D.M. Ceperley, G.V. Chester, and M.H. Kalos, Phys. Rev. B **16**, 3081 (1977); D.M. Ceperley, Phys. Rev. B **18**, 3126 (1978); D.M. Ceperley, and B.J. Alder, Phys. Rev. Lett. **45**, 566 (1980); J. Chem. Phys. **81**, 5833 (1984); Phys. Rev. B **36**, 2092 (1987).
9. P.J. Reynolds, D.M. Ceperley, B.J. Alder, and W.A. Lester, J. Chem. Phys. **77**, 5593 (1982); J.W. Moskowitz, K.E. Schmidt, M.A. Lee, and M.H. Kalos, J. Chem. Phys. **77**, 349 (1982); *ibid.* **81**, 5833 (1984)
10. W.A. Lester, Jr., and B.L. Hammond, Annual Rev. Phys. Chem. **41**, 283 (1990).
11. J.B. Anderson, J. Chem. Phys. **63**, 1499 (1975); *ibid.* **65**, 4121 (1976); *ibid.* **86**, 2839 (1987); *ibid.* **96**, 3702 (1992).
12. J. Vrbik, and S.M. Rothstein, J. Comput. Phys. **63**, 130 (1986); J. Vrbik, D.A. Legare, and S.M. Rothstein, J. Chem. Phys. **92**, 1221 (1990).

13. M.M. Hurley and P.A. Christiansen, J. Chem. Phys. **86**, 1069 (1987); B. L. Hammond, P.J. Reynolds, and W.A.Lester, Jr., *ibid.* **87**, 1130 (1987).
14. P.A. Christiansen, J.Chem. Phys. **88**, 4867 (1988); J. Phys. Chem. **94**, 7865 (1990).
15. S. Fahy, X.W. Wang, and S.G. Louie, Phys. Rev. Lett. **61**, 1631 (1988); Phys. Rev. B **42**, 3503 (1990)
16. G.B. Bachelet, D.M. Ceperley, and M.G.B. Chiocchetti, Phys. Rev. Lett. **62**, 1631 (1988).
17. L. Mitáš, E.L. Shirley, and D.M. Ceperley, J. Chem. Phys. **95**, 3467 (1991).
18. B.L. Hammond, P.J. Reynolds, and W.A. Lester,Jr., Phys. Rev. Lett. **61**, 2312 (1988).
19. L. Mitáš and R.M. Martin, Phys. Rev. Lett. **72**, 2438 (1994).
20. C.J. Umrigar, K.G. Wilson, and J.W. Wilkins, Phys. Rev. Lett. **60**, 1719 (1988); in *Computer Simulation Studies in Condensed Matter Physics: Recent Developments*, edited by D.P. Landau, K.K. Mon, and H.B. Schuttler (Springer, New York, 1988).
21. C.J. Umrigar, Int. J. Quantum Chem.: Quantum Chem. Symp. **23**, 217 (1989).
22. C.J. Umrigar, M.P. Nightingale, and K.J. Runge, J. Chem. Phys. **99**, 2865 (1993).
23. D.M. Ceperley, J. Stat. Phys. **43**, 815 (1986) and private communication.
24. G.B. Bachelet, D.M. Ceperley, M.G.B. Chiocchetti, and L. Mitáš, in *Progress on Electron Properties of Solids*, edited by R. Girlanda et al. (Kluwer Academic, Dordrecht, 1989); L. Mitáš and G.B.B. Bachelet, Preprint University of Trento UTF174, March 1989, (unpublished work).
25. W.M.C. Foulkes and M. Schluter, Phys. Rev. B **42**, 11505 (1990).
26. A. Bosin, V. Fiorentini, A. Lastri, and G.B. Bachelet, submitted to Phys. Rev. B
27. X.-P. Li, D.M. Ceperley, and R.M. Martin, Phys. Rev. B **44**, 10929 (1991).
28. W.E. Pickett, Comput. Phys. Rep. **9**, 115 (1989); M. Krauss and W.J. Stevens, Annual Rev. Phys. Chem. **35**, 357 (1984).
29. L. Mitáš, in *Computer Simulation Studies in Condensed Matter Physics V*, edited by D.P. Landau, K.K. Mon and H.B. Schüttler (Springer, Berlin, 1993).
30. K.E. Schmidt and J.W. Moskowitz, J. Chem. Phys. **93**, 4172 (1990).
31. G. Rajagopal, R.J. Needs, S. Kenny, W.M.C. Foulkes, and A. James, submitted to Phys. Rev. Lett.
32. D.M. Ceperley, Phys. Rev. B **18**, 3126 (1978).
33. L. Mitáš, to be published
34. C. Mailhiot, L.H. Yang, and A.K. McMahan, Phys. Rev. B **46**, 14419 (1992).
35. W.J. Stevens, H. Basch, and M. Krauss, J. Chem. Phys. **81**, Pt.II, 6026 (1984).
36. D.M. Ceperley and B. Bernu, J. Chem. Phys. **89**, 6316 (1988).
37. G. Engel and Y. Kwon, to be published.
38. D.M. Ceperley, private communication.

LOCALIZED-SITE CLUSTER EXPANSIONS

M.A. Garcia-Bach

Departament de Física Fonamental
Facultat de Física
Universitat de Barcelona
Av. Diagonal, 647
E-08028 Barcelona

I. INTRODUCTION

Two important topics in Solid State Physics are those related to the occurrence and characterization of magnetic materials and higher T_c superconductors. Often both properties are present to some extent in several families of compounds such as Chevrel Phases, heavy fermions, ternary perovskites and spinel compounds and, very specially, quaternary compounds, among which high-T_c superconductors are found.[1] Also, after Little's[2] suggestion in 1964, that long one-dimensional (1D) systems could present excitonic high T_c superconductivity, quasi-one-dimensional donor-acceptor and charge transfer organic compounds have been extensively investigated.

Traditionally, electronic properties of solids have been treated through two well differentiated theoretical schemes: a) Bloch-Wilson independent-electron extended states formulation, applicable to conductors and semiconductors, and b) Localized description, based on the atomic limit and crystal field, giving rise to a Heisenberg Hamiltonian, applied to magnetic insulators.

However, the aforementioned compounds lie close to the metal-insulator transition border where the role of correlation is important, and to a certain extent their properties lie between these two limiting descriptions. Since *ab initio* full CI calculations are limited to small clusters, effective Hamiltonians and their solutions are of interest for extended systems.

To illustrate the problem, let us focus our attention on the Hubbard model. The Hubbard Hamiltonian is the simplest model capable of mapping the behavior from a metal to an insulator, through correlated systems. Furthermore, it is claimed by some to be the appropriate model for describing high T_c superconductors. For simplicity, let us consider a half filled band system, with as many electrons as sites

$$H = t \sum_{\langle ij \rangle} c^{+}_{i\sigma} c_{j\sigma} + U \sum_{i} n_{i\alpha} n_{i\beta}, \tag{1}$$

where $\langle ij \rangle$ indicates summation over nearest neighbors, $c^{+}_{i\sigma}$ ($c_{i\sigma}$) creates (annihilates) an

Electronic Properties of Solids Using Cluster Methods
Edited by T.A. Kaplan and S.D. Mahanti, Plenum Press, New York, 1995

electron with spin $\sigma(\sigma = \alpha, \beta)$ on site i and $n_{i\sigma}$ is the number operator of electrons with spin σ on site i . U and t are system-dependent parameters.

In the limit $U = 0$ (see Fig. 1), this Hamiltonian can be solved exactly. Its solutions are products of Bloch extended states and the system is a conductor for most band fillings. For small U values Perturbation Theory (PT) may be reasonably used to extend the independent electron description, and $t = 0$ corresponds to the independent atom limit. For $|t|/U < 1/2\nu$, where ν is the coordination number, a Heisenberg Hamiltonian can be obtained at half filling by Degenerate Perturbation Theory (DPT) upon $t = 0$ as the zeroth order, the system being described as a magnetic insulator with the exchange parameter $J_0 = -2t^2/U$. In this sense $|t|/U = 1/2\nu$ is considered a rough limit of applicability of Heisenberg Hamiltonians. The question remains for intermediate values of $|t|/U$, where PT based on either one or the other limit diverges.

We will not consider here extensions from the independent electron limit that remain efficient in the strongly correlated domain, such as is the Gutzwiller-type *Ansatz*.[3-5] Our attention will focus on efforts in treating systems beyond the $|t|/U < 1/2\nu$ Hubbard domain by using Heisenberg Hamiltonians which incorporate delocalization through effective exchanges, differing from the exchange parameter that would be obtained through DPT, and not necessarily based on the separated atomic limit.[6-14]

II. LOCALIZED-SITE CLUSTER EXPANSIONS FORMALISM FOR THE HAMILTONIAN

In this section we are going to focus our attention on a cluster expansion formalism[8] proposed by Poshusta and Klein to obtain an effective Hamiltonian with a *dressed* local interaction, in a way reminiscent of cluster expansions proposed long ago in Statistical Mechanics.

Let $H(G)$ be the associated effective Hamiltonian of a system G spanned by a basis of products of site states. Here G will be thought of as a graph consisting of vertices and edges, vertices being understood as sites and edges as bonds. Sites can be atoms, groups of

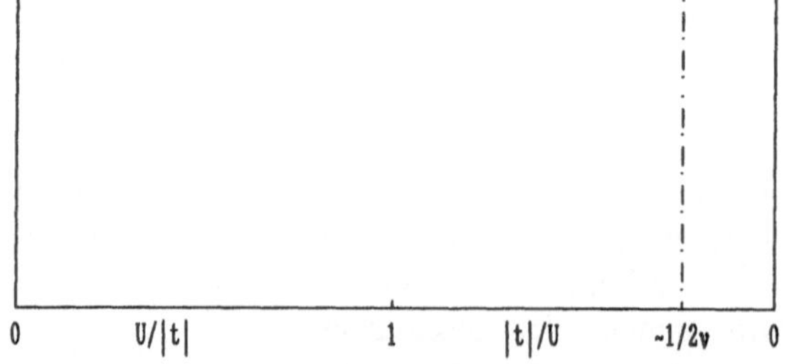

Figure 1. Schematic representation, as a function of $U/|t|$ or $|t|/U$, of different regions of Hubbard Hamiltonian. This Hamiltonian can be solved exactly for the limits $U/|t| = 0$ (metals) and $|t|/U = 0$ (atomic limit). $|t|/U = 1/2\nu$, where ν is the coordination number, separates the region where DPT upon the atomic limit is applicable, giving rise to Heisenberg Hamiltonian. This critical value is interpreted as separating the metallic and magnetic insulator regimes. There is a large zone around $|t|/U = 1$, where perturbative solutions are not acceptable.

atoms, bonds or even interstitial positions. *Irreducible cluster* interactions $V(G')$ acting only on the site states of a subsystem G' contained in G, are introduced such that

$$H(G) = \sum_{G'} V(G'). \tag{2}$$

Because the largest G' is G itself, the relation (2) can be inverted to give

$$V(G) = \sum_{G'} \mu(G,G')H(G') \tag{3}$$

where $\mu(G,G')$ is the so-called Möbius function that only depends on the site ordering and does not depend on details of $H(G)$. Finally the nth order cluster approximant to $H(G)$ is defined as

$$H_n(G) = \sum_{G'}^{s(G')\leq n} V(G') \tag{4}$$

where $s(G)$ is the size of G. Then, if more complete or *ab initio* Hamiltonians $H(G')$ for smaller size systems G'; $s(G')\leq n$, are solved with high accuracy, the appropriate parameters of the effective Hamiltonians can be obtained as those which reproduce the eigenvalues with the same symmetry of the more complete Hamiltonian. From parameters of effective Hamiltonians for smaller size systems, those of irreducible cluster interactions $V(G)$, $s(G')\leq n$, may be obtained, via Eq. (3), and the nth order cluster approximants constructed for an arbitrarily large system.

In Ref. 8 appropriate irreducible parameters for bipartite clusters of diameter up to two (see Fig. 2), which are used to generate $s(G)$ Heisenberg Hamiltonian approximants to Hubbard Hamiltonians with $0 \leq |t|/U \leq 1$, have been obtained. In this case the effective Hamiltonian and the irreducible cluster interactions can be spanned in terms of permutations

$$H(G) = \sum_{P} J_P(G)\,P, \tag{5}$$

$$V(G) = \sum_{P} j_P(G)\,P, \tag{6}$$

where the summations are restricted[15] to permutations such that $P^2 = 1$.

Using these irreducible parameters, the parameters of Heisenberg approximants for different extended systems can be computed very easily. In Table 1, the ratio $J_{(ij)}/J_0$, with i and j nearest neighbors, are given for different extended systems and $|t|/U$ ratios up to 1. For comparison, we have added previous results[12] for a 1D infinite system and the 2D graphite lattice, using a coherent state derivation of Heisenberg Hamiltonians. It can be observed that the *dressed* $J_{(ij)}$ is close to the value yielded by DPT for ratios $|t|/U$ small, but differs dramatically when the ratio $|t|/U$ increases.

III. CLUSTER EXPANSION HEISENBERG MODEL FOR CONJUGATED HYDROCARBONS

Easily interpreted in the spirit of the preceding section and supported by other available procedures,[9-13] a Heisenberg Hamiltonian was obtained[16] as a 2-site cluster approximant for conjugated hydrocarbons, extracted from an *ab initio* extended basis set plus CI calculations on the lowest state of ethylene.

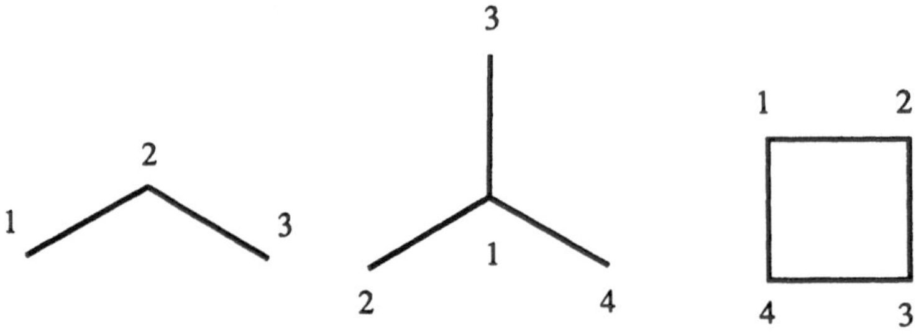

Figure 2. The three diameter $n = 2$ graphs.

For the conjugated hydrocarbons the Hamiltonian so obtained involves a scalar term characterizing the interaction between two σ-bonded carbon atoms with parallel spin in the π systems, and a spin operator part,

$$H = \sum_{\langle ij \rangle} \{R(r_{i,j}) + g(r_{i,j})[(i,j) - 1]\}, \tag{7}$$

where (i,j) denotes transposition of spin i and j and is related to the usual spin operator via the Dirac identity,

$$(i,j) = 2\bar{S}_i \cdot \bar{S}_j + \frac{1}{2} \,. \tag{8}$$

Both $R(r)$ and $g(r)$ are distance-dependent. Their values can be obtained from the lowest states of the ethylene molecule, which are

$$E(^1A_g) = R - 2g \tag{9}$$

$$E(^3B_u) = R \,. \tag{10}$$

Then the functions *g(r)* and *R(r)* may be readily obtained from the potential energy surfaces of both states resulting from either *ab initio* calculations or experimental data. In Ref. [16] a polynomial development of *g* and *R* has been used,

Table 1. Values of $J_{(ij)}$ of Eq. (5) for i and j nearest neighbors, in units of J_0.

$\lvert t \rvert / U$	1D, n=1	1D, n=2 2D, n=1	2D, n=2 sq. lat.	Coh. St. 1D	Coh. St. graphite
0.05				0.9842	0.9802
0.1	0.9629	0.9623	0.8697	0.9460	0.9313
0.2	0.8770	0.8717	0.6574	0.8352	0.8023
0.3	0.7806	0.7683	0.4899	0.7240	0.6833
0.4	0.6928	0.6742	0.3781	0.6305	0.5824
0.5	0.6180	0.5954	0.3021	0.5550	0.5140
0.6	0.5556	0.5304	0.2487	0.4941	0.4551
0.7	0.5034	0.4771	0.2098	0.4444	0.4078
0.8	0.4595	0.4328	0.1806	0.4035	
0.9	0.4223	0.3957	0.1581		
1.0	0.3904	0.3642	0.1401		

$$g(r) = \sum_{n=0}^{5} a_n(g) r^n \tag{11}$$

and

$$R(r) = \sum_{n=0}^{5} a_n(R) r^n \tag{12}$$

in fitting the *ab initio* extended-basis set CI calculations. The coefficients are given in Table 2. Eq. (11) at $r = 1.40\,\text{Å}$ yields a value of $g = 1.8573\,eV$. This value compares very well with $J_{(ij)} = 1.8597\,eV$, as obtained from column 1 of Table 1, by extrapolation to the suggested[8] appropriate value of $|t|/U = 0.7287$ for hydrocarbons.

Nevertheless, for regular or slightly distorted systems, useful information can be extracted even without knowing the appropriate parametrization of this lower order Hamiltonian approximant, with only nearest neighbor interactions. For small distortions δr_{ij}, the modification of the interaction between pairs of sites can be expressed as

$$g(r_{ij}) \equiv g_0(1 + \delta_{ij}). \tag{13}$$

Then the Heisenberg Hamiltonian is easily written as

$$H_T \equiv H - F(\{r_{ij}\}) = g_0 \sum_{\langle ij \rangle} (1 + \delta_{ij})\,(i,j) \tag{14}$$

where

$$F(\{r_{ij}\}) \equiv \sum_{\langle ij \rangle} [R(r_{ij}) - g(r_{ij})] \tag{15}$$

is fixed by the interatomic distances. H_T allows comparison among different distortion symmetries when the corresponding $F(\{r_{ij}\})$ take the same value.

IV. LOCALIZED-SITE CLUSTER EXPANSIONS OF THE WAVE FUNCTION

Cluster expansions of the wave function has long been recognized as a size-consistent means to introduce many-body correlation effects. Development of cluster expansions within a real-space localized viewpoint is natural when dealing with Hamiltonians with *dressed* dominant interactions that are local in nature.

Nevertheless, there are different ways to implement a localized viewpoint cluster expansion of the wave function, even without considering further classification acording to

Table 2. The coefficients of polynomial expansion of $g(r)$ and $R(r)$ functions of Eqs. (11) and (12) in eV, for r given in Å.

	g	R
a_0	22.5192336	$5.28692998\ 10^{2}$
a_1	-30.34483144	$-1.43624353\ 10^{3}$
a_2	14.97588275	$1.58780103\ 10^{3}$
a_3	-2.6987315	$-8.9538503\ 10^{2}$
a_4	-0.146733	$2.5791813\ 10^{2}$
a_5	0.08128577	$-3.0275671\ 10^{1}$

the method of computing matrix elements. One way is focused on the cluster expansion operator acting on a selected zeroth order wave function, as for example those of Refs. 17 and 18. In this case, the zeroth order wave function and the expansion operator selection represents an additional degree of freedom to construct a variational wave function. A second way is to focus on the expansion of the configuration coefficients (as those of Refs. 19-22) on the model space, spanned by a basis of products of site-states, or one of their subspaces.

Zeroth Order Selection

When performing an operator type cluster expansion, a zeroth order wave function is needed. For the antiferromagnetically signed $s = \frac{1}{2}$ Heisenberg Hamiltonian two main locally described zeroth-order states are of interest. First is the Néel state,

$$[\Psi_0^N\rangle \equiv \prod_i^{i\in A} \alpha(i) \prod_j^{j\in B} \beta(j) \tag{16}$$

where A and B denote the two sets of sites in which a system may be partitioned such that each member of one set is a nearest neighbor solely to sites of the other set, and $\alpha(i)$ and $\beta(i)$ represents the spin-up and spin-down states for site i. The second is a Kekülé state, also termed a bond singlets product state or a nearest-neighbor valence-bond (VB) state,

$$|K\rangle = \prod_{\langle ij\rangle}^{K} [\alpha(i)\beta(j) - \beta(i)\alpha(j)] \tag{17}$$

where K is a VB subgraph structure (also called the dimer covering) with exactly one edge into each vertex of G.

The expected value per site, in units of g_0, of the Heisenberg Hamiltonian of Eq. (14) for a Kekülé state can be readily obtained as

$$E_K = \frac{\nu - 3}{4}, \tag{18}$$

where ν is the coordination number, while in the case of the Néel state its value is $E_N = 0$. The energy of a Kekülé state is lower for $\nu\langle 3$ while Néel state yields lower energy for $\nu\rangle 3$. In fact, Néel-state-based approaches are usually applied to inorganic antiferromagnets which typically involve three-dimensional structures of relatively high coordination number. In the literature, Kekülе states and Resonance, in the lower level, appeared limited to small finite organic systems until Anderson's suggestion[23] of Resonating Valence Bond (RVB) state applied to high-T_c superconducting compounds. At a coordination number around 3, both zeroth order wave functions yield comparable energy values, so any correction, as inclusion of next-nearest-neighbor interaction, could introduce frustration and energy ordering inversion.

V. EXAMPLE APPLICATIONS

As some example of applicability of Heisenberg Hamiltonians, we have used the Hamiltonian of Eq. (7) to obtain the ground state geometry of polyacetylene (PA)[24] and polyacene (PCE).[25] Other conjugated polymers as polyacenacene (PAA), polyperylene (PP), polyphenantrene (PPH) and poly(benz[m,n]anthracene), have also been investigated.[26] Also the simpler Hamiltonian of Eq. (14), with δ_{ij} set to zero has been investigated for square lattice (SQL) strips.[26]

Néel Based *Ansätze*

Néel-based cluster expansion *ansätze* are defined by

$$|\Psi^N\rangle \equiv U_0(e^{T^+})|\Psi_0^N\rangle \tag{19}$$

so the zeroth order function is selected to be the Néel state of Eq. (15). T^+ is defined as

$$T^+ = \sum_{i(m)} T^+_{i(m)}, \tag{20}$$

U_0 indicates that only the unlinked portion of $\exp\{T^+\}$ is to be retained, i.e., a product as $\Pi_{i(m)}T^+_{i(m)}$ is to be retained in the Taylor series expansion of $\exp\{S^+\}$ only if none of the sets $i(m) \equiv i_1, i_2, \ldots, i_m$ of sites have a site in common.

For PA the non-zero $T^+_{i(m)}$ of Eq. (19) are chosen.[24]

$$T^+_{i,i+1} \equiv x_i\,(i,i+1), \tag{21}$$

$$T^+_{i,i+1,i+2,i+3} \equiv y_i\,(i,i+1)(i+2,i+3) + z_i\,(i,i+3), \tag{22}$$

where x_i, y_i and z_i are variational parameters with alternating translation symmetry conditions. The energy per carbon atom as a function of variational parameters and interatomic distances is found to be

$$E_N = \frac{1}{L}\sum_i \left\{ R_{i,i+1} + \frac{2g_{ii+1}}{f_0 f_{i+1}}\left[f_{i+1} - 1 + \frac{z_i^2}{f_{i+1}f_i} + \frac{x_i(f_{i+1}f_i + y_i)}{f_{i+1}f_i - y_i} + \frac{y_{i+1}z_{i+1}}{f_i^2} + \frac{x_{i+1}^2 z_{i+1} f_{i+1}^2}{(f_{i+1}f_i)^2} \right]\right\} \tag{23}$$

with

$$f_i \equiv 1 + \frac{y_i^2 + z_i^2}{f_{i+1}^2 f_i} + \frac{x_i^2(f_i f_{i+1} + y_i)}{f_{i+1}(f_i f_{i+1}) - y_i} \tag{24}$$

$$f_0 \equiv 1 + \frac{2(y_i^2 + z_i^2)}{f_{i+1}^2 f_i} + \frac{x_i^2 y_i(2f_i f_{i+1} - y_i)}{f_{i+1}(f_i f_{i+1} - y_i)^2}$$

and

$$+\frac{2\left(y_{i+1}^2+z_{i+1}^2\right)}{f_{i+1}f_i^2}+\frac{x_{i+1}^2 y_{i+1}\left(2f_if_{i+1}-y_{i+1}\right)}{f_i\left(f_if_{i+1}-y_{i+1}\right)^2} \tag{25}$$

For the aforementioned more complicated conjugated polymers or square lattice strips, the non-zero $T_{i(m)}$ are restricted[25, 26] to nearest-neighbors pair excitations

$$T^+ = \sum_{<ij>}^{i\in A} x_{ij}\, S_i^- S_j^+ , \tag{26}$$

where S_j^+ and S_i^- are the spin raising and lowering operators, and A stands for the sublattices with *site spin* α at zeroth order. Then the energy is obtained using a powerful transfer matrix technique.[25]

Resonating Valence Bond *Ansätze*

For PA the RVB *ansatz* we have used is of the cluster expansion of the configuration coefficients type. It is defined as

$$|\Psi^{RVB}\rangle \equiv \sum_s \prod_j \prod_i x_i^{m(jis)}|s\rangle , \tag{27}$$

where s is restricted to all the linearly independent overall singlets made as a product of spin-pairings with no more than M = 4 pairings crossing any plane perpendicular to the chain. Here x_i is the variational parameter associated with the occurrence of i sites preceding a site j + 1 being spin-paired to i sites succeeding site j, $m(jis)$ is one when the

Table 3. Geometry of polyacetylene. The results of the three first rows[24] are obtained using the Heisenberg Hamiltonian of Eq. (7) and different approaches for the wave function. VLSC higher level stands for Neel-based ansatz of Eqs. (23-25), lower level results are obtained when Eq. (22) is set to zero. RVB-VLSCE stands for ansatz of Eq. (27). For comparison we have added CEPA results of König and Stollhoff[29] and experimental values.[28]

Ψ_0	Method		r_c	r_s	r_l	E
	PT 4th order		1.43	1.401	1.401	-0.616
Néel State	VLSCE:	lower level	1.46	1.400	1.400	-0.661
		higher level	1.32	1.360	1.443	-0.704
Product of bond singlets	PT 2nd order			1.361	1.438	-0.720
	RVB-VLSCE			1.364	1.436	-0.724
	CEPA *ab initio*			1.343	1.436	
	Experimental			1.36	1.44	

spin-pairings in s crossing the plane perpendicular to the chain midway to j and $j + 1$ is exactly i and zero otherwise. The energy is computed through a powerful transfer matrix technique[25] as a function of interatomic distances and variational parameters.

In the case of PCE and PAA, different *ansätze* of a cluster expansion operator type, acting on symmetry differentiated Kekülé structures, have been used,[25, 26]

$$|\Psi_X\rangle \equiv U_0(e^{T^+})|K_X\rangle, \tag{28}$$

where X labels different symmetry Kekülé structures, and

$$T^+ \equiv \sum_{\langle ef\rangle} x_{ef} T^+_{ef} \quad . \tag{29}$$

T^+_{ef} constructs a non-Kekülé four site singlet when acting on a product of two adjacents bond singlets e and f in zeroth order Kekülé structure, and a variational parameter x_{ef} associated to the recoupling of singlets is introduced. For the more complicated conjugated polymers and the wider square lattice strips, a linear combination of Kekülé structures have been used.

VI. RESULTS AND DISCUSSION

Because of space limitation we present here only some characteristic results that we have obtained, using the Heisenberg Hamiltonian of Eq. (7), on PA and PCE. Also, using the simpler Hamiltonian of Eq. (14) SQL strips have been investigated,[26] but only long range spin order is presented along with those for the rest of the polymers. We explicitly omit here our work related with the study of excited states[27] from a variational localized-site cluster expanded wave function.

Polyacetylene

PA is usually described by an independent electron model Hamiltonian. This description is supported by the fact that the appropriate value[8] of $|t|/U \sim 0.7$ is far beyond the $|t|/U \langle \frac{1}{4}$ DPT limit of applicability of Heisenberg Hamiltonians. Nevertheless we have used the Heisenberg Hamiltonian of Eq. (7). Using either the Néel-based *ansatz* energy of Eqs. (23-25) or the RVB *ansatz* of Eq. (27), when geometry is optimized an alternating geometry is obtained[24] with interatomic distances that compare very well with experimental values,[28] as presented in Table 3. The lower order Néel-based *ansatz* is not able to yield an alternating geometry for this system. CEPA *ab initio* calculation of König and Stollhoff[29] and the experimental value[28] are included in the Table for comparison.

When energy is computed as a function of the alternation parameter, δ , for different mean interatomic distances $\bar{r}$

$$r_1 \equiv \bar{r}(1 - \delta); \quad r_2 \equiv \bar{r}(1 + \delta), \tag{30}$$

it is observed (see Fig. 3), that the minima lay at values of $\delta = 0$ for $\bar{r} \langle \bar{r}_c$ shown in Table 3, which implies a regular system, while an alternating geometry has lower energy for longer $\bar{r}$, though the whole energy surface is very flat. The energy gain brought by the

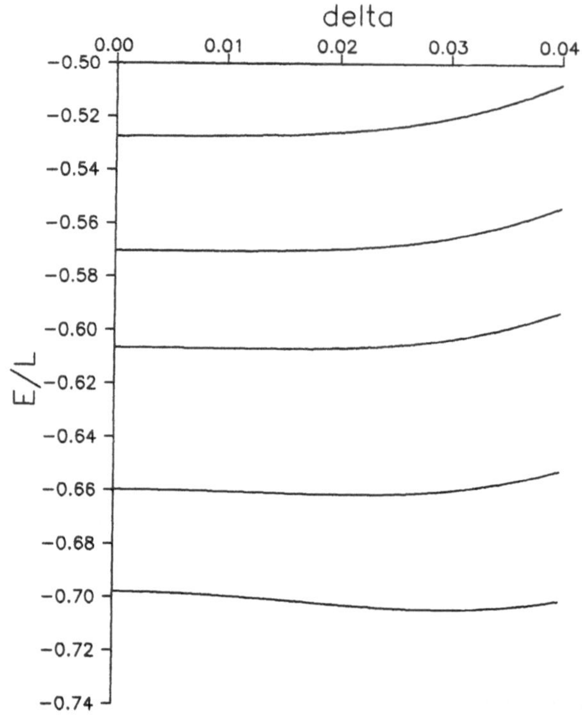

Figure 3. Néel-based *ansatz* energy per atom (in eV) as a function of δ for $\bar{r}$ = 1.32, 1.33, 1.34, 1.36, and 1.40 (in Å), from top to bottom, taking the energy of the Néel state with 1.40 Å equal bond lengths as zero of energy.

dimerization is 0.0118 eV per C_2H_2 unit, when the Néel-based *ansatz* is used, and 0.0391 when the RVB *ansatz* is used, as presented in Table 4.

Polyacene

The PCE polymer should exhibit some deviation from a perfect chain of regular hexagons. There are four possible translationally symmetric bond stretching irreducible modes of distortion (Fig. 4) orthogonal to the *breathing* mode; they are associated to A_1, A_2, B_1, and B_2 irreducible representation of C_{2v}. We have used the Heisenberg Hamiltonian of Eq. (7) to describe this system. In this case (with cyclic boundary conditions) there really are just two types of orthogonal and noninteracting Kekülé states, with A_2 and B_2 symmetries of the group C_{2v}, so two RVB *ansätze* of Eq. (28) are constructed. Using these two RVB *ansätze* and the lower order Néel-based one, along with a transfer-matrix technique to compute matrix elements, the optimal geometry of Table 5 are obtained. As in the case of PA, it can be seen that the lower order Néel-based *ansatz* yields a regular chain of hexagons. RVB *ansätze* yield degenerate A_2 and B_2 distortions.

Table 4. Optimized regular interatomic distances (in Å) and gain in energy per C_2H_2 unit (in eV) of regular PA by report to the alternating ground state.

	r	ΔE
Néel based *ansatz*	1.399	0.0118
RVB-VLSCE	1.398	0.0391

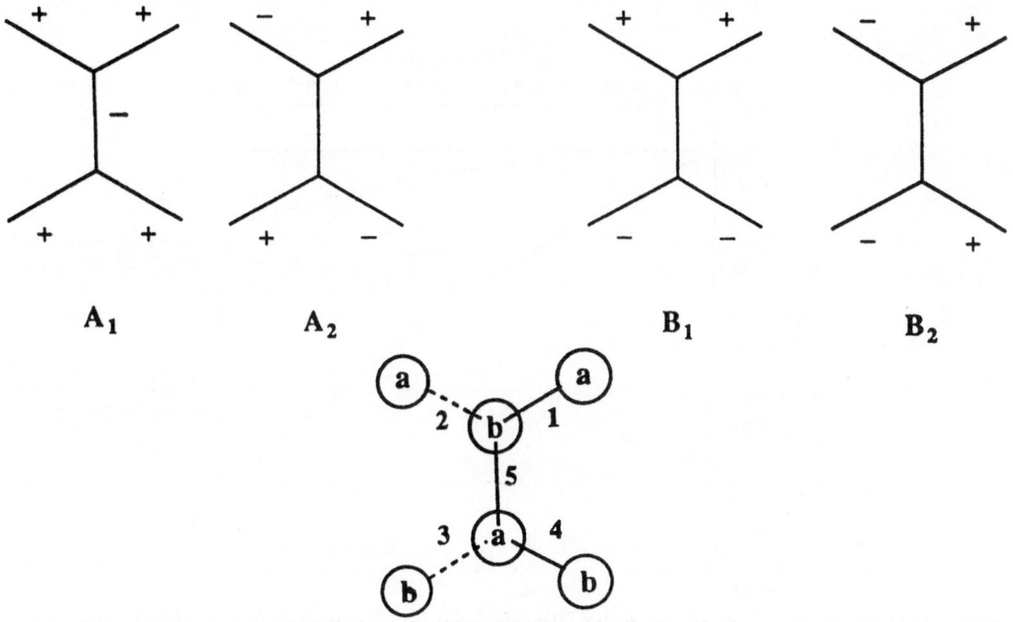

Figure 4. (a) A unit cell of PCE with the irreducible distortion modes. (b) The labeling convention for the bonds and sites of a PCE unit cell.

We have also computed the energy per unit cell as a function of δ for the different bond stretching irreducible distortion modes, using the simpler Hamiltonian of Eq. (14) and Néel-based and RVB *ansätze*. The ordering of the energy is the same for both the Néel-based and RVB *ansätze* (see Figs. 5,6), though the RVB results show a cusp at $\delta = 0$, since they are broken symmetry *ansätze*.

Site Spin and Long Range Spin Ordering

Using the lower order Néel-based *ansatz*, we have computed the expected value of S_z for different sites of the conjugated polymers. Results are presented in Table 6 as a function of the number of nearest neighbors, ν, and the next-nearest-neighbors, ν'. It can be seen that comparable surroundings yield similar site S_z, and its value increases with the number of neighbors, in agreement with Néel order being more favorable for higher coordination number. Nevertheless, it is interesting to notice that values in column $\nu = 2$ and $\nu' = 4$ are higher than one would expect for a strictly increasing order from left to right. These correspond to edge sites. Their S_z values compare with those of highest coordination number column, presumably closer to Néel order. These results could be interpreted as a

Table 5. The minimum varationally optimized total energies for each proposed *ansatz* and associated values for r_i, where the labeling is shown in Fig. 4.

Ansätz	E(cell) (eV)	r_1	r_2	r_3 (Å)	r_4	r_5
Néel-based	-6.23174	1.40	1.40	1.40	1.40	1.43
A-RVB	-6.05387	1.36	1.45	1.36	1.45	1.45
B-RVB	-6.05387	1.36	1.45	1.45	1.36	1.45

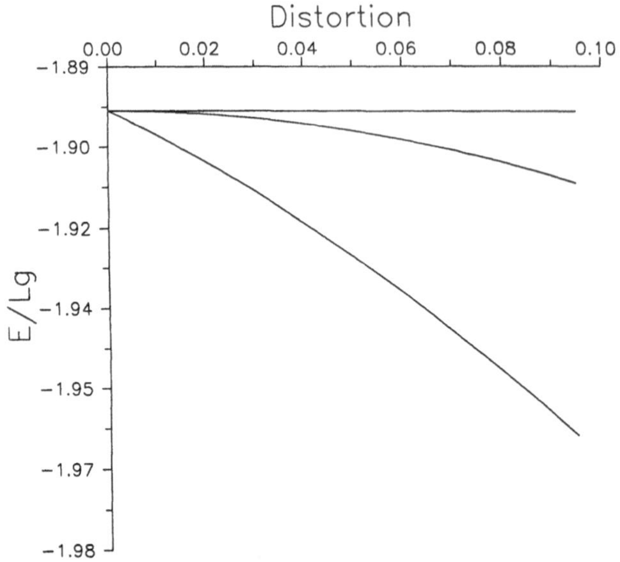

Figure 5. Néel-based *ansatz* for the irreducible distortions orthogonal to the *breathing* mode. The steep curve is $E(A_1)$, the flat one is $E(B_1)$, and the parabolic one is $E(A_2) = E(B_2)$.

manifestation of the edge-localized unrestricted band-orbitals suggested very recently.[30] When long range spin order,

$$\rho = \lim_{|ij| \to \infty} \left|\left\langle \bar{S}_i \bar{S}_j \right\rangle\right|, \tag{31}$$

is computed for edge sites we obtain

$$\rho_{edge}(PCE) = 0.1334, \quad \text{and} \quad \rho_{edge}(PAA) = 0.1352, \tag{32}$$

which are high in comparison with long range order mean values shown in Table 7 for several other systems.

Table 6. Site spin z expected value, $\left|\left\langle S_z^i \right\rangle\right|$, for site i with ν nearest neighbors and ν' next nearest neighbors.

	$\nu = 2$			$\nu = 3$		
	$\nu' = 2$	$\nu' = 3$	$\nu' = 4$	$\nu' = 4$	$\nu' = 5$	$\nu' = 6$
PA	0.2893					
PCE			0.3653	0.3263		
PAA			0.3677	0.3296		0.3665
PPH		0.3153			0.3461	
PP		0.3165			0.3496	0.3617

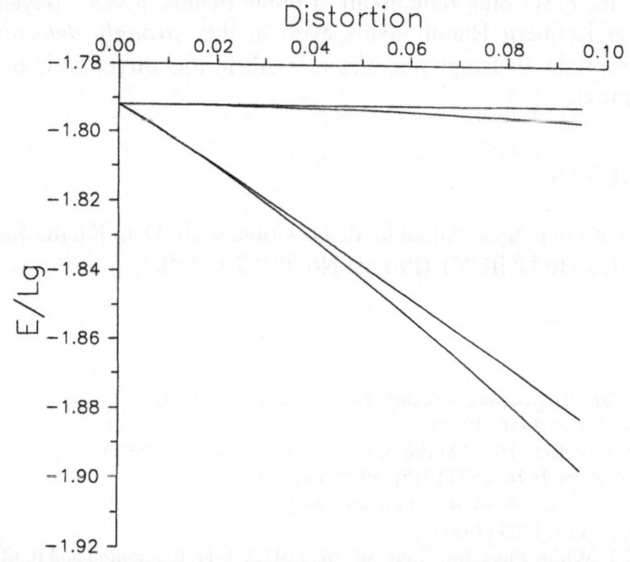

Figure 6. RVB *ansätze* for the irreducible distortions orthogonal to the *breathing* mode. If the parabolic curve is not considered, there is a correspondence of curves with those of Fig. 5. The parabolic curve is obtained when using an A-symmetric RVB *ansatz* for a B-symmetric distortion and the oposite.

VII. CONCLUSIONS

The treatment of conjugated polymers through a carefully parametrized r-dependent Heisenberg Hamiltonian proved to be useful for studying the ground state distortion modes. For PA and PCE this Hamiltonian proved to give reliable results for the mean bond length and the existence and extent of bond alternation. It is also able to give information regarding the transition from alternating to regular structures, or low lying excited states.[27] Edge states in correspondence with band results[30] are also present using this method.

Since for PA $|t|/U \sim 0.7$, the success of these results illustrates the possibility of treating systems beyond the $|t|/U\langle 1/2\nu$ Hubbard domain by using Heisenberg Hamiltonians which incorporate delocalization through effective exchanges, which can be obtained using high accuracy cluster results. It is worthwhile to point out that this is not an isolated fact. Recently a very similar problem of phosphorus metal-insulator transition has been treated,[31] using an r-dependent Heisenberg Hamiltonian, based on the spectroscopy of P and P_2, with excellent results.

While the cluster method for obtaining effective Hamiltonians of Heisenberg type could be seen as a mathematical artifact, the recent suggestion of interstitial singly occupied

Table 7. Long range spin orientation order. It can be notice PCE and PAA present higher Long range order values due to edge states.

	ρ	$\bar{\nu}$	$\bar{\nu}'$
PA	0.0837	2.0	2.0
PPH	0.1094	2.5	3.5
PCE	0.1196	2.5	4.0
PP	0.1148	2.6	4.4
PAA	0.1257	2.7	4.7
SQL $w = 2$	0.1008	3.0	4.0
SQL $w = 3$	0.1274	3.3	5.3
SQL $w = 4$	0.1340	3.5	6.0

localized orbitals,[32,14] for describing bulk alkali or noble metals, gives a physical picture to support the use of Heisenberg Hamiltonians even in this *strongly delocalized* domain, though with the appropriate exchange parameter. Work in this direction is being pursued[33] in a more general context.

ACKNOWLEDGMENTS

The author acknowledges valuable discussions with D.J. Klein. Support of the research is acknowledged to DGICYT (Project No. PB92-0868).

REFERENCES

1. J.C. Phillips, *Physics of High-T_C Superconductors* (Academic Press, 1989).
2. W.A. Little, Phys. Rev. A **134**, 1416 (1964).
3. M.C. Gutzwiller, Phys. Rev. Lett. **10**, 159 (1963); Phys. Rev. A **134**, 993 (1964).
4. G. Stolhoff, P. Fulde, Z. Phys. B **26**, 257 (1977); **29**, 231 (1978).
5. T.A. Kaplan, P. Horsch, P. Fulde, Phys. Rev. Lett. **49**, 889 (1982).
6. L.F. Matheiss, Phys. Rev. **123**, 1209 (1961).
7. E.N. Economou and C.T. White, Phys. Rev. Lett. **38**, 289 (1977); E.N. Economou and P. Mihas, J. Phys. C **10**, 5017 (1977).
8. R.D. Poshusta and D.J. Klein, Phys. Rev. Lett. **48**, 1555 (1982); R.D. Poshusta, T.G. Schmalz and D.J. Klein, Mol. Phys. **66**, 317 (1989); R.D. Poshusta and D.J. Klein, J. Mol. Struc **229**, 103 (1991).
9. S. Kuwajima, J. Chem. Phys. **77**, 1930 (1982).
10. J.P. Malrieu and D. Maynau, J. Amer. Chem. Soc. **104**, 3021, 3029 (1982).
11. Ph. Durand, J. Phys. Lett. **43**, 1461 (1982); Phys. Rev. A **28**, 3184 (1983); D. Maynau, Ph. Durand, J.P. Daudey, and J.P. Malrieu, Phys. Rev. A **26**, 3193 (1983).
12. D.J. Klein and D.C. Foyt, Phys. Rev. **A8**, 2280 (1973); D.J. Klein, W.A. Seitz, M.A. Garcia-Bach, J.M. Picone, and D.C. Foyt, Int. J. Quant. Chem. Sym. **17**, 555 (1983).
13. J.-P. Malrieu, D. Maynau and J.P. Daudey, Phys. Rev. B **30** 1817 (1984); D. Maynau, M.A. Garcia-Bach and J.P. Malrieu, J. Phys., Paris, **47**, 207 (1986).
14. M.B. Lepetit, J.P. Malrieu, and F. Spiegelmann, Phys. Rev. B **41**, 8093 (1990).
15. D.J. Klein, J. Phys. A **13**, 3141 (1980).
16. M. Said, D. Maynau, J.P. Malrieu, and M.A. Garcia-Bach, J. Amer. Chem. Soc. **106**, 571 (1984).
17. R.R. Bartkowski, Phys. Rev. B **5**, 4536 (1972).
18. D.J. Klein, J. Chem. Phys. **64**, 4868 (1976).
19. P.W. Kasteleyn, Physica **28**, 104 (1952).
20. D.A. Huse and V. Elser, Phys. Rev. Lett. **60**, 2531 (1988).
21. S. Liang, B. Doucot, and P.W. Anderson, Phys. Rev. Lett. **61**, 365 (1988).
22. M.A. Garcia-Bach, R. Valentí, S.A. Alexander, and D.J. Klein, Croatica Chemica Acta, **64**, 415 (1991).
23. P.W. Anderson, Science **235**, 1196 (1987).
24. M.A. Garcia-Bach, P. Blaise, and J.P. Malrieu, Phys. Rev. B **46**, 15645 (1992).
25. M.A. Garcia-Bach, A. Peñaranda, and D. J. Klein, Phys. Rev. B **45**, 10891 (1992).
26. R. Valentí, Ph.D., Universitat de Barcelona, 1989; A. Peñaranda, Ph.D., Universitat de Barcelona, 1991; M.A. Garcia-Bach, R. Valentí, A. Peñaranda, and D.J. Klein, to be published.
27. D.J. Klein, M.A. Garcia-Bach, and W.A. Seitz, J. Mol. Struct. **185**, 275 (1989); M.A. Garcia-Bach, R. Valentí, and D.J. Klein, ibid p. 287; D.J. Klein, M.A. Garcia-Bach, and R. Valentí, Int. J. Mod. Phys. B **3**, 2159 (1989); M.A. Garcia-Bach, R. Valentí, S.A. Alexander and D.J. Klein, Croatica Chemica Acta **64**, 415 (1991).
28. C.S. Yannoni and T.C. Clarke, Phys. Rev. Lett. **51**, 1191 (1983). H. Kalhert, O. Lertner, and G. Leising, Synthetic Metals **17**, 467 (1987).
29. G. König and G. Stollhof, Phys. Rev. Lett. **65**, 1239 (1990).
30. D.J. Klein, Chem. Phys. Lett. **217**, 261 (1994).
31. A. Pellegatti and J.P. Malrieu, Phys. Rev. B **46**, 9946 (1992).
32. M.H. McAdon and W.A. Goddard, Phys. Rev. Lett. **55**, 2563 (1985); J. Phys. Chem. **91**, 2607 (1987).
33. F. Illas, J. Casanovas, M.A. Garcia-Bach, R. Caballol, and O. Castell, Phys. Rev. Lett. **71**, 3549 (1993).

GENERATION AND SOLUTION OF EFFECTIVE MANY-BODY HAMILTONIANS FOR RARE EARTH AND TRANSITION METAL COMPOUNDS

A. K. McMahan

Lawrence Livermore National Laboratory
University of California
Livermore, CA 94550

I. INTRODUCTION

The high-T_c superconductors have brought about greatly renewed interest in the subject of strong electron-electron correlation, since the parent cuprates are believed to belong to the class of Mott insulators which are prototypical examples of strong correlation.[1] The challenge for such materials is to calculate electronic excitations such as ionization and affinity energies, whose least difference determines the insulating gap, and whose dependence on crystal momentum provides the dispersion of quasiparticle states bounding the gap. Local density approximate (LDA) and Hartree-Fock (HF) theories provide relatively accurate ground electronic state properties, for example, the dependence of total energy on atomic position,[2,3] and successful LDA applications of this capability have been noted for the high-T_c cuprates.[1] The one-electron spectra of both of these mean field theories, however, give poor approximations to the gaps in Mott insulators.[1,4-6] To properly calculate electronic excitations in these systems one must in principle use a many-body basis capable of providing correlated wavefunctions, the size of which unfortunately grows combinatorially with system size. Two strategies may be simultaneously employed to face this challenge. The first is to reduce the number of orbitals required per site by mapping the important low-lying excitations onto an effective Hamiltonian which ignores those degrees of freedom associated with higher lying excitations. The second is the focus of the present workshop, namely the use of finite clusters to restrict system size. By using periodic boundary conditions, such cluster calculations can still have well characterized values of crystal momentum (**k** vector), and therefore make contact with the familiar solid-state concept of dispersion.

While the parameters defining effective Hamiltonians have traditionally been defined in an empirical fashion,[7-11] it has become appreciated in recent years that the LDA total energy capability may also provide reasonable values of these parameters, by carrying out a series of ground state calculations for appropriately constrained orbital occupations.[12-21] In contrast to empirical determinations usually limited to a small number of parameters, such approaches are capable of providing far more detailed description (e.g., any number of orbital types, multineighbor interactions, etc.) providing more chemical realism than such

Edited by T.A. Kaplan and S.D. Mahanti, Plenum Press, New York, 1995

models as the one- or even three-band Hubbard Hamiltonian.[20] Quantum chemistry calculations provide yet a third approach. On the whole, these calculations have been used to make direct comparison to experimental data; however, there are some recent determinations of effective Hamiltonian parameters which provide for valuable comparison with the LDA values.[22-24]

It is the purpose of this paper to give an overview of the LDA generation of such effective Hamiltonians for a series of *d* and *f* electron Mott insulators, and review their approximate solution for periodic clusters. The materials considered are CeO_2 and PrO_2,[16] the high-T_c parent La_2CuO_4,[18,25,26] and three isostructural compounds, K_2CuF_4, K_2NiF_4, and La_2NiO_4.[25] These compounds are more properly termed charge-transfer insulators, since their gaps are generally bounded on one side by predominantly ligand *p* states; and on the other, by transition metal *d* or rare earth *f* states.[27] Beyond the mechanics of this approach, there are important physical conclusions illustrated by this body of work, both in regard to the large-energy-scale charged excitations associated with adding electrons or holes to a charge-transfer insulator, and to the low energy scale associated with the spin degrees of freedom. The most important correlation effect required for the large energy scale is seen to be simply screening of the electron-electron interaction, a fact which comes out simply from the constrained-occupation LDA calculations which yield reduced, solid-state screened values of the Hubbard repulsion parameter, *U*, that in turn result in reasonable insulating gaps. It might be noted that screening of the electron-electron interaction is also one important function of the GW method which has been used to calculate accurate gaps in semiconductors.[28] At the smaller magnetic energy scale there is a persistent tendency for *d* or *f* electrons in these materials to enter into local complexes with the surrounding ligand spins which approximate eigenstates of the local total spin. The structure of these spin clouds is seen to have an important impact on quasiparticle dispersion.

The organization of this paper will be to introduce the effective Hamiltonians in stages by considering first the Coulomb operator for a single atom or ion in Sec. II, and then introducing the additional hopping parameters required for a collection of atoms in Sec. III. Sec. IV will deal with improved treatment of screening effects, noting that HF calculations with screened interactions do indeed provide approximately correct insulating gaps for charge-transfer insulators. Section V will discuss essential spin correlations and their impact on quasiparticle dispersion, based on configuration interaction (CI) calculations for clusters. An assessment of the range of validity of the present effective Hamiltonians is discussed in Sec. VI, and brief conclusions are given in Sec. VII.

II. COULOMB OPERATOR

This section considers the Coulomb operator and a simple effective Hamiltonian which describes the electronic excitations of one atom or ion. It is seen that mean field total energies for constrained occupations give a relatively accurate representation of these excitations, and that the importance of screening of the electron-electron interaction is already apparent for this example.

For simplicity, consider first a fictitious atom which supports only, say, *d* states characterized by the magnetic quantum number *m* and by spin σ. Within this manifold the Coulomb operator may be written

$$\hat{U} = \frac{1}{2} \sum_{m_1,m_2,m_3,m_4} \sum_{\sigma,\tau} U_{m_1,m_2,m_3,m_4} c^+_{m_1\sigma} c^+_{m_2\tau} c_{m_4\tau} c_{m_3\sigma} \tag{1}$$

where presumably

$$U_{m_1,m_2,m_3,m_4} = \int d\mathbf{r} \int d\mathbf{r}' \phi^*_{m_1}(\mathbf{r}) \phi_{m_3}(\mathbf{r}) \phi^*_{m_2}(\mathbf{r}') \phi_{m_4}(\mathbf{r}') \frac{e^2}{|\mathbf{r}-\mathbf{r}'|}. \quad (2)$$

Three numbers, the Slater integrals F^0, F^2, and F^4, characterize the Coulomb operator in this manifold of d states.[29] The monopole integral F^0 is the largest contributor to the overall energy difference between d^1 and d^2 configurations and is the dominant component of the Hubbard U parameter. The multipole integrals F^2 and F^4 create the term structure, e.g., split the 45 two-electron d^2 states into ^3F, ^1D, ^3P, ^1G, and ^1S terms, and serve to define various exchange parameters. The monopole contribution to Eq. (2) is $F^0\delta_{m_1m_3}\delta_{m_2m_4}$, so that Eq. (1) may be expanded

$$\hat{U} = \frac{1}{2}F^0\,\hat{n}_d\,(\hat{n}_d - 1) + \text{multipole terms}, \quad (3)$$

where $\hat{n}_d = \Sigma_{m\sigma} c^+_{m\sigma} c_{m\sigma}$ is the operator which counts the total number of d electrons.

Now consider a more realistic case, the Cu ion, and in particular an effective Hamiltonian which might describe low-lying configurations $d^ms^n \equiv 3d^m4s^n$ near the $d^{10}s^1$ atomic ground state.

$$H = \varepsilon_d \sum_{m,\sigma} d^+_{m\sigma} d_{m\sigma} + \frac{1}{2}U_d \sum_{m_1,m_2} \sum_{\sigma,\tau} d^+_{m_1\sigma} d^+_{m_2\tau} d_{m_2\tau} d_{m_1\sigma} + \varepsilon_s \sum_{\sigma} s^+_\sigma s_\sigma$$
$$+ \frac{1}{2}U_s \sum_{\sigma,\tau} s^+_\sigma s^+_\tau s_\tau s_\sigma + U_{ds} \sum_{\sigma,m,\tau} s^+_\sigma s_\sigma d^+_{m\tau} d_{m\tau} \quad (4)$$

Here ε_d and ε_s are site energies for adding $3d$ and $4s$ holes, respectively, to the full-shell $d^{10}s^2$ configuration, while $d_{m\sigma}$ and s_σ are the respective destruction operators for holes. For simplicity, only the monopole parts of the Coulomb interaction are considered, so that the Hubbard U values designate appropriate monopole Slater integrals. The number operators for the total number of d and s holes are good quantum numbers in terms of which the eigenvalues of Eq. (4) are

$$E(n_d,n_s) = \varepsilon_d n_d + \tfrac{1}{2}U_d n_d(n_d - 1) + \varepsilon_s n_s + \tfrac{1}{2}U_s n_s(n_s - 1) + U_{ds} n_d n_s \quad (5)$$

The total energy differences given by Eq. (5) may be mapped onto spectroscopic data by using configurational averages over the experimental spectra (e.g., over all 45 states for d^8) in order to assign a specific energy to each configuration. We consider six configurations corresponding to the neutral atom ($d^{10}s^1$ and d^9s^2) and to singly (d^{10}, d^9s^1, and d^8s^2) and doubly (d^9) charged ions. The five linearly independent total energy differences may be used to fix the five parameters in Eq. (5), and the resultant *experimental* values are shown in Table 1, where only the more physically meaningful difference between the two site energies is given. The use of configuration averages here actually yields[29] $F^0(3d,3d)-(2/63)[F^2(3d,3d)+F^4(3d,3d)]$, $F^0(4s,4s)$, and $F^0(3d,4s)-G^2(3d,4s)/10$, for U_d, U_s, and U_{ds}, respectively, so that 0.6 and 0.1 eV have been added to the experimental U_d and U_{ds} values obtained by this procedure, respectively, in order that the corresponding numbers in Table 1 more closely approximate true monopole integrals. These estimated corrections were taken from the multipole integrals of Mann,[30] since there is evidence that even in the solid there is relatively little screening of the effective multipole integrals, in sharp distinction to the monopole case,[31] as will be discussed subsequently.

Standard LDA calculations for atoms and ions spherically symmetrize the one electron potential, and therefore retain (for non spin polarized calculations) only monopole contributions to the Coulomb energy. Our LDA total energies obtained for the same six configurations, using the von Barth-Hedin exchange-correlation potential,[32] are reproduced by the LDA parameter values in Table 1. The agreement is on the whole good, better than 1 eV, which is the general experience in such calculations. This illustrates one conclusion of the present section, namely that constrained-occupation LDA total energies can give reasonable electronic excitation energies for atoms and ions. More generally, this appears to be true of mean field calculations, as the HF value for the Cu U_d obtained by Martin and Hay is only 1.1 eV larger than experiment.[23] Note that their 17.4 eV value corresponds to the ^3F state of the d^8 configuration, and the corresponding monopole value should be about 2 eV larger, ~19.4 eV. Our LDA calculations, on the other hand, utilized a d^8s^2 configuration, and a comparable calculation for U_d using only d^8, d^9, and d^{10} configurations gives U_d =17.7 eV, or 2.5 eV larger than the LDA value in Table 1. In principle, our five parameter fit should have unraveled the interplay of the Cu *4s* and *3d* interactions, and the lesson here is that such mappings as in Eq. (5) apply to a restricted configurational range. Evidently, the Cu *3d* wavefunctions are somewhat more localized in the absence of Cu *4s* electrons, which leads to the larger U_d. This anticipates the next point of discussion.

Both experiment and LDA total energies suggest U_d = 15–18 eV whereas a direct evaluation of the monopole counterpart of Eq. (2) with our ground state ($d^{10}s^1$ configuration) wavefunctions yields U_d = 26.9 eV, while Mann gives 26.0 eV based on his HF wavefunctions.[30] The reason for this 18 to 27 eV difference is effectively screening in the atom as has been discussed in detail by Gunnarsson and coworkers.[17,21] It is generally assumed that HF does not do a good job of screening the electron-electron interactions,[33] and one might presume the same for LDA because of its Hartree-like wavefunction. The observations of Gunnarsson and coworkers,[17,21] however, suggest that electron-electron screening effects are still reflected in an average way by the radial extent of the various orbitals involved. In particular, when separate self consistent calculations are carried out for, say, differing Cu(*3d*) occupations, the resultant *3d*, *4s*, and *4p* orbitals ϕ are found to be expanded radially for the larger *3d* occupancies. This orbital relaxation serves to reduce the increase in the total energy brought about by the increased *3d* occupancy, which appears intuitively to be a screening effect, and which results in the reduced and relatively accurate values of U_d obtained from the total energy differences.

A dramatic example of Cu(*3d*) electrons screening one another is provided by the Cu ion with only *n 3d* electrons outside the Ar core, total energy $E(n)$, and $U_d(n) \equiv E(n+1) + E(n-1) - 2E(n)$. As noted above, the LDA value for $U_d(9)$ = 17.7 eV; however, $U_d(1)$ = 34.8 eV. This follows from a negative n^3 term in $E(n)$, which if included in the form $(1/2)(U_0 + n\Delta U)n(n-1)$, leads to the value $\Delta U \sim -0.7$ eV. It is easy to see, then, that $U_d(9) = U_0 + 26\Delta U$, while $U_d(1) = U_0 + 2\Delta U$, creating a 17 eV difference between the two values. Thus because of the large number of *d–d* interactions for large *n*, a small reduction

Table 1. Parameters describing the $d^{10}s^1$, d^9s^2, d^{10}, d^9s^1, d^8s^2, and d^9 configurations of the Cu atom. Hole conventions are used, i.e., ε_s and ε_d are the energies to add one hole to full *4s* and *3d* shells, respectively. All values are in eV.

	$\varepsilon_s - \varepsilon_d$	U_d	U_s	U_{ds}
Expt	−1.5	16.1	8.3	9.8
LDA	−2.0	15.2	8.3	9.8

in each interaction, caused by a small relaxation in *all* the *3d* orbital, has a big impact on the total energy.

It therefore appears that screened Coulomb interactions are essential to an accurate representation of the electronic excitations. Such screening is trivially included in the second quantized effective Hamiltonians simply by using reduced values for the Slater monopole integrals. One interesting implication of this procedure is that the operator $d^+_{m\sigma}$ creates orbitals of differing spatial extent depending on existing *3d* occupation. This should not be disturbing, however, for the function of an effective Hamiltonian is to reproduce a many-electron spectra just as a tight-binding fit is used to represent a one-electron spectra, in each case without regard to what the actual orbitals might be.

III. HYBRIDIZATION

This section introduces the hybridization or hopping parameters which greatly complicate the generation of parameters defining effective many-body Hamiltonians. Nevertheless, reasonable, although far from rigorous, procedures do exist for the calculation of these quantities, and in particular generic LDA decoupling schemes are reviewed. An important difference in philosophy of approach is noted between methods which seek to build most of the solid-state screening effects into highly renormalized parameters defining the effective Hamiltonians, versus those which seek to obtain these screening effects from configuration interaction solutions of Hamiltonians characterized by bare or less-renormalized parameters.

For a collection of atoms, as in a cluster or solid, the one-body terms in the electronic Hamiltonian are no longer in general diagonal, and an appropriate generalization of Eq. (4) becomes

$$H = \sum_{i,\sigma} \varepsilon_i \, c^+_{i\sigma} c_{i\sigma} + \sum_{i,j,\sigma}^{i\neq j} t_{ij} \, c^+_{i\sigma} c_{j\sigma} + \frac{1}{2} \sum_{i,j,\sigma,\tau} U_{ij} \, c^+_{i\sigma} c^+_{j\tau} \, c_{j\tau} c_{i\sigma} \, . \tag{6}$$

As before σ and τ designate spin; however, *i* and *j* indicate atomic site in addition to the usual atomic (principle, angular and magnetic) quantum numbers. The new *hopping* term allows electrons or holes to move from one atomic site to another. In a mean field solution, this term is responsible for the self-consistent one-electron wavefunctions ψ being mixtures or linear combinations of the various basis orbitals ϕ_i, which is the familiar concept of hybridization. Equation (6) now exhibits the main types of parameters found in effective many-body Hamiltonians: site energies, ε_i, hopping parameters, t_{ij}, and Coulomb integrals, U_{ij}. For simplicity we continue to display only monopole contributions to the direct Coulomb integrals, $U_{ij} = F^0(i,j)$, and more generally there are in addition a variety of exchange integrals as dictated by the full set of Slater F^k and G^k integrals.[29]

A given system is considered to be weakly or strongly correlated depending in some average sense on whether *U/t* is small or large. Both limits are straightforward, as HF is exact for *U=0*, and *t=0* uncoupled atoms may be treated (less term structure) as in the previous section. It is the competition between hopping and Coulomb terms when neither is negligible which makes the many-body solutions of Eq. (6) such a challenge for bulk-like systems. The parameters in the effective Hamiltonians must then either be determined by utilizing relatively rigorous solutions (e.g., HF plus CI) for small clusters, or by turning to approximate mean field solutions in the case of large or bulk systems. The former category includes cluster solutions of the effective Hamiltonian which are then compared either to experiment[8-11] or to *ab initio* quantum chemistry calculations[22-24] in order to extract the parameters. The latter category includes a mapping of HF solutions of the effective Hamiltonian onto *ab intio* LDA calculations,[19] and the more common LDA *decoupling* methods which will now be briefly described.[12-18,20]

The LDA decoupling methods assume that the many-body hybridization or hopping matrix elements in Eq. (6) may simply be approximated by the corresponding one-body matrix elements in the LDA one-electron Hamiltonian.

$$t_{ij} \approx t_{ij}^{LDA} \ . \tag{7}$$

If, furthermore, *both* sets of *t* 's were set to zero, one can in this limit decouple the atoms and recover the simple mapping of the previous section. Specifically, let $\tilde{E}$ and $\tilde{E}^{LDA}$ be the corresponding total energies for $t_{ij} = t_{ij}^{LDA} = 0$. The eigenstates of modified Eq. (6) are then

$$\tilde{E}(\{n_i\}) = \sum_i \varepsilon_i n_i + \frac{1}{2}\sum_i U_i n_i(n_i - 1) + \frac{1}{2}\sum_{i,j}^{i\neq j} U_{ij} n_i n_j \tag{8}$$

while the mapping becomes

$$\begin{aligned}\varepsilon_i &= \tilde{E}(1) - \tilde{E}(0) \approx \tilde{E}^{LDA}(1) - \tilde{E}^{LDA}(0) \\ &\approx \tilde{\varepsilon}^{LDA}(\tfrac{1}{2}) + \cdots\end{aligned} \tag{9}$$

$$\begin{aligned}U_i &= \tilde{E}(2) + \tilde{E}(0) - 2\tilde{E}(1) \approx \tilde{E}^{LDA}(2) + \tilde{E}^{LDA}(0) - 2\tilde{E}^{LDA}(1) \\ &\approx \tilde{\varepsilon}^{LDA}(\tfrac{3}{2}) - \tilde{\varepsilon}^{LDA}(\tfrac{1}{2}) + \cdots\end{aligned} \tag{10}$$

Here, only the occupations n_i are indicated, with the others being assumed zero. Also advantage is taken of the definition of the LDA one-electron eigenvalues in Taylor expanding $\tilde{E}^{LDA}(n_i)$.

This approach may be simply implemented if those states exhibiting large Coulomb interactions are sufficiently localized so as to permit core-like treatment in the required *self-consistent* calculation of the $\tilde{\varepsilon}^{LDA}(n)$ for use in Eqs. (9) and (10), while a separate calculation including these same states amongst the valence bands provides the matrix elements for Eq. (7). There has been discussion of the optimal one-electron potential to be used in the latter calculation,[34] related to the occupation-dependence of the *t*'s, but usually the standard, converged LDA potential has been taken. Applications include *4f* states in rare-earth metals[12,21] and compounds,[14-16] and *3d* states in NiO,[13] Mn doped CdTe,[17] and in the high-T_c cuprates.[18] For more extended states which still have significant Coulomb interactions, such as O(*2p*), generalization of the decoupling approach requires the use of Wannier functions as employed to obtain the results in Fig. 1.[20]

Figure 1 shows a plot of the LDA eigenvalues $\tilde{\varepsilon}^{LDA}(n)$ as a function of occupation, as obtained for Cu(*3d*), in-plane O(*2p*), and apical O(*2p*) sites in La_2CuO_4. These are averages over magnetic quantum number and are presented as *hole* eigenvalues. The truncated Taylor expansions given by Eqs. (9) and (10) yield quite accurate values for the total energy differences here, and the latter gives $U_d = F^0(3d,3d) = 7.4$ eV for one hole (9 *d* electrons). Comparison to $U_d = 15$–18 eV for the isolated atom shows significant additional screening in the solid by the surrounding ligand electrons. The curvature seen in the results in Fig. 1 is precisely the orbital relaxation effect discussed in the previous section. Since these are hole eigenvalues, the negative n^3 term in $E(n)$ discussed in Sec. II appears here as

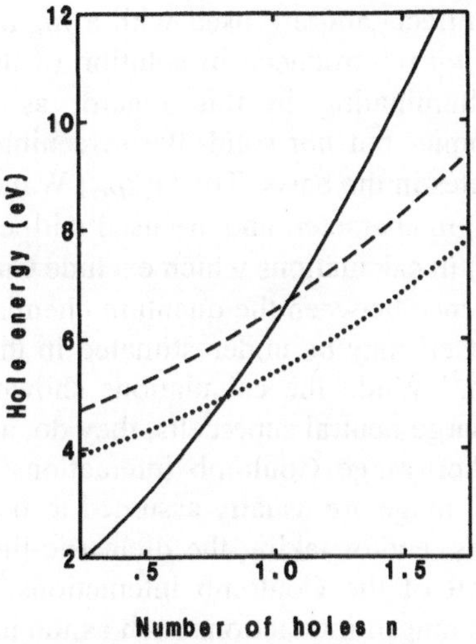

Figure 1. Hole LDA eigenvalues $\tilde{\varepsilon}^{LDA}(n)$ as a function of occupation *n* for Cu(*3d*) (solid), in-plane O(*2p*) (dashed), and apical O(*2p*) (dotted) states in La_2CuO_4.

a positive n^2 term in $-\partial E(n)/\partial n$, i.e., the smallest *U*'s occur for the least number of holes on a given site.

The parameter determinations by Eqs. (7), (9), and (10) are reasonable, although obviously not rigorous. As previously noted, comparisons with spectroscopic data for isolated atoms and ions suggest that such site energy differences and Coulomb interactions are reliable to about ±1 eV. All LDA values for the La_2CuO_4 solid are in similar agreement with empirical determinations based on the analysis of photoemission, Auger, and optical data for the solid,[8-11] with differences occasionally as large as 2 eV, but usually within ~1 eV (see Table VI of Ref. 20). The approximation, Eq. (7) for the hopping parameters has one obvious shortcoming, namely that the true many-body integrals include nearby *spectator* orbitals whose differences between initial and final states might be expected to reduce the true value relative to the one-body matrix element. Analysis of experimental valence spectra and static susceptibilities for rare earth metals and compounds suggests overestimation of the *t*'s by 13–30% for these materials,[35] which is also consistent with the kind of reduction of the Eq. (7) calculated t_{pd} which would yield a superexchange frequency *J* in agreement with experiment for La_2CuO_4.[20] Quantum chemistry determinations, on the other hand, suggest 10–40% larger values of t_{pd} which are compensated by significantly larger Coulomb interactions so that predictions for *J* are also reasonable.[22,24]

The large differences between quantum chemistry and LDA determinations of the Coulomb interactions has been emphasized in the context of Pariser-Parr-Pople model fits to Cu_2O_{11} cluster calculations.[23] The parameter combination $U-4|t|$ in this case provides a rough approximation to the insulating gap associated with an infinite CuO_2 layer. Two quantum chemistry calculations, one *ab initio*[23] and the other an incomplete neglect of differential overlap (INDO) approximation,[24] give 8.4 and 4.8 eV, respectively, for this quantity. For comparison, a calculation[23] for the same Cu_2O_{11} cluster using LDA parameters[19] gives 2.5 eV. Extracluster polarization effects are known to reduce the bulk NiO gap as determined from NiO_6 cluster calculations by ~5 eV,[36] and corrections of the same order of magnitude could bring these quantum chemistry calculations into agreement with, e.g., the La_2CuO_4 gap of ~2 eV. The Cu_2O_{11} calculation with LDA parameters, on the other hand, requires no further screening, since the parameters themselves are renormalized to include solid-state screening effects and are used with a highly restricted basis so that additional screening effects are *not* reintroduced in solution of the effective Hamiltonian. The INDO calculations are illuminating in this regard, as they assume monopole interactions characteristic of atomic- but *not* solid-like screening (e.g., $F^0(2p,2p)$ = 13.0 eV), yet also include Cu(*4sp*) states in the basis. The $O(2p_\sigma)$ Wannier functions of Ref. 20, by comparison, are 22% Cu(*4sp*) in character, and are used with effective, further screened interactions ($F^0(2p,2p)$ = 3.6 eV) in calculations which exclude Cu(*4sp*) states.

A second potential difference between the quantum chemistry and LDA parameters is that the degree of screening itself may be underestimated in the former case, and over-estimated in the latter approach.[23] While the calculations shown in Fig. 1 did not treat La_2CuO_4 as a metal, by using charge-neutral supercells, they do, as do LDA calculations in general, measure on-site and short-range Coulomb interactions relative to more distant interactions, which beyond some range are usually assumed to be zero. Actual calculation of these more distant interactions, and/or taking the dielectric-limit values for them, will generally increase the size of all of the Coulomb interactions.[20] These changes can be significant for some intermediate ranged interactions such as the in-plane near neighbor O–O repulsion (U_{pp} = 0.1 → 0.6 eV), with much smaller percentage increases for the on-site values (U_d = 7.1 → 7.7 eV, U_p = 3.1 → 3.7eV).

IV. IMPROVED ONE-ELECTRON SPECTRA

Since CI calculations are in general expensive and numerically challenging, it would be advantageous to improve mean field treatments so that their one-electron spectra might better approximate, to the extent possible, the true many-body spectra. The simplest measure of such improvement for Mott-Hubbard and charge-transfer insulators would be the value of the gap, E_g. Various modified one-electron theories are reviewed here in light of their E_g results for NiO and La_2CuO_4, materials for which the experimental gaps are well known.[37-40] The discussion first considers bulk calculations (Table 2), and then turns to insights provided by cluster calculations. In this context it is then observed that HF solutions for effective Hamiltonians whose Coulomb interactions reflect *solid-state* screening also provide respectable gaps for charge-transfer insulators, as is demonstrated for La_2CuO_4 and three isostructural compounds.[25]

Table 2. Bulk calculations of the insulating gaps for NiO and La_2CuO_4 as obtained from the one-electron eigenvalues. References are given to the right of the gap values.

	NiO		La_2CuO_4	
	E_g (eV)	Refs.	E_g (eV)	Refs.
Expt	4.3, 4.0	[37,38]	1.8–2.0	[39,40]
LSDA	0.3	[4]	0	[1]
LSDA–OP	1.4	[41]	0.2	[41]
LSDA–SIC	2.54	[42]	1.04	[43]
L(S)DA+U	3.1	[44]	1.65	[45]
HF	14	[6]		
HF–U_{eff}			2.0, 2.2	[25,26]

The predilection of LDA and local spin-density approximate (LSDA) calculations to give gaps which are too small is well known.[28] Often, as for La_2CuO_4, LSDA incorrectly predicts metallic character.[1] Sometimes, as for NiO, LSDA correctly predicts insulating character, however, with a gap an order of magnitude too small.[4] It is now generally appreciated that such overly small gaps arise from the LSDA's approximate treatment of exchange interactions, e.g., the Slater F^2 and F^4 contributions to the Coulomb operator, whereas the general size of the gap for these materials should reflect the monopole terms F^0. LDA and LSDA fail to acknowledge this impact of F^0 because they use the same one-electron potential to calculate both occupied valence and empty conduction states.[41-45] While orbital polarization (OP) [46] is seen in Table 2 to be an improvement, this analog of open shell HF provides better treatment of the multipole parts of the Coulomb operator, but still doesn't confront the fundamental F^0 aspect of the problem. The self-interaction corrected (SIC) generalization of LSDA does face this issue, and is seen in the table to provide significant increases in the calculated E_g.[42,43] The numerically simpler LDA+U[44] and LSDA+U[45] methods address the same problem in a more *ad hoc* way, by adding to the total energy functional diagonal approximations to Eqs. (3) and (1), respectively, less appropriate averages.

While HF has no self-interaction problem, its one-electron excitations are characterized by unscreened Coulomb interactions which generally result in insulating gaps that are far too large.[47] Recent HF calculations for bulk NiO, for example, exhibit a gap of about 14 eV.[6] An examination of the one-electron partial densities of states suggests a ~26 eV separation between valence and conduction band centroids of Ni(*3d*) character, consistent with Mann's unscreened $F^0(3d,3d)$ = 26.3 eV.[30] The top of the valence band is of predominant O(*2p*) character, with the centroids of valence Ni(*3d*) and O(*2p*) character separated by about 5 eV. These results suggest a charge transfer gap of $E_g \sim \varepsilon_d + U_d - \varepsilon_p \sim$ 21 eV, which is then reduced down to ~14 eV by band width (i.e., *t*) effects. The same E_g expression follows from Koopmans' theorem[48] total energy differences, in which case ε_d and ε_p are the energies to add one hole to full Ni(*3d*) and O(*2p*) shells, respectively. Bulk HF calculations for $CaCuO_3$ give a similarly large 13.5 eV separation between occupied and empty Cu–O *dps** antibonding bands.[5] For both NiO and $CaCuO_3$, one expects that screened Coulomb interactions would lead to improved gaps.

A careful analysis of the insulating gap in NiO has been presented based on the NiO_6 cluster, where the charge-transfer gap $E_g=\Delta$ was defined as the sum of the smallest O(*2p*) ionization and the smallest Ni(*3d*) affinity energies.[36] The initial ΔSCF value of 14.1–15.5 eV was found to be reduced by 4–5 eV due to intracluster polarization and correlation effects, and by ~5 eV due to extracluster polarization contributions, bringing the final 4.4–5.2 eV result for Δ in line with experiment. Since the ΔSCF value is a total-energy difference obtained from separate self consistent HF calculations for the two configurations, it should already include some relaxation effects relative to the gap in the one-electron spectrum of the bulk results above. Nevertheless, these cluster results clearly point to the omission of large screening effects as the major problem in the HF one-electron spectra. Note, that the separation between bonding and antibonding $O(p_\sigma)$–$Cu(d_{x^2-y^2})$ hybrids in the HF one-electron spectrum for a CuO_6^{11-} ($d^{10}p^6$) cluster has been found to be 7.75 eV,[23] which is close to the $[(\varepsilon_p-\varepsilon_d)^2+16t^2]^{1/2}$ = 6.3–6.9 eV range predicted for this cluster using the screened LDA parameters,[20] where these site energies also correspond to the addition of one hole to the respective filled shells. It is significant that such HF energy differences between like-charge states appear reliable, whereas those between dissimilar charge states do not, again pointing to inadequate treatment of screening in the latter case.

It is quite simple to carry out HF calculations with screened Coulomb interactions when the starting point is the effective Hamiltonian given by Eq. (6) — one simply assumes appropriately reduced values of the Slater integrals which define the Coulomb operator. The last entry in Table 2 shows the bulk-calculated gap for La_2CuO_4 obtained using six-[25] and eight-band[26] effective Hamiltonians for this material, in which all intrasite exchange interactions relevant to the restricted bases were included. Table 3 compares equivalent six-band results for three other K_2NiF_4-structure insulators.[25] The calculated gaps in Table 3 appear to be within ~1 eV of experiment, and their trend follows that evident in the charge transfer energy difference for holes, $\varepsilon(p_\sigma)-\tilde{\varepsilon}(d_{x^2-y^2})$, where $\tilde{\varepsilon}$ is the energy to add the first (second) *d* hole for the Cu (Ni) materials. The origin of the trend is simply that *3d* states are more tightly bound (more positive hole energies ε_d) in Cu than Ni, and similarly for the *2p* states of O and F. Since $\varepsilon_p > \varepsilon_d$, this implies the largest gap for the Ni–F compound, and the smallest, for the cuprate. Experimental gaps for the two fluorides are unknown, beyond that these materials are transparent (E_g^{expt}> 3.5 eV).

V. ESSENTIAL CORRELATIONS

There are limits to improvements that can be made in mean field theories, and it seems obvious that the true many-body excitation spectra contain some features which can only be reproduced by correlated wavefunctions. Cluster calculations play a pivotal role here, first in making the required CI calculations manageable, and secondly in elucidating the nature of the correlations in the localized limit. It is suggested here that the essential correlations for charge-transfer insulators are often associated with localized structures which approximate eigenstates of the local total spin, a statement so obvious to quantum chemists that it might not bear mentioning. Nevertheless, the participation in such local-spin complexes by an electron or hole added to a charge-transfer insulator implies a significant response of the environment to the added particle, which makes contact with the solid state concept of a quasiparticle. These ideas are illustrated here first from impurity-model perspective for CeO_2, PrO_2,[16] and hole-doped La_2CuO_4[18] and then the results of CI calculations for periodic La_2CuO_4 clusters are reviewed in order to comment on the dispersion of quasiparticle states bounding the insulating gap in this material.[26]

Impurity Model Perspective

An impurity model is considered here to be a cluster with only one *d*- or *f*-electron site surrounded by an assembly of ligand sites, which may range from just near neighbors to an infinite sea as in the Anderson impurity model.[49] As the localized electron site is the center of symmetry, the corresponding irreducible representations, μ, and partner, i ($i=1,N_\mu$),

Table 3. Comparison to experiment of theoretical gaps calculated with Hartree-Fock theory using screened Coulomb interactions.[25] The last row gives the charge-transfer energy difference for holes. All quantities are in eV.

	La_2CuO_4	La_2NiO_4	K_2CuF_4	K_2NiF_4
E_g^{expt}	1.8–2 ^{a}	4 ^{b}	>3.5	>3.5
E_g^{HF}	2.0	5.1	5.0	8.3
$\varepsilon(p_\sigma)-\tilde{\varepsilon}(d_{x^2-y^2})^c$	3.3	5.6	5.7	8.8

^{a} Refs. 39 and 40.
^{b} H. Eisaki et al., Phys. Rev. B **45**, 12513 (1992).

^{c} $\tilde{\varepsilon}(d_{x^2-y^2}) = \varepsilon(d_{x^2-y^2}) + nU^{S=1}(d_{x^2-y^2}, d_{3z^2-r^2})$, where n=0 (1) for Cu (Ni).

labels of its *d* or *f* orbitals become important indices. One familiar irreducible representation for a *d* spin-orbital in a tetragonal environment, for example, is $\mu = b_{1g} = x^2-y^2$, for which the partner index *i* runs over the two spin possibilities. The most interesting ligand orbitals are molecular orbitals centered about the localized site which transform according to the same irreducible representations, as these can then couple with the localized orbitals. The counterpart of the example just given is the b_{1g} molecular orbital formed from four $O(2p_\sigma)$ orbitals, one on each of the four near-neighbor O sites surrounding a given Cu in the cuprate CuO_2 layers. Depending on the size of the cluster, additional indices beyond μi may be required for these coupling ligand molecular orbitals, and in the spirit of the Anderson impurity model, an energy index, ε, will be added.

With this background in mind, consider now the ground state of CeO_2, a charge-transfer insulator with a nominal $Ce^{4+}(4f^0)$, $O^{2-}(2p^6)$ configuration consistent with its nonmagnetic behavior, yet at first glance apparently inconsistent with photoemission experiments which indicate a ~0.5 $4f$ occupancy.[50] An approximation to the impurity model singlet ground state easily reconciles these facts[16]

$$\Psi_0 \approx A\left|f^0 p^6\right\rangle + \sum_{\mu}\int d\varepsilon\, a_\mu(\varepsilon)\left[\frac{1}{\sqrt{N_\mu}}\sum_i f^+_{\mu i} p_{\varepsilon\mu i}\right]\left|f^0 p^6\right\rangle . \quad (11)$$

Here $\left|f^0 p^6\right\rangle$ is a singlet closed-shell Slater determinant, while the operator combination in the square brackets creates *multiconfigurational* (many Slater determinant) singlets out of $Ce(4f)$–$O(2p)$ particle-hole pairs. Equation (11) therefore mixes f^0p^6 and f^1p^5 configurations, however, in the latter case, surrounding $O(2p)$ spins form a singlet complex with the *f* electron reminiscent of the Kondo effect.[49] The result is a wavefunction of precise symmetry, e.g., a singlet as if f^0, yet with a nonintegral *f* occupation of ~0.5.

PrO_2 has a nominal $Pr^{4+}(4f^1)$, $O^{2-}(2p^6)$ configuration for which LDA incorrectly predicts metallic character,[51] a familiar signature of charge-transfer insulators. To a rough approximation its ground state may be written $\Psi_{\nu j} \approx f_{\nu j}^{+}\Psi_0$, although more properly one should use the appropriate Clebsch-Gordan coefficients to combine the two *f* electrons and one *p* hole into an f^1-symmetry multiconfigurational combination. In analogy with the Ce case, one has precise f^1 symmetry here with a ~1.5 *f* occupancy. Based on the total energies obtained from the wavefunctions $\Psi_{\nu j}$, impurity model calculations[16] correctly predict a $\nu=\Gamma_8$ ground state, with a 0.157 eV $\Gamma_7 - \Gamma_8$ crystal field splitting, as compared to 0.13 eV measured in neutron scattering experiments.[52]

The local singlet[53] in hole-doped La_2CuO_4 has precise meaning within the context of the impurity model, as indicated by Eq. (12), where only details of the more important d^9p^5 configurations are indicated.

$$\Psi_0 \approx \int d\varepsilon a(\varepsilon)\left[\frac{1}{\sqrt{2}}\left(d_{\mu\uparrow}p_{\varepsilon\mu\downarrow} - d_{\mu\downarrow}p_{\varepsilon\mu\uparrow}\right)\right]\left|d^{10}p^6\right\rangle + \left|d^8p^6\right\rangle + \left|d^{10}p^4\right\rangle . \quad (12)$$

Here $\mu=b_{1g}$, and the intrinsic $Cu(3d)$ x^2-y^2 hole forms an ${}^1A_{1g}$ singlet with a doped $O(2p_\sigma)$ hole in a molecular orbital of the same b_{1g} irreducible representation. Although this is the lowest energy two-hole state for the cluster, a triplet ${}^3B_{1g}$ state, which pairs an intrinsic $Cu(3d)$ x^2-y^2 hole with an a_{1g} $O(2p)$ molecular orbital, has been found in many calculations to lie close by in energy.[54-56] It has been argued on the basis of cluster calculations that the nature of $O(2p)$ holes doped into the cuprates changes from b_{1g} to a_{1g} symmetry in the doping range near optimal T_c.[57] Thus the precise ${}^3B_{1g}$ to ${}^1A_{1g}$ energy separation is of some

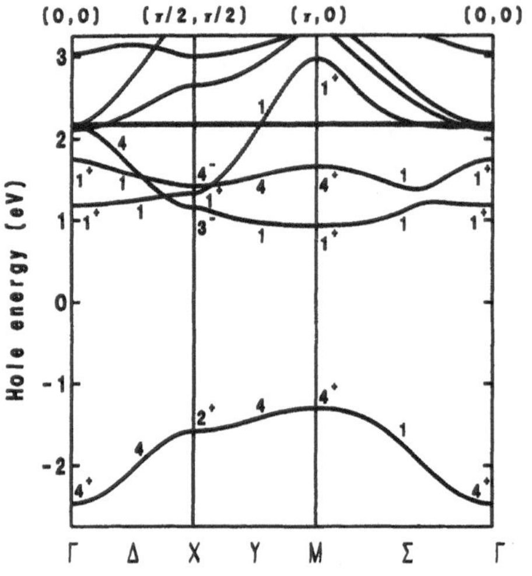

Figure 2. Spin-up hole ands from the unrestricted Hartree-Fock solution of an eight-band effective Hamiltonian for La_2CuO_4.[26] The intrinsic Cu(*3d*) holes completely fill the single branch below the gap, while the states above the gap are candidates for O(*2p*) hole doping.

importance, and it is to be emphasized that this depends critically on the total spin of these states and their associated multiconfigurational structure.

Periodic Clusters

The use of periodic boundary conditions with finite clusters allows contact to be made with the familiar solid-state parameter of crystal momentum, **k**. It is noted that analogs of the ${}^1A_{1g}$ and ${}^3B_{1g}$ two-hole states in the impurity models correspond to the lowest energy states of one-hole-doped La_2CuO_4 for different values of **k**, so that the issue of relative ${}^1A_{1g}$ and ${}^3B_{1g}$ ordering is really one of quasiparticle dispersion. The results of Grant and McMahan[26] in this regard are reviewed here, namely that HF even with screened interactions gives qualitatively the wrong quasiparticle dispersion, whereas CI, which appears to restore some of the local configurational balance of Eq. (12), agrees with experiment.

Figure 2 gives the up-spin unrestricted HF bands for the eight-band effective Hamiltonian which led to the 2.2 eV gap for La_2CuO_4 listed in Table 2.[26] These are hole bands, so that the intrinsic Cu(*3d*) holes completely fill the predominantly $b_{1g} = x^2{-}y^2$ band (Γ_4^+ - X_2^+ - M_4^+ - Γ_4^+) below the gap. The mostly O(*2p*) states at X_3^- and M_1^+ above the gap are the lowest candidates for doped holes, and are of primary interest in the remaining discussion. The **k** vectors corresponding to the *X* and *M* points are ($\pi/2$, $\pi/2$) and (π, 0), respectively, in units of $1/a$ where the coordinate frame and the lattice constant, *a*, are those of the single (non magnetic) unit cell.

Figure 3 shows a schematic of the local symmetry implicit in the HF one-electron wavefunctions within the CuO_2 layers, for both the intrinsic Cu holes and the doped-hole states. The symmetry is assessed relative to Cu sites which lie at the center of each small square, where shaded and unshaded squares designate the two sublattices created by the Néel-ordered HF solution. Note in Fig. 3 that the X_3^- state is incompatible with local a_{1g} symmetry, whereas the O(*2p*) phases associated with the M_1^+ states appear locally to be of b_{1g} symmetry relative to the Cu's on one sublattice; and a_{1g}, relative to those on the other.

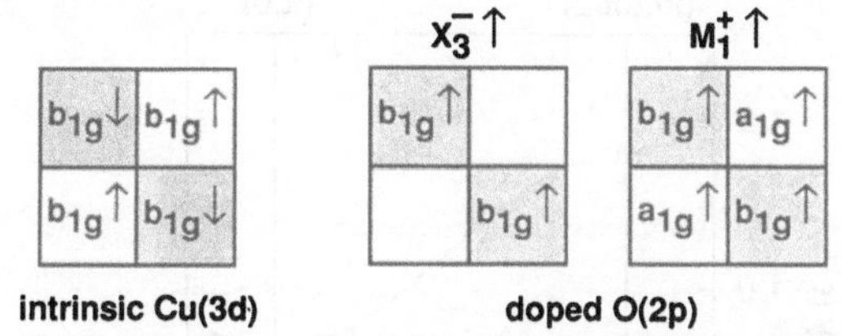

Figure 3. Schematic of the local symmetries about each Cu (center of each little square) for the intrinsic Cu(*3d)* holes (all states), and O(2p) holes doped at just the *X* and *M* points of Fig. 2.

Figure 3 indicates intrinsic-doped hole relations of $b_{1g\downarrow}\ b_{1g\uparrow}$ for the X_3^- doped hole, suggestive of the ${}^1A_{1g}$ two-hole state; and both $b_{1g\uparrow}\ a_{1g\uparrow}$ and $b_{1g\downarrow}\ b_{1g\uparrow}$, relative to the different sublattices, for the M_1^+ doped hole, suggestive of ${}^3B_{1g}$ and ${}^1A_{1g}$ two-hole states, respectively. The ${}^3B_{1g}$ component in the latter case is by far the dominant one, so that doped O(*2p*) holes at X_3^- and M_1^+ are seen to be the analogs of the local singlet, ${}^1A_{1g}$, and triplet, ${}^3B_{1g}$, encountered in the impurity calculations.

Quasiparticle energies are properly defined in terms of peaks in the one-particle spectral functions. It has been suggested for the class of Mott insulators that such functions should show relatively sharp peaks at the edge of the insulating gap, followed by broad tails of incoherent spin excitations at energies away from the gap.[58] Numerical simulations for the strong-coupling limit of the one-band Hubbard model are consistent with this picture for states along the line between *X* and *M*,[59] suggesting that these quasiparticle energies may be approximated by

$$\varepsilon_{\mathbf{k}} = \left\langle \Psi_{\mathbf{k}}^{N+1} \middle| H \middle| \Psi_{\mathbf{k}}^{N+1} \right\rangle - \left\langle \Psi_0^N \middle| H \middle| \Psi_0^N \right\rangle . \tag{13}$$

Here, Ψ_0^N is the ground state of the N-hole antiferromagnetic insulator, while $\Psi_{\mathbf{k}}^{N-1}$ is the ground state of the system with one added hole and total momentum **k**. CI representations for these wavefunctions may be obtained for periodic clusters,[26] where the size of the cluster determines the specific grid of **k** points that may be accessed. For the CuO_2 layers, clusters of 8 or more formula units are required to access both the *X* and *M* points of interest to this discussion. Very similar results were found for both 8 and 16 formula units in Ref. 26, suggesting convergence with cluster size.

Figure 4 provides a summary of the limited CI results of Ref. 26 for the lowest *X–M* branch of $\varepsilon_{\mathbf{k}}$ above the gap, which will include the lowest-energy states available for a single hole doped into insulating La_2CuO_4 The curves show four different levels of approximation to this single branch, and those labeled "2"–"4" are second-neighbor tight-binding fits to the *X* and *M* point CI results shown, assuming a value at Γ of about 2 eV. The curve labeled "1" is the X_3^- to M_1^+ branch of Fig. 2, and corresponds via Koopmans' theorem[48] to the HF Néel solution for Ψ_0^N with $\Psi_{\mathbf{k}}^{N+1} = c_{\mathbf{k}\uparrow}^+ \Psi_0^N$. Here $c_{\mathbf{k}\uparrow}^+$ adds an up-spin hole which does not disturb the underlying Néel order as shown in Fig. 3, suggesting the name *doped* Néel for these *N+1* hole Slater determinants. In sharp contrast, the spin flipped counterpart (↑↓ to ↓↑), needed to complete any local singlet or triplet relation with an intrinsic Cu hole, requires flipping one of the Cu spins relative to the Néel-ordered HF background. This breaks translational symmetry, yielding localized *N+1* hole *spin-bag* Slater determinants.[60] Nevertheless, an appropriately phased sum of translated spin-bag determinants provides a multiconfigurational state of well characterized momentum. Curves "1"– "3" in Fig. 4 were obtained[26] from Eq. (13) taking (1) the doped Néel, (2) this

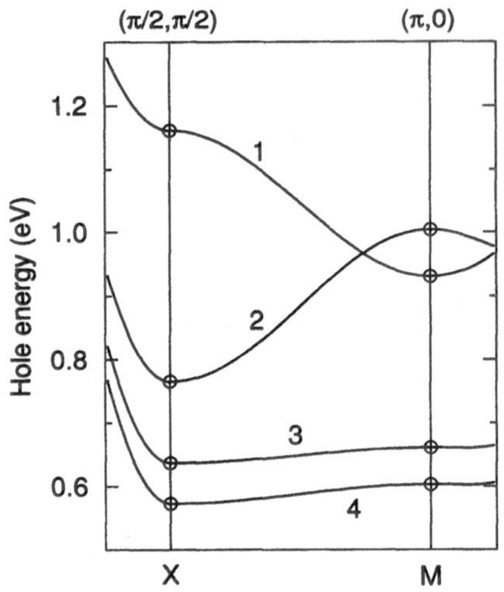

Figure 4. Periodic cluster configuration interaction results for the single X_3^- to M_1^+ branch shown in Fig. 1. Four levels of approximation are shown (see text).

Bloch projected spin bag, and (3) both states in limited CI representations of $\Psi_{\mathbf{k}}^{N+1}$; where in all cases the HF Néel solution was used for Ψ_0^N. By comparison, curve "4" represents far more extensive, but still limited, CI for both Ψ_0^N and $\Psi_{\mathbf{k}}^{N+1}$.

The local total spin is not a well defined quantity, nevertheless, an essential conclusion from Fig. 4 is that some representation of both spin flipped components in the square brackets of Eq. (12) is essential to an accurate treatment of the quasiparticle dispersion in La_2CuO_4. In the present case these two components (doped Néel and Bloch projected spin bag states) are significantly different in nature only because of bias introduced by the HF Néel-ordered background, as opposed to the true singlet ground state of the antiferromagnetic insulator. Additional CI (curve "4") has much larger impact on the individual terms in Eq. (13) than on $\varepsilon_{\mathbf{k}}$, suggesting further improvements in this background which don't significantly impact the relatively localized quasiparticle. Finally, while screened Coulomb interactions can improve the gap in the HF one-electron spectrum, curve "1" in Fig. 4 shows qualitatively incorrect results for the dispersion of the quasiparticle states bordering this gap. This curve would have the first doped O(*2p*) holes going into states near the M point, which are far too rich in apical O($2p_z$) character to be in agreement with experiment.[61,62] Note also that the overall size of correlation corrections to E_g are not small, ~ 1 eV, if a consistent comparison is made between HF and CI, taking in each case the indirect gap from the valence *M* to the conduction *X* points. The most converged CI gap is 1.5 eV,[26] suggesting the La_2CuO_4 Coulomb interactions used in this calculation are too small.[45]

VI. DISCUSSION

This paper has described the generation and solution of effective many-body Hamiltonians for an unusual class of strongly correlated rare earth and transition metal compounds, namely charge-transfer insulators.[27] In the absence of exact yet feasible methods for the calculation of electronic excitations in bulk samples of these materials, it is argued that the present approach is reasonable and provides agreement with experiment over a range of energy scales. An essential aspect of this approach is the premise that the problem of charged excitations in these materials may be folded into renormalized parameters in reduced-basis effective Hamiltonians, leaving the primary correlation problem to be that associated with the spin degrees of freedom. These points are discussed below in the context of the range of validity anticipated for these Hamiltonians, with specific numbers cited for La_2CuO_4.

The range of validity sought for the present effective Hamiltonians is a description of the low-lying electronic excitations of the charge-transfer insulator (*N* electrons) and of the insulator doped with one electron and one hole ($N\pm1$ electrons). There are two different energy scales here. Charged excitations govern the overall placement of the ground state energies of the *N* and $N\pm1$ electron systems relative to one another, while spin excitations determine the lowest-lying neutral excitations above these ground states. The large energy scale is built into the present effective Hamiltonians by using *solid*-state screened values of the Coulomb interactions and site energies, and an assessment of the present values for La_2CuO_4 is provided by the HF+CI gap value of 1.5 eV[26] as compared to the 1.8–2.0 eV experimental value.[39,40] Similarly the low energy scale obtained from the same parameters is indicated by the superexchange frequency J~0.24 eV[25] as compared to the experimental 0.13 eV value.[63] Since J is notoriously sensitive to the parameters defining the effective Hamiltonian, modest (~20%) tuning of these parameters could easily provide excellent values of both E_g and J. The quasiparticle dispersion of the $N-1$ electron system is another measure of the spin interactions, and it would appear that their role in favoring initial O($2p$) doping at $\mathbf{k} = (\pi/2, \pi/2)$ is fundamentally correct and in agreement with experiment.

Higher lying neutral excitations are more problematic. These include the $d(x^2-y^2) \rightarrow d(3z^2-r^2)$ exciton in the La_2CuO_4 insulator, which simple cluster calculations with LDA parameters put in the range ~1 eV,[11,56] compared to 0.4 eV as indicated by the most recent experiments.[64] At still higher energies, closer to the ~2 eV gap, would come Cu($3d$)–O($2p$) electron-hole pairs. It would seem logical that the use of solid-state screened parameters in the effective Hamiltonian might imply projection onto a Hilbert subspace which precludes accessing these and higher-lying states.

VII. CONCLUSIONS

Effective Hamiltonians offer a realistic and practical approach to the representation of low-lying electronic excitations in strongly correlated charge transfer insulators, such as the high-T_c parent La_2CuO_4. Not only do such Hamiltonians minimize the required number of orbital degrees of freedom, but it is suggested that the fundamental problem of charged excitations in these materials may be folded into renormalized Hamiltonian parameters which already reflect solid state screening, and that such parameters may easily be obtained from constrained local density functional calculations. Beneficial consequences of the renormalization include reasonable insulating gaps already from the Hartree-Fock one-electron spectra, and a restriction of configuration interaction to those essential correlations associated with the spin degrees of freedom. In large periodic clusters where states have well characterized crystal momentum, such spin correlations in the vicinity of particles doped into the parent insulator are still reminiscent of the total spin eigenstates found in impurity model solutions.

ACKNOWLEDGMENTS

This work was performed under the auspices of the U.S. Department of Energy by Lawrence Livermore National Laboratory under Contract No. W-7405-Eng-48. The author gratefully acknowledges his collaborators in various parts of the work reviewed here, especially J. B. Grant and R. M. Martin.

REFERENCES

1. W. E. Pickett, Rev. Mod. Phys. **61**, 433 (1989)
2. See, e.g., R. M. Martin, in *Electronic Structure, Dynamics, and Quantum Structural Properties of Condensed Matter*, edited by J. T. Devreese and P. Van Camp (Plenum, New York, 1985), p. 175.
3. See, e.g., R. Orlando, C. Pisani, C. Roetti, and E. Stefanovich, Phys. Rev. B **45** (1992); M. I. McCarthy and N. M. Harrison, *ibid.* **49**, 8574 (1994).
4. K. Terakura, A. R. Williams, T. Oguchi, and J. Kübler, Phys. Rev. Lett. **52**, 1830 (1984).
5. S. Massidda, M. Posternak, and A. Baldereschi, Phys. Rev. B **46**, 11705 (1992).
6. W. C. Mackrodt, N. M. Harrison, V. R. Saunders, N. L. Allan, M. D. Towler, E. Aprà, and R. Dovesi, Phil. Mag. A **68**, 653 (1993).
7. J. W. Allen, S.-J. Oh, O. Gunnarsson, K. Schönhammer, M. B. Maple, M. S. Torikachivili, and I. Lindau, Adv. in Phys. **35**, 275 (1986)
8. E. B. Stechel and D. R. Jennison, Phys. Rev. B **38**, 4632 (1988); **38**, 8873 (1988)
9. F. Mila, Phys. Rev. B **38**, 11358 (1988).
10. H. Eskes, G. A. Sawatzky, and L. F. Feiner, Physica C **160**, 424 (1989).
11. H. Eskes, L. H. Tjeng, and G. A. Sawatzky, Phys. Rev. B **41**, 288 (1990).
12. B. I. Min, H. J. F. Jansen, T. Oguchi, and A. J. Freeman, Phys. Rev. B **33**, 8005 (1986).
13. M. R. Norman and A. J. Freeman, Phys. Rev. **B** 33, 8896 (1986).
14. A. J. Freeman, B. I. Min, and M. R. Norman, in *Handbook on the Physics of Rare Earths*, edited by K. A. Gschneidner, L. Eyring, and S. Hafner (Elsevier, New York, 1987), Vol. 10, p. 165.
15. J. M. Wills and B. R. Cooper, Phys. Rev. B 36, 3809 (1987).
16. A. K. McMahan and R. M. Martin, in *Narrow-Band Phenomena*, edited by J. C. Fuggle, G. A. Sawatzky, and J. W. Allen (Plenum, New York, 1988), p. 133.
17. O. Gunnarsson, O. K. Andersen, O. Jepsen, and J. Zaanen, Phys. Rev. B **39**, 1708 (1989).
18. A. K. McMahan, R. M. Martin, and S. Satpathy, Phys. Rev. **B** 38, 6650 (1988).
19. M. S. Hybertsen, M. Schlüter, and N. E. Christensen, Phys. Rev. B **39**, 9028 (1989).
20. A. K. McMahan, J. F. Annett, and R. M. Martin, Phys. Rev. B **42**, 6268 (1990).
21. V. I. Anisimov and O. Gunnarsson, Phys. Rev. B **43**, 7570 (1991).
22. R. L. Martin, Physics B **163**, 553 (1990).
23. R. L. Martin, in *Cluster Models for Surface and Bulk Phonomena, NATO ASI Series*, edited by G. Pacchioni and P. S. Bagus (Plenum, New York, 1992), p. 485; R. L. Martin and P. J. Hay, J. Chem. Phys. **98**, 8680 (1993); R. L. Martin, *ibid* 8691 (1993).
24. Y. J. Wang, M. D. Newton, and J. W. Davenport, Phys. Rev. B **46**, 11935 (1992).
25. J. B. Grant and A. K. McMahan, Phys. Rev. Lett. **66**, 488 (1991).
26. J. B. Grant and A. K. McMahan, Phys. Rev. B **46**, 8440 (1992).
27. J. Zaanen, G. A. Sawatzky, and J. W. Allen, Phys. Rev. Lett. **55**, 418 (1985).
28. M. S. Hybertsen and S. G. Louie, Phys. Rev. Lett. **55**, 1418 (1985).
29. J. C. Slater, *Quantum Theory of Atomic Structure* (McGraw-Hill, New York, 1960), Vol. 1, p. 324 .
30. J. B. Mann, "Atomic Structure Calculations, I. Hartree-Fock Energy Results for the Elements Hydrogen to Lawrencium," Los Alamos Report LA-3690, July 28, 1967.
31. E. Antonides and G. A. Sawatzky, in *Transition Metals (1977)*, Proceedings of the International Conference on the Physics of Transition Metals, edited by M. J. G. Lee, J. M. Perz, and E. Fawcett, I. P. Conf. Ser. No. 39 (IP, London, 1979), p. 134; D. van der Marel and G. A. Sawatzky, Phys. Rev. B **37**, 10674 (1988).
32. U. von Barth and L. Hedin, J. Phys. C **5**, 1629 (1972).
33. S. Doniach and E. H. Sondheimer, *Green's Functions for Solid State Physicists* (Benjamin, Reading, 1974), p. 124.
34. O. Gunnarsson and O. Jepsen, Phys. Rev. B **38**, 3568 (1988).
35. O. Gunnarsson, N. E. Christensen, and O. K. Andersen, J. of Mag. Mag. Mat. **76** & **77**, 30 (1988)

36. G. J. M. Janssen and W. C. Nieuwpoort, Phys. Rev. B **38**, 3449 (1988).
37.G. A. Sawatzky and J. W. Allen, Phys. Rev. Lett. **53**, 2339 (1984)
38. S. Hüfner, J Osterwalder, T. Riesterer, and F. Hullinger, Solid State Commun. **52**, 793 (1984)
39. S. L. Cooper, G. A. Thomas, A. J. Millis, P. E. Sulewski, J. Orenstein, D. H. Rapkine, S-W. Cheong, and P. L. Trevor, Phys. Rev. B **42**, 10785 (1990).
40. T. Thio, R. J. Birgeneau, A. Cassanho, and M. A. Kastner, *ibid.* **42**, 10800 (1900).
41. M. R. Norman, Phys. Rev. B **44**, 1364 (1991).
42. A. Svane and O. Gunnarsson, Phys. Rev. Lett. **65**, 1148 (1990).
43. A. Svane, Phys. Rev. Lett. **68**, 1900 (1992).
44. V. I. Anisimov, J. Zaanen, and O. K. Andersen, Phys. Rev. B **44**, 943 (1991).
45. M. T. Czyżyk and G. A. Sawatzky, Phys. Rev. B **49**, 14211 (1994).
46. O. Eriksson, M. S. S. Brooks, and B. Johansson, Phys. Rev. B **41**, 7311 (1990).
47. C. Pisani, R. Dovesi, and C. Roetti, *Hartree-Fock Ab Initio Treatment of Crystalline Systems* (Springer-Verlag, Berlin, 1988).
48. T. A. Koopmans, Physica **1**, 104 (1933).
49. O. Gunnarsson and K. Schönhammer, Phys. Rev. B **28**, 4315 (1983).
50. E. Wuilloud, B. Delley, W.-D. Schneider, and Y. Baer, Phys. Rev. Lett. **53**, 202 (1984); J. W. Allen, J. Mag. Magn. Mater. 47,-48, 168 (1985).
51. D. D. Koelling, A. M. Boring, and J. H. Wood, Solid State Commun **47**, 227 (1983).
52. S. Kern, C.-K. Loong, and G. H. Lander, Phys. Rev. B **32**, 3051 (1985).
53. F. C. Zhang and T. M. Rice, Phys. Rev. B **37**, 3759 (1988).
54. H. Eskes and G. A. Sawatzky, Phys. Rev. Lett. **61**, 1415 (1988).
55. A. Fujimori, Phys. Rev. B **39**, 793 (1989).
56. J. F. Annett, R. M. Martin, A. K. McMahan, and S. Satpathy, Phys. Rev. B **40**, 2620 (1989).
57. H. Kamimura and M. Eto, J. Phys. Soc. Jpn. **59**, 3053 (1990).
58. S. Schmitt-Rink, C. M. Varma, and A. E. Ruckenstein, Phys. Rev. Lett. **60**, 2793 (1988).
59. S. A. Trugman, Phys. Rev. B **41**, 892 (1990); K. J. von Szczepanski *et al.*, *ibid.* 2017 (1990); E. Dagotto *et al.*, *ibid.* 2585 (1990); J. Song and J. F. Annett, Europhys. Lett. **18**, 549 (1992).
60. J. R. Schrieffer, X.-G. Wen, and S.-C. Zhang, Phys. Rev. Lett. **60**, 944 (1988).
61. H. Romberg *et al.*, Phys. Rev. B **41**, 2609 (1990).
62. C. T. Chen *et al.*, Phys. Rev. Lett. **68**, 2543 (1992).
63. G. Aeppli *et al.*, Phys. Rev. Lett. **62**, 2052 (1989); R. R. P. Singh *et al.*, *ibid.* **62**, 2736 (1989).
64. J. D. Perkins, J. M. Graybeal, M. A. Kastner, R. J. Birgeneau, J. P. Falck, and M. Greven, Phys. Rev. Lett. **71**, 1621 (1993).

NUMERICAL STUDIES OF STRONGLY CORRELATED ELECTRONIC SYSTEMS

Adriana Moreo

Department of Physics
and National High Magnetic Field Lab
Florida State University
Tallahassee, FL 32306

I. INTRODUCTION

One of the simplest models often used to describe the physics of the high temperature cuprate superconductors is defined by the two dimensional (2D) Hubbard Hamiltonian. Several normal state properties of these materials, such as antiferromagnetism and the behavior of the optical conductivity are qualitatively reproduced by this model.[1] Although superconductivity has not been observed in Monte Carlo simulations,[2] self-consistent techniques suggest its existence at low temperatures in the $d_{x^2-y^2}$ channel.[3] Superconductivity in the high temperature copper-oxides[4] appears in the vicinity of antiferromagnetic order. This may have important implications for the pairing mechanism. Studies of the uniform magnetic susceptibility, χ in $La_{2-x}Sr_xCuO_4$[5,6] have shown that as the doping x increases from zero at a fixed temperature, so does χ. Fixing the electronic density at half-filling and changing the temperature, the susceptibility reaches a maximum value at a finite temperature, T_m, that eventually goes to zero as the material is doped with holes. This behavior, in turn, leads to an anomalous temperature dependence of the Knight shift and the nuclear spin relaxation rates $1/TT_1$.[7]

It is widely believed that the interesting low temperature magnetic behavior in the cuprate high T_c superconductors comes predominantly from the copper-oxide planes. As stated above, some of the experimental properties of these planes have been successfully qualitatively reproduced using two-dimensional purely electronic models, like the *t-J* and the one band Hubbard models.[8,9] Since the qualitative behavior of some of these properties does not depend strongly on *U/t*, weak coupling techniques have successfully described them, but more quantitative comparisons with experimental results suggest that $U/t \approx 10$ is more realistic to phenomenologically describe the cuprate materials.[10] For the *t-J* model, on the other hand, qualitative changes in the ground state, like phase separation and d-wave superconductivity[11] are observed as *J/t* increases. Other properties of the copper-oxide compounds are not theoretically understood. Among them are the behavior of the uniform magnetic susceptibility as a function of temperature and doping, the shape of the Fermi surface and, of course, superconductivity. The behavior of the magnetic susceptibility will be presented in Sec. II; in Sec. III the shape of the Fermi surface will be discussed; superconductivity will be addressed in Sec. IV by introducing a simplified Hamiltonian without free parameters. Sec. V is devoted to conclusions.

II. UNIFORM MAGNETIC SUSCEPTIBILITY

The magnetic susceptibility[12] has been analyzed in the context of the Hubbard model using weak coupling methods like RPA.[13] Under this approximation T_m is always zero, and in addition the susceptibility decreases with doping in disagreement with experiments on $La_{2-x}Sr_xCu0_4$. However, recent high temperature expansions (HTE) up to ninth order in $1/T$, and extrapolated to low temperatures by Padé approximants were carried out for the *t-J* model.[14] The observed behavior is in qualitative agreement with experiments for $J/t \approx 0.5$. The discrepancy between RPA and HTE is difficult to explain.[13,15] One possibility is that the behavior of the susceptibility is different in the weak coupling and strong coupling limit and that RPA is capturing only the weak coupling behavior. Another possibility is that the behavior that resembles experiments is characteristic of the *t-J* rather than the Hubbard model and appears at some finite value of *J/t*. One of our goals is to solve this paradox. Analytical approaches such as mean-field, RPA and high temperature expansions have proved to be very useful in describing bulk properties of Hamiltonians, but are somewhat uncontrolled. RPA sums an arbitrary set of Feynman diagrams while HTE relies on analytic continuation to extract low temperature properties. On the other hand, there are numerical methods, like exact diagonalization and quantum Monte Carlo, that provide exact results on finite lattices. Hence, in order to obtain bulk results with these techniques finite size effects have to be considered. Thus, numerical and analytical techniques are complementary to each other. Our goal is to study the uniform magnetic susceptibility of the Hubbard model for *U/t* ranging from weak to strong coupling, using an unbiased method like quantum Monte Carlo.[16] Finite size effects will be analyzed and comparisons with experiments and previous analytical results will be made. The Hubbard model is defined by the standard Hamiltonian

$$H=-t\sum_{\langle \mathbf{ij}\rangle,\sigma}\left(c^{+}_{\mathbf{i},\sigma}c_{\mathbf{j},\sigma}+h.c.\right)+U\sum_{\mathbf{i}}n_{\mathbf{i},\uparrow}n_{\mathbf{i},\downarrow}\ , \tag{1}$$

where $c^{+}_{\mathbf{i},\sigma}$ creates an electron at site $\mathbf{i}$ with spin projection σ, and $n_{\mathbf{i},\sigma}$ is the number operator, and the sum $\langle\mathbf{ij}\rangle$ runs over pairs of nearest neighbor lattice sites. U is the on-site coulombic repulsion, and t the hopping parameter. The *t-J* model is defined by

$$H_{tJ}=-t\sum_{\langle \mathbf{ij}\rangle,\sigma}\left(\bar{c}^{+}_{\mathbf{i},\sigma}\bar{c}_{\mathbf{j},\sigma}+h.c.\right)+J\sum_{\langle \mathbf{ij}\rangle}\left(S_{\mathbf{i}}\cdot S_{\mathbf{j}}-\frac{1}{4}n_{\mathbf{i}}n_{\mathbf{j}}\right) \tag{2}$$

where $c^{+}_{\mathbf{i},\sigma}$ creates a hole at site $\mathbf{i}$ with spin projection σ, and $S_{\mathbf{i}}$ are spin operators. As the Hubbard coupling U increases, double occupancy of a lattice site becomes less likely and the Hubbard Hamiltonian approximately maps on the $t-J$ model with $J=4t^2/U$. The two are strictly equivalent in the limit $J/t=4t\ \ U=0$. Then, the $t-J$ model can be regarded as the strong coupling limit of the Hubbard model, although Zhang and Rice derived it directly from the copper oxide planes.[17] It is important to notice that as *J/t* increases the mapping is no longer accurate. Actually, exact diagonalizations on 4×4 lattices show that the one band Hubbard and the $t-J$ models are no longer equivalent for $J/t \gtrsim 0.4$.[18]

The magnetic susceptibility is given by

$$\chi = \lim_{\mathbf{q}\to 0} \frac{1}{N} \sum_{\ell} e^{i\mathbf{q}\cdot\ell} \int_0^\beta d\tau \left\langle \left[n_{\mathbf{i}+\ell,\uparrow}(\tau) - n_{\mathbf{i}+\ell,\downarrow}(\tau) \right] \left[n_{\mathbf{i},\uparrow}(0) - n_{\mathbf{i},\downarrow}(0) \right] \right\rangle , \tag{3}$$

where β=1/T and T is the temperature, **q** denotes momentum and N is the number of lattice sites. Before presenting our results, let us discuss the expected behavior of χ in several limits. For a Fermi liquid χ is finite at low temperature (proportional to the density of states) and as T increases it decreases to zero following the Curie-Weiss law. We should expect to observe this behavior at very low fillings where double occupancy of a site is unlikely, and certainly also at U/t=0. When U/t is large and the system is half-filled, Eq. (1) maps onto the Heisenberg model. In this case χ reaches a maximum at $T\approx J$ and then drops by about a factor two at T=0.[19] As the system is doped away from half-filling the maximum value of χ is expected to increase. This occurs because $\chi \propto 1/J$ and we can mimic the effect of doping by replacing J by an effective coupling J_{eff} with $J_{eff}<J$. Neither the increase of χ with doping starting at half-filling nor the increase with temperature starting at T=0, have been reproduced by RPA calculations.

In Fig. 1 we show χ vs. T for U/t=4 at different fillings obtained with quantum Monte Carlo. We present results on 4×4 and 8×8 lattices in order to discuss finite size effects. It can be seen that they are small at low hole doping including half-filling. As the doping increases they become important at low temperature. Let us try to understand why this occurs and what we should expect at larger values of the coupling. At U=0, finite size effects are significant since we expect that $\chi \propto \beta\, e^{-\beta\Delta_{sw}}$ where Δ_{sw} is the gap between the ground state and the first excited state with total spin 1. For U=0 there are degeneracies that in small lattices will cause χ to diverge as β at low temperature for some dopings. In small systems there are finite gaps between states and therefore certain fillings χ will go to 0 below temperatures of the order of these gaps. As U/t increases we expect less degeneracies and smaller gaps, and thus finite size effects should be less important.

At half-filling (Fig. 1a) Heisenberg-like behavior is observed showing the existence of long range antiferromagnetic order. Of course, the position of the peak in χ is not well predicted by the relation valid in the Heisenberg model, $T_m\approx J=4/U$ since U/t is small. As the filling decreases to 0.75 (Fig. 1b) the susceptibility *decreases* and becomes flatter. This behavior is in agreement with RPA calculations but not with that observed experimentally in the cuprate superconductors. At quarter filling finite size effects are strong (Fig. 1c) since even on an 8×8 lattice a spurious divergence in the susceptibility is observed at low temperature. At low density, $\langle n\rangle$=0.25 (Fig. 1d) Monte Carlo results for U/t=4 are compared with U=0 results. We had to use a 32×32 lattice to obtain a behavior without finite size effects for U=0 at T as low as 0.1. Similar behavior, though, was obtained for U/t=4 on an 8×8 lattice in agreement with our earlier remark stating that finite size effects are expected to diminish as the value of the coupling increases. Clearly at this low filling Fermi-liquid like behavior is observed.

Our results for U/t=4 indicate that RPA successfully reproduces the physics of the Hubbard model in this limit. What happens when the coupling increases? In Fig. 2a we present the behavior of χ as a function of temperature for different values of U/t at half-filling on 4×4 clusters. For U/t=4 finite size effects are not strong for $\langle n\rangle$ >0.5 and, as shown before, we expect them to become smaller as U/t increases. χ increases with the coupling because the magnetic correlations become stronger. At large U, when the Heisenberg model is a good approximation to Eq. (1) the susceptibility is expected to peak at $T\approx 4/U$, as it occurs for U/t=10 and 20. In Figs. 2b and 2c we show the behavior of χ at

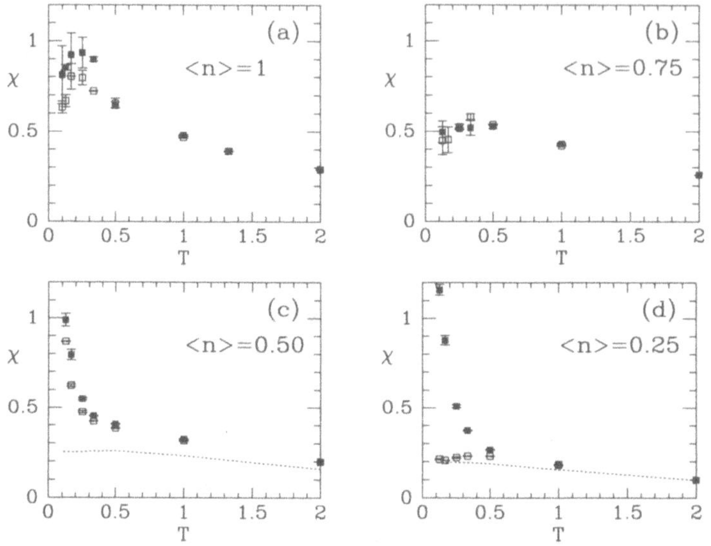

Figure 1. Uniform magnetic susceptibility χ as a function of temperature for *U/t*=4 on a 4×4 (full squares) and on an 8×8 lattice (open squares) for <n>=1.0 (a); 0.75 (b); 0.50 (c) and 0.25 (d). The dotted line denotes the U=0 susceptibility.

T=0.25 for *U/t*=4 and 10, respectively, as a function of doping. In Fig. 2b we see that χ clearly decreases with doping for *U/t*=4. The behavior of χ changes qualitatively for *U/t*=10 as shown in Fig. 2c. As the system is doped away from half-filling, the susceptibility increases reaching a maximum at <*n*>≈0.83. Fig. 2c is in qualitative agreement with experiments. These results suggest that the RPA approach successfully describes the behavior of the one band Hubbard model for *U/t*≈4 but fails for stronger couplings. On the other hand, comparisons with the results for the *t-J* model imply that for *U/t*=4*t/J*≥10 Hubbard and *t-J* models are very similar. It is important to notice that the behavior of χ for the Hubbard model in 2D appears to be different than in 1D[20] where a maximum at finite T is present for all values of *U/t* and χ increases slightly with doping at very small *U*.

III. FERMI SURFACE[21]

As stated in the introduction, self-consistent techniques suggest the existence of superconductivity in the Hubbard model. In order to understand the pairing mechanism leading to such superconducting instability it is important to study the shape of the Fermi surface in this model, and compare it against angle-resolved photoemission experiments for the cuprates. Theories based on antiferromagnetism (AF) explain in a natural way the apparent d-wave symmetry of the condensate, but they also suggest the existence of *hole pockets* in $n(\mathbf{k})$.[22] Since for some time evidence of a large electron-like Fermi surface has been reported in experiments,[23] ideas based on AF have been questioned. Moreover, in agreement with these experiments, theoretical studies of the Hubbard[8] and t-J Hamiltonians,[24] reported electron-like Fermi surfaces very similar in shape to the non-interacting case, even at low hole densities where antiferromagnetic correlations should still be strong in the system. Possible many-body effects have been invoked to explain such a behavior.

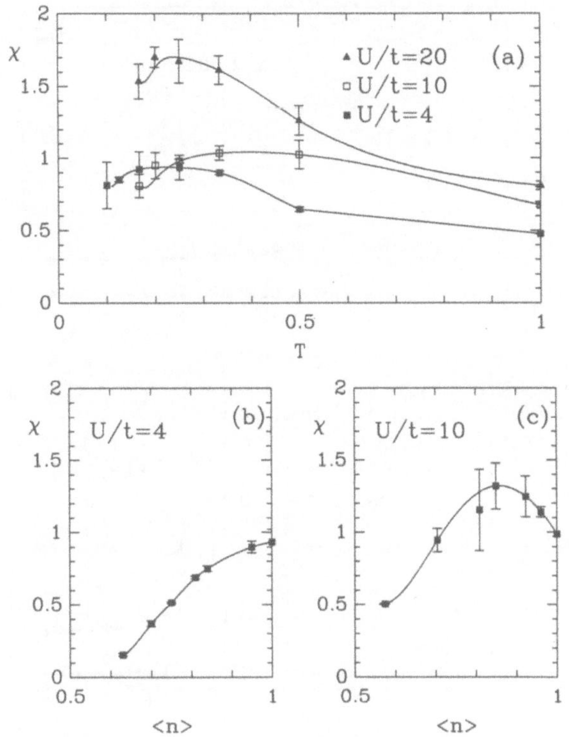

Figure 2. Uniform magnetic susceptibility as a function of temperature at half-filling on a 4×4 lattice for different values of *U/t*; (b) Uniform magnetic susceptibility as a function of density at temperature 0.25 on a 4 ×4 lattice for *U/t*=4; (c) same as (b) for *U/t*=10.

However, on the experimental side, these results have been recently challenged by Aebi et al.[25] Using a photoemission technique that allows the mapping of the whole Fermi surface, these authors have shown the existence of hole pockets in their data for Bi2212 consistent with a rigid band filling of the hole states predicted by antiferromagnetically based mean-field approximations.[22] In parallel to the experimental results, recent theoretical arguments by Dagotto et al.[26] conjectured that the large Fermi surface of the 2D Hubbard model observed at finite temperature can be made compatible with the existence of hole pockets. Their reasoning is that unless the temperature of the simulation is smaller than the energy difference of holes states at momenta $\mathbf{k}=(\pm\pi/2,\pm\pi/2)$ and $(0,\pi)$, $(0,\pi)$ (which they found to be very small at *U/t*=10), the hole pocket effect is washed out for thermodynamical reasons. These authors suggested that studies at stronger couplings, lower temperatures and larger lattices should provide evidence for the existence of hole pockets. Motivated by these challenging experimental and theoretical ideas, we perform a careful analysis of Quantum Monte Carlo data for the 2D Hubbard model defined in Eq. (1). We conclude that numerical evidence indeed supports the existence of hole pockets at $\mathbf{k}=(\pm\pi/2,\pm\pi/2)$ at low hole density. Thus, our results reopen the possibility that the normal state properties of the high temperature superconductors at half-filling can be qualitatively approximated by a band filling of the states obtained in the mean-field spin-density-wave approximation. As in the previous section, the Hubbard Hamiltonian will be analyzed using standard Quantum Monte Carlo techniques.[16] Clusters with up to 8×8 sites will be studied

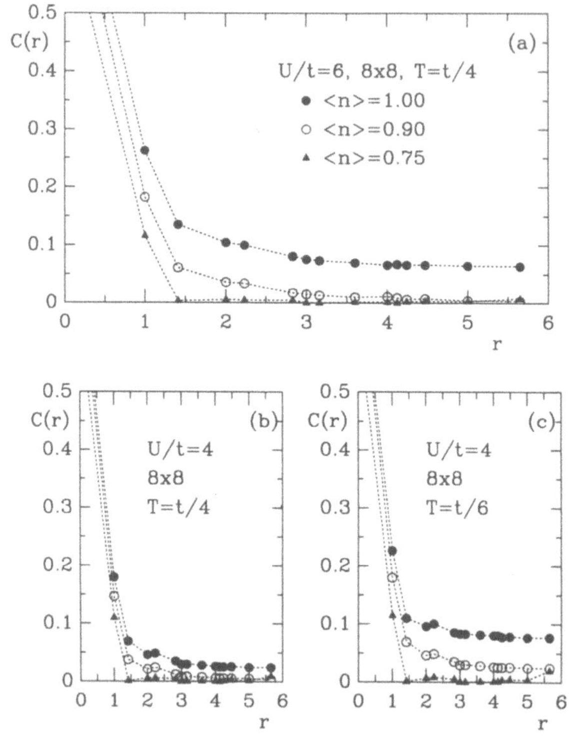

Figure 3. (a) Spin-spin correlation $C(r)=\langle S_i^z S_{i+r}^z \rangle (-1)^r$ for U/t=6, T=t/4 on an 8×8 lattice at different fillings. The error bars are of the size of the dots; (b) Same as (a) but for U/t=4; (c) Same as (b) but for T=t/6.

varying the temperature *T*, filling ⟨n⟩, and coupling *U/t*. The notorious sign-problem prevents us from working at very low temperatures and large couplings. Those used here correspond to the lowest *T* and largest *U/t* that can be reliably reached with confidence using the QMC algorithm.[27] We will show below that they are enough to observe hole pockets in the Hubbard model.

The first issue to analyze is whether antiferromagnetic correlations are strong in the system away from half-filling, since this is a necessary condition for the existence of hole pockets induced by AF. In Fig. 3a, the spin-spin correlation function $C(r)=(-1)^{|r|}\langle S_i^z S_{i+r}^z \rangle$ is presented for *U/t*=6 on an 8×8 lattice at *T*=*t*/4. At half-filling, ⟨*n*⟩=1 (filled circles) the correlations are strong and remain finite even at the largest possible distance clearly indicating the expected antiferromagnetic long range order (LRO) in the system. More interesting for our analysis is that at density ⟨*n*⟩=0.9 (open circles) where antiferromagnetic LRO does not exist, it is clear that the spin correlations are strong up to a distance of about 3 to 4 lattice spacings. Increasing further the hole density to ⟨*n*⟩=0.75 (triangles), antiferromagnetism is further suppressed but it remains strong between nearest-neighbor spins showing that the moments are still well-formed even at this fairly large hole doping. When the coupling is reduced to the value that has been more widely analyzed by QMC techniques namely *U/t*=4, a qualitatively similar behavior is found at the same temperature. The spin correlations at this coupling (shown in Fig. 3b) are less developed, but LRO is still observed at half-filling. Working at ⟨n⟩=0.9, short range antiferromagnetism is also robust as it occurs at higher couplings. Decreasing the temperature to T=t/6 (shown in Fig. 3c) the spin correlations at U/t=4 become as strong as for U/t=6 at T=t/4. Then, Figs. 3a-c have shown that at a realistic density of $\langle n\rangle$=0.9, the antiferromagnetic correlations are sufficiently developed that they may originate hole-pockets in the Fermi surface. Eventually, these pockets should evolve into a noninteracting-like Fermi surface as the electronic density is further decreased from half-filling.

Since the antiferromagnetic correlations are robust, then the natural question to discuss is why hole pockets have not been observed in previous studies (which were only carried out at U/t=4). To address this question let us concentrate on the analysis of the momentum distribution $n(\mathbf{k})=\sum_\sigma c^+_{\mathbf{k},\sigma}c_{\mathbf{k},\sigma}$, where $c^+_{\mathbf{k},\sigma}=\frac{1}{\sqrt{N}}\sum_{\mathbf{j}} e^{i\mathbf{k}\cdot\mathbf{j}}c^+_{\mathbf{j},\sigma}$. Note that n(**k**) has been previously measured[8,24] but in these and some other papers, the shape of the Fermi surface was determined by calculating the position of the momenta where n(**k**)=0.5. In principle, this procedure is correct. However, using such a convention here we show that the pockets could be overlooked due to finite temperature and/or small lattice effects. To understand this point, in Fig. 4a n(**k**) versus **k** is shown along the k_x=k_y direction for U/t=6 at T=t/4 on an 8×8 cluster and density $\langle n\rangle$=0.9 (open squares). Comparing the interacting results with noninteracting U/t=0 data for the same parameters (filled squares) we observe that, in both cases, n(**k**) becomes 0.5 at approximately the same momentum. Repeating the calculation along other directions in momentum space, and *if* the n(**k**) ≈ 0.5 criterion is still used, it is thus understandable why a Fermi surface resembling a non-interacting system was obtained in previous studies. However, this does not rule out the existence of pockets. The crosses shown in Fig. 4a correspond to spin-density-wave mean-field (SDW-MF) results.[22] In this approximation, an antiferromagnetic state is used which effectively produces a 2×2 unit cell. The mean-field Hamiltonian is diagonalized producing conduction and valence bands separated by the antiferromagnetic gap. The energy levels are given by $E_k=\pm\sqrt{\varepsilon_k^2+\Delta^2}$, where $\varepsilon_k=-2t(\cos k_x+\cos k_y)$ and Δ is found using a self-consistent equation. At half-filling, the valence band is filled, and $n(\mathbf{k})=(1-\varepsilon_\mathbf{k}/E_\mathbf{k})/2$ is in very good agreement with the numerical data[28] as shown in Fig. 4b. Now, let us assume that a robust antiferromagnetism survives the introduction of hole doping, as was suggested by the results of Fig. 3a-c. Quasiparticles are removed from the top of the valence band to mimic the presence of doping. $n(\mathbf{k})$ now is given by,

$$n(\mathbf{k})=\frac{1}{2}\left(1-\frac{\varepsilon_\mathbf{k}}{E_\mathbf{k}}\right)\left(\frac{1}{e^{-\beta(|E_\mathbf{k}|-\mu)}+1}\right), \tag{4}$$

where μ is selected such that the density is $\langle n\rangle$. In Figs. 4a and 4b the SDW mean-field results are compared against the Monte Carlo data. The agreement along the k_x=k_y direction both at half-filling and at finite hole density is remarkable. We have explicitly checked that only at densities where the antiferromagnetic correlations become of one lattice spacing or less ($\langle n\rangle$ ≈ 0.75), the comparison between the numerical data and the SDW results deteriorates. Reducing further the density, the QMC data converges smoothly to a weakly interacting gas of electrons.

The agreement between a theory based on strong AF correlations and the numerical data is consistent with the strong AF correlations shown in Fig. 3. However, the main issue that we are addressing remains paradoxical, i.e. if the numerical data and the SDW approximation are in good agreement, why is there no hole pocket in the occupation number mean value of Fig. 4a? To understand this point, it is instructive to study the effects that high temperatures and a finite lattice produce in n(**k**) in the mean-field approximation.

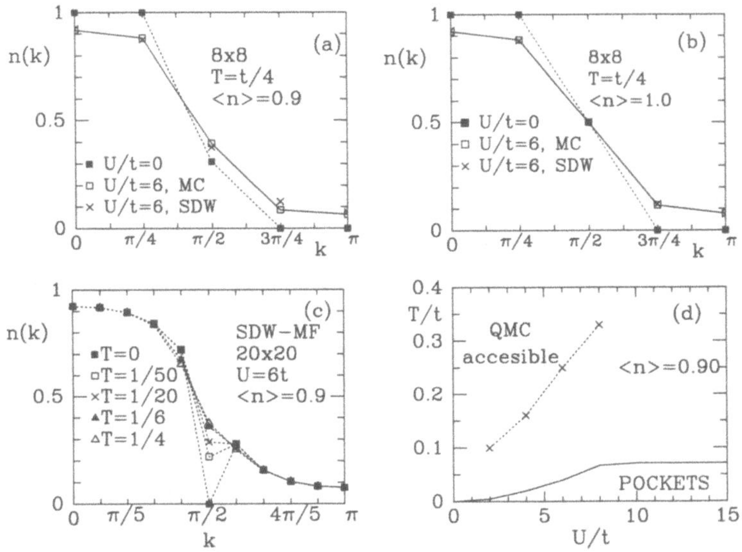

Figure 4. (a) n(**k**) as a function of momentum along the $k_x = k_y$ diagonal direction on an 8×8 lattice for U/t=6, T=t/4 at ⟨n⟩ =0.9. The open squares are Monte Carlo results, the filled squares are U/t=0 results, and the crosses correspond to the SDW mean field approximation (see text); (b) Same as (a) for ⟨n⟩ =1.0 (half-filling); (c) Mean field values of n(**k**) vs momenta along the diagonal $k_x = k_y$ direction using a 20×20 lattice, U/t=6, ⟨n⟩ =0.9 for T/t=0, 1/50, 1/20, 1/6 and 1/4; (d) Mean field determination of the region where pockets, defined as a relative minimum along the $k_x = k_y$ direction, are observed in the U/t vs T/t plane for ⟨n⟩ =0.9 on a 20×20 cluster. The crosses indicate the lowest temperatures that can be reached with our Monte Carlo technique for different values of the coupling, due to the sign problem.

In Fig. 4c we show the mean field n(**k**) along the $k = k_x = k_y$ direction on a 20×20 lattice, for U/t=6 and ⟨n⟩=0.9. At T=0 there is a clear pocket-like feature at $(\pi/2, \pi/2)$, very different from that observed numerically in Fig. 4a. However, as the temperature increases the pocket becomes less pronounced and for temperatures T=t/6 and t/4, which are the values most commonly studied with Monte Carlo techniques, it has all but disappeared. In Fig. 4a the crosses are the mean-field results and it is clear that they are in much better agreement with the Monte Carlo data than the U/t=0 points. This shows that the existence of pockets along the $k = k_x = k_y$ direction is a feature very difficult to see and quickly washed out by high temperature effects[29] due to the rapid change in n(**k**) where the pockets are expected. In Fig. 4d, we show the region on the U/t-T/t plane where a pocket, defined as a local *minimum* in the momentum distribution, is expected at $(\pi/2, \pi/2)$, according to the SDW mean-field results at ⟨n⟩=0.9. In the same plot the crosses indicate the lowest temperatures that can be presently reached for each U/t value with the QMC algorithm, the limitation being caused mainly by the well-known sign problem. Clearly, all the crosses lie in a region where pockets are not observed in the mean-field approximation. Thus, the absence of a local minimum in our results of Fig. 4c are not incompatible with hole pockets, and it is caused by finite temperature and small lattice effects. The most striking evidence showing that the interacting system is different from the non-interacting one is obtained analyzing the line from **k** = (0,π) to (π,0), (i.e. from Y to X). The SDW-MF predicts a constant momentum distribution along $cosk_x + cosk_y$=0 even at finite coupling, but this is a spurious degeneracy of this simple approximation and thus it is not useful to contrast against the

QMC data. The noninteracting system U/t=0, also has $n(\mathbf{k})$ constant along this line at any density and temperature, and it is given by

$$n(\mathbf{k}) = \frac{1}{e^{\beta(\varepsilon_{\mathbf{k}}-\mu)}+1} = \frac{1}{e^{-\beta\mu}+1} \tag{5}$$

At half-filling $\mu = 0$ and $n(\mathbf{k})$ along this line is 0.5. This symmetry allows us to study pockets in the interacting system by looking for a reduction along the X-Y line of the momentum distribution as the coupling is increased. If $n(\pi/2, \pi/2)$ is the smallest along this line, it would be an evidence in favor of hole pockets. In Fig. 5a $n(\mathbf{k})$ vs. $\mathbf{k}$ along the $(\pi,0)$-$(0,\pi)$ line is presented, using an 8×8 cluster, U/t=6, and $T=t/4$. In the half-filled case (filled circles) $n(\mathbf{k})$ is constant even for an interacting system. However, at density $\langle n\rangle$=0.9 (open circles) there is a clear minimum (pocket) at $\mathbf{k} = (\pi/2, \pi/2)$. The minimum lasts as long as antiferromagnetism is present in the system; i.e. at density $\langle n\rangle \leq 0.75$ or smaller the minima are no longer observed giving support to our interpretation of this feature as hole pockets caused by AF correlations.

In Fig. 5b, similar results are presented for U/t=4. Here the pockets are smaller than at U/t=6 because the AF order is not so strong but the qualitative behavior is the same. The maximum depth occurs again for $\langle n\rangle$ =0.9. In Fig. 5c we show the pocket at several couplings U/t ranging from 0 to 6 at a fixed density $\langle n\rangle$=0.9. As the coupling increases the pocket becomes deeper which is intuitively understandable using the AF picture. Finally, in Fig. 5d the dependence with temperature at U/t=4 and $\langle n\rangle$=0.9 is presented. As with the coupling dependence, the pocket becomes deeper as the temperature is lowered from $T=t/2$ to $t/6$.[30] It is clear that the behavior of the pocket is correlated with the behavior of the magnetic correlations presented in Fig. 3. In other words, when the spin correlations are strong, the pockets appear in the spectra. When the correlations increase, due to an increase in coupling or decrease in temperature, the pockets become deeper. But when the AF correlations are of the order of one lattice spacing, as for $\langle n\rangle < 0.75$, the pockets are no longer observed, and n(**k**) becomes constant along the $(\pi, 0)$,$(0, \pi)$ direction as in the noninteracting case.

IV. SUPERCONDUCTING STATE

Recent experimental results for the high temperature cuprate superconductors[31] have suggested that the pairing state is highly anisotropic, probably a $d_{x^2-y^2}$ singlet. In theories where the pairing mechanism is produced by antiferromagnetic fluctuations, superconductivity in the d-wave channel appears naturally, as has been shown in the context of the 2D Hubbard model using self-consistent techniques,[32] phenomenologically using the nearly-antiferromagnetic Fermi liquid state,[33] and using various techniques in the 2D t-J model.[34] While these approaches seem successful in the prediction of the superconducting symmetry, some phenomenological details of the cuprates remain somewhat hidden, like the existence of an *optimal* doping where the d.c. resistivity is linear. A different family of theories for the cuprates makes extensive use of the concept of van Hove (vH) singularities in the carriers density of states (DOS) which are believed to be caused by band structure features of these materials.[35] In this context, the quasiparticle dispersion is extracted from angle-resolved photoemission (ARPES) experiments or band structure calculations, an ad-hoc vertex interaction is proposed, and predictions for

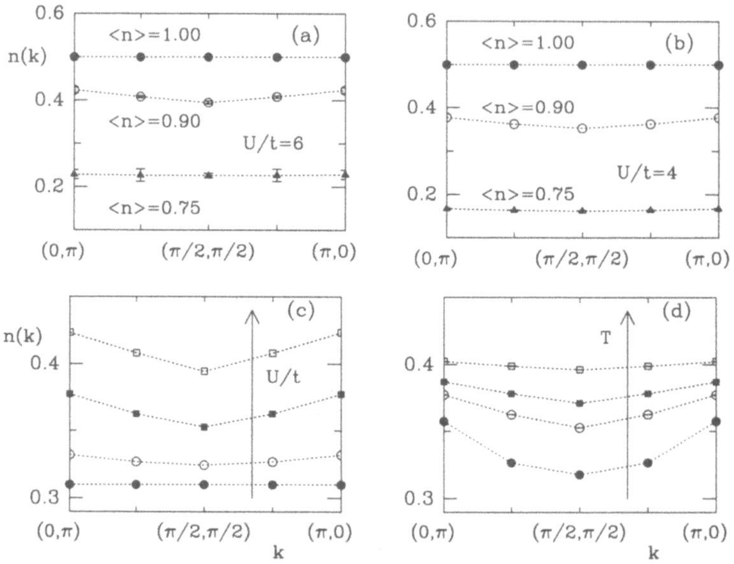

Figure 5. (a) Momentum distribution n(**k**) as a function of momenta along the (π, 0) to (0, π) direction on an 8×8 cluster for U/t=6, T=t/4 at ⟨n⟩ =1, (filled circles), ⟨n⟩ =0.9 (open circles) and ⟨n⟩ =0.75 (filled triangles); (b) Same as (a) for U/t=4; (c) Momentum distribution n(**k**) as a function of momentum along the (π, 0) to (0, π) direction on an 8×8 lattice at ⟨n⟩ =0.9 and T/t=1/4 for (from above to below) U/t=6, 4, 2 and 0; (d) Momentum distribution n(**k**) as a function of momentum along the (π, 0) to (0, π) direction on an 8×8 lattice at ⟨n⟩ =0.9 and U/t=4 for temperatures (from above to below) T=t/2, t/3, t/4 and t/6.

superconductivity are made using standard many-body techniques. The strong point of this approach is the natural existence of an optimal doping, i.e. when the chemical potential reaches the vH singularity, reproducing satisfactorily the experimental result $\rho_{dc} \sim T$. However, these vH theories should be considered phenomenological rather than microscopic since they make extensive use of dispersions and interactions which are not calculated from a microscopic Hamiltonian but taken mostly from experimental data.

In this section, we will describe a microscopically-based theory of the cuprates that combines the strong features of the antiferromagnetic and van Hove scenarios. The first step in the construction of such a theory is the observation that the ARPES quasiparticle dispersion may be caused by holes moving in a local antiferromagnetic environment, rather than by band structure effects. This idea is motivated by the existence of universal flat bands near the X and Y points (in the 2D square lattice language) in the spectrum of hole-doped cuprates, which are difficult to understand unless caused by strong correlation effects in the CuO_2 planes.[36,37] Recently it has been shown[26,38] that models of correlated electrons can produce such flat bands in a natural way, and here we further elaborate on this idea showing that the agreement with ARPES is *quantitative*.

Using the well-known two dimensional (2D) t-J model defined in Eq. (2) the dispersion of one hole in an antiferromagnet can be calculated accurately[38] with numerical or analytical techniques. Working at J/t=0.3, and taking J=0.125 eV as the scale, the hole dispersion can be approximated by

$$\varepsilon_{\mathbf{k}} = 1.33J \cos k_x \cos k_y + 0.37J\left(\cos 2k_x + \cos 2k_y\right) \qquad (6)$$

(valid at small J/t) which was obtained using Green's Function Monte Carlo and Lanczos methods.[38] Holes move within the same sublattice to avoid distorting the AF background. To improve the agreement with experiments described below, here a small hopping amplitude $t' = 0.05t$ has been included in the Hamiltonian (hopping along the plaquette diagonals). The dispersion is plotted in Fig. 6a against momentum with the Fermi level at the flat band corresponding to a hole density of x=0.15 (in Fig. 6a and 6c, $\varepsilon_{\mathbf{k}}$ is inverted i.e. the electron language is used rather than the hole language to facilitate the comparison with experiments). $\varepsilon_{\mathbf{k}}$ contains a saddle-point located close to the X and Y points, which induces a large DOS in the spectrum. In addition, $\varepsilon_{\mathbf{k}}$ is nearly degenerate along the $\cos k_x + \cos k_y = 0$ line increasing the DOS in the vicinity of the Fermi level of Fig. 6a. All these qualitative features are common to several models of correlated electrons, and should not be considered as exclusively produced by the t-J model.

What is the influence of a finite hole density on this dispersion? Recent experiments by Aebi et al. [25] have shown the presence of strong antiferromagnetic (AF) correlations in the normal state of BSSCO, leading to the formation of hole pockets in their photoemission spectra. $\varepsilon_{\mathbf{k}}$ reproduces qualitatively their results, which are shown in Fig. 6b. The most important issue to consider at a finite (but small) hole density is that the quasiparticle weight, Z, is smaller for the bands centered at momentum (π, π) than those at (0, 0) as represented pictorially in Figs. 6a and 6b, with dashed lines). These are the *shadow bands* which were originally introduced in the context of the spin-bag approach.[39]

To further elaborate on whether our dispersion (rigorously derived at $x \to 0$) can be used at finite x, we have carried out extensive QMC numerical simulations of the 2D Hubbard model at several electronic densities, a large coupling U/t=8 (i.e. in the regime of the t-J model with J/t=0.5), and temperature T=t/4, where $t \approx 0.4$eV. The results (Fig. 7a) show that for a given spin at an arbitrary site of the lattice, their nearest and next-to-nearest neighbor spins tend to be antiparallel even with hole densities as large as x=0.25.[40] At smaller temperatures, the correlations would clearly be even stronger than those in Fig. 7a. An even more convincing proof in favor of our hole dispersion is produced by a contrast with experiments. In Fig. 6c we compare $\varepsilon_{\mathbf{k}}$ along the Γ-Y direction against ARPES results obtained by the Argonne group[36,41] for YBCO. The agreement is excellent. It is worth emphasizing that the theoretical curve of Fig. 6c is derived from a microscopic Hamiltonian, and it is *not* a fit of ARPES data. This is a major difference between the present approach and previous vH scenario calculations. Thus, the assumption that the dispersion Eq. (6) holds near half-filling is justified and natural. The intuitive picture to remember is that as long as the antiferromagnetic correlation length is larger than the typical size of a spin-polaron, the quasiparticles behave as if moving in a nearly perfect AF background.

As a second step in building up a model for the cuprates, the interaction among the hole quasiparticles is necessary, and our intuition will be based again on results obtained for the 2D t-J model. In this case, it is well-known that an effective attractive force exists in an antiferromagnet leading to the bound state of two holes in the d-wave channel.[1] Since this problem is non-trivial (it is the analog of the Cooper problem for low temperature superconductors), we will simplify its analysis by studying the potential in the atomic limit (large J/t) where the attraction is induced by the minimization of the number of missing antiferromagnetic links in the problem[1] (a process similar to the attraction among spin-bags in a spin-density-wave background[39]). Assuming that the link spin-spin correlation is not much distorted by the carriers in this limit, the binding energy of two holes is $\Delta_B = e_{2h} - e_{1h} = J(\langle S_i \cdot S_j \rangle - 0.25)$ where e_{nh} is the energy of n holes with respect to the antiferromagnetic ground state energy, and **i** and **j** are nearest neighbors. Accurate numerical simulations[42] have shown that $\langle S_i \cdot S_j \rangle \approx -0.3346$, and thus $\Delta_B \approx$-0.6 J. To mimic

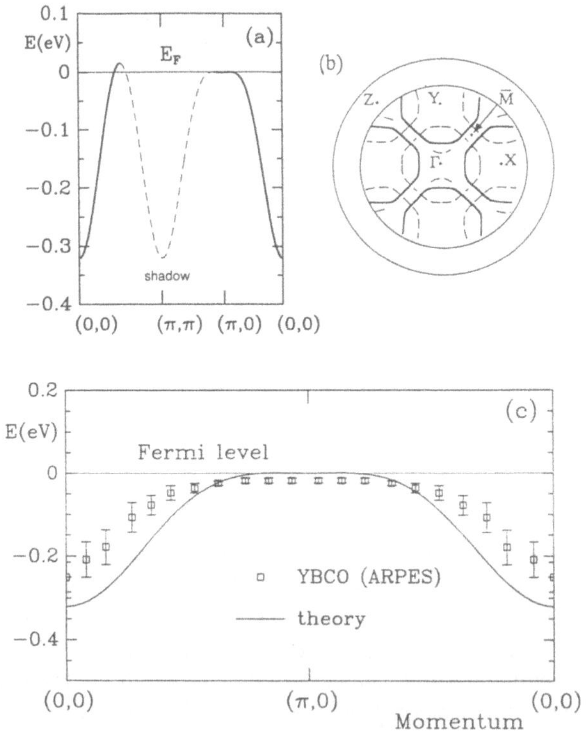

Figure 6. (a) Quasiparticle dispersion of the 2D t-J model (Eq. (6)) using J = 0.125 eV and J/t=0.3. The electron notation is used, thus all levels below E_F are occupied. The thick lines are the position of the quasiparticles that should be easily observed in ARPES experiments at a finite hole concentration. The dashed line represents the ``shadow dispersion", caused by the remnant AF at finite density. The calculation was done using a Monte Carlo technique,[38] (b) Experimental results for Bi2212.[25] The $\overline{M}$(Y) point corresponds to **k** = (π, 0)((0,π)) in the 2D square lattice language. The thick line is a strong experimental signal, while the dashed line is weak; (c) Direct comparison between experiments and theory: the solid line is our dispersion Eq. (6), while the experimental results are taken from Ref. 41.

this effect, we introduce an attractive term that reduces the energy when two quasiparticles share the same link. Thus, the Hamiltonian proposed here is

$$H = -\sum_{\mathbf{k},\alpha} \varepsilon_{\mathbf{k}} c^{+}_{\mathbf{k}\alpha} c_{\mathbf{k}\alpha} - |V| \sum_{\langle \mathbf{ij} \rangle} n_{\mathbf{i}} n_{\mathbf{j}} \tag{7}$$

where $c_{\mathbf{k}\alpha}$ is an operator that destroys a (point-like) quasiparticle with momentum **k** and in sublattice α = A, B; $n_{\mathbf{i}}$ is the number operator at site **i**; $|V| = 0.6J$, and $\varepsilon_{\mathbf{k}}$ is given in Eq. (6).[43] Since quasiparticles with spin-up(down) move in sublattice A(B), the interaction term can also be written as a spin-spin interaction. This Hamiltonian has been deduced based on strong AF correlations, and it has a vH singularity in the DOS; thus we will refer to it as the *antiferromagnetic-van Hove* (AFVH) model. Pictorially, it is shown in Fig. 7b: quasiparticles move within the A or B sublattices and interact when they share a link. Note that in constructing the low energy AFVH Hamiltonian, retardation effects have been neglected. It will be shown below that this model leads to d-wave superconductivity, which is mainly induced by the shape of the real space potential among carriers.[44] Thus, we do not

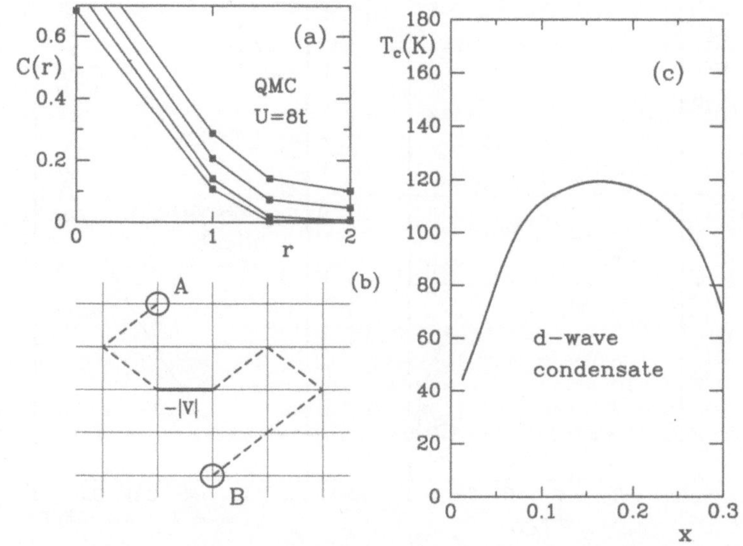

Figure 7. (a) Spin-spin correlation $C(\mathbf{r})=\langle S_i^z S_{i+r}^z \rangle (-1)^{|r|}$ vs distance for the 2D Hubbard model calculated with QMC at U/t=8 on a 8×8 cluster. The *electronic* densities, $\langle n \rangle$, starting from above are 1.0, 0.90, 0.83, and 0.74. The temperature is T = 0.1 eV, $\Delta\tau$ = 0.0625, and the error bars are typically about 0.02; (b) a pictorial representation of the AFVH model. Quasiparticles (large circles) move within sublattice A or B, and they interact when they share a link; (c) Critical temperature T_c of the AFVH model as a function of hole density x (=1-$\langle n \rangle$), obtained with the standard gap equation and a d-wave order parameter.

believe that retardation effects will change our conclusions, and their consideration will be postponed for a future publication. The excellent results described below make this argument plausible.

The AFVH model can be studied using different many-body techniques. Computational techniques are the most reliable ones, but their implementation is non-trivial.[45] Here, the analysis of the AFVH Hamiltonian will be carried out with the standard BCS formalism. Since the ratio |V|/W is about 0.2, where W is the bandwidth of the quasiparticles, the gap equation will produce a reliable estimation of the critical temperature since we are effectively exploring the *weak* coupling regime of the AFVH model. Solving the gap equation on 200×200 grids, we consistently observed that the free energy is minimized (or critical temperature maximized) using a $d_{x^2-y^2}$ order parameter. After some tedious but straightforward algebra it can be shown that the equation to be used for the thermodynamical properties in this channel is

$$\frac{1}{0.6J} = \frac{1}{2N}\sum_{\mathbf{k}} \frac{\cos k_x (\cos k_x - \cos k_y) \tanh(E_{\mathbf{k}}/2T)}{E_{\mathbf{k}}} \tag{8}$$

where $E_{\mathbf{k}} = \sqrt{(\varepsilon_{\mathbf{k}} - \mu)^2 + \Delta_{\mathbf{k}}^2}$ and $\Delta_{\mathbf{k}} = \frac{\Delta_0}{2}(\cos k_x - \cos k_y)$, Δ_0 is a parameter, and T is the temperature. Note that here the temperature dependence of the chemical potential μ cannot be neglected. In Fig. 7c, T_c against the hole density is shown.[46] Two features need to be remarked: i) an optimal doping exists at which T_c is maximized. This is a direct consequence of the

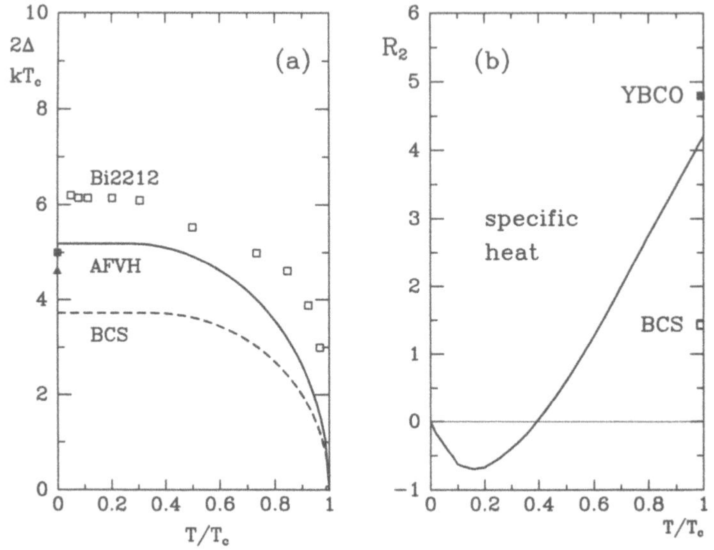

Figure 8. $2\Delta_{max}(T)/kT_c$ against T/T_c. The solid line corresponds to the AFVH model at the optimal doping, x ≈ 0.15. The results are weakly hole density dependent. The open squares correspond to tunneling data for $Bi_2Sr_2CaCu_2O8$ from Ref. 47. The full triangle at T = 0 corresponds to PES data by Ma et al. [48] while the full square is Batlogg's summary of experimental data.[49] The dotted line is the BCS prediction (i.e. attractive Hubbard model at half-filling and weak coupling); (b) R_2 as defined in the text vs T/Tc for the AFVH model at density x= 0.15 (i.e. optimal doping). The dot corresponds to experimental results for YBCO.[51] The BCS result is shown for reference.

presence of a large peak in the DOS of the quasiparticles; ii) both the optimal doping (15%) and the optimal T_c of about 100K are in excellent agreement with the cuprates phenomenology.[3] Although in the effective AFVH Hamiltonian the natural scale of the problem is the exchange J~1000K, since the ratio of coupling to bandwidth is small, T_c is further reduced (as is standard in the weak coupling BCS formalism) to about 100K. It is remarkable that this quantitative agreement with experiments is deduced without the need of ad-hoc fitting parameters. It is also very important to remark that the presence of a finite optimal density of holes *cannot* be achieved if a simpler $(\cos k_x + \cos k_y)^2$ dispersion is used. Thus, as noticed in Ref. 38 the small energy difference between $k = (\pi/2, \pi/2)$ and $(\pi,0)$ in the hole dispersion is crucial for the quantitative success of this approach.

From the gap Eq. (8), the ratio $R(T) = 2\Delta_{max}(T)/kT_c$ can be calculated, where for a d-wave condensate at temperature $T, \Delta_{max}(T)$ is defined as the maximum value of the gap. In Fig. 8a, R(T) is shown as a function of temperature, compared with recent tunneling data by Vedeneev et al.[47] At zero temperature, the AFVH model predicts R(0) = 5.2 and the tunneling experiment gives 6.2. While the agreement is already encouraging, note that other experiments have reported a smaller value for R(0). For example, PES data by Ma et al. [48] obtained R(0) = 4.6, while an average over the pre-1992 literature[49] suggested R(0) = 5 ± 1 supporting the results of the AFVH model. At T_c we have also calculated the ratio $R_2 = \Delta C/C_n$, where $\Delta C = C_s - C_n$ is the difference between the specific heat of the normal state C_n (which we calculate turning off the interaction |V|), and the superconducting state

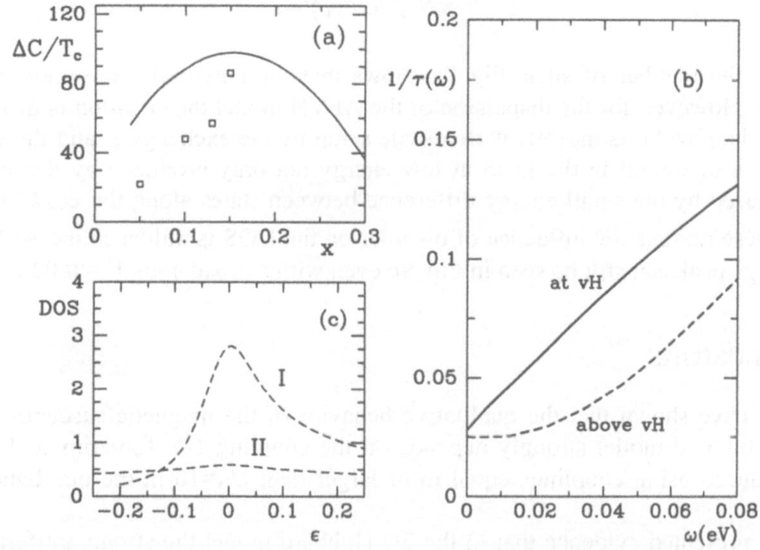

Figure 9. (a) $\Delta C/T_C$ vs hole density x using the AFVH model. The open squares correspond to experimental data[55] for $YBa_2Cu_3O_{7-\delta}$, using the convention that the result at the optimal δ(approximately 0.02) is plotted at our optimal doping x ~ 0.15, and that δ ≈ 0.02+(0.15-x) in presenting the rest of the data. The vertical axis is in units of mJ/K^2/mol for the experiment and mJ/K^2/ site for the theory; (b) Quasiparticle lifetime against energy, ω, at zero temperature calculated using the quasiparticle dispersion Eq. (2) as described in the text. The solid line corresponds to the case where the chemical potential, μ, is at the saddle-point, while for the dashed line μ is below the saddle-point. The units in the vertical axis are arbitrary; (c) DOS as defined in Eq. (5) corresponding to the AVFH model (I), and a tight-binding model (II) For more details see the text. The broadening is Γ = 0.02*e*V.

C_s. The result is shown in Fig. 8b.[50] At T_c, $R_2 \approx 4.2$ which is again in good agreement with YBCO experiments reported by Phillips et al., and Loram and Mirza[51] which found R_2 = 4.8 and 4.1, respectively. It is also gratifying that not only dimensionless ratios but also absolute values are in reasonable agreement with experiments. For example, in Fig. 9a $\Delta C/T_c$ is shown as a function of density compared with results for YBCO. Although the relation between hole carriers and the excess of oxygen in this material is unknown, we can easily see that the general trends between theory and experiment are similar, as was observed in other vH calculations.[52] We have also verified that an important feature of previous vH scenarios also exists in our model, namely a quasiparticle lifetime linear with frequency at the optimal doping. To carry out the calculation we switched off the attractive term leading to superconductivity, introduced particle-particle repulsion within the *same* sublattice at a distance of one lattice spacing,[53] and carried out the standard one bubble approximation to get the imaginary part of the self-energy. The result is shown in Fig. 9b for a density below the vH singularity and at this singularity. Clearly, a linear behavior is observed at the optimal doping.[54] Finally, let us address the stability of the van Hove singularity against effects not explicitly considered in the Hamiltonian. In some van Hove theories the singularity in the DOS is rapidly removed by inhomogeneities or 3D effects. As a simple illustration, the DOS of a 2D tight-binding model with dispersion $\varepsilon_k = -2t(\cos k_x + \cos k_y)$ is plotted in Fig. 9c with an energy-independent broadening Γ= 0.02 *e*V using

$$N(\omega)=\frac{2}{\pi N}\sum_{\mathbf{k}}\frac{\Gamma}{(\omega-\varepsilon_{\mathbf{k}})^2+\Gamma^2}, \tag{9}$$

where N is the number of sites. Fig. 9a shows that the singularity is removed with this broadening. However, for the dispersion of the AFVH model the situation is different since the carriers bandwidth is narrow, with a scale setup by the exchange J, and there is a large accumulation of weight in the DOS at low energy not only produced by the saddle-point but also caused by the small energy difference between states along the $\cos k_x + \cos k_y = 0$ line. For these reasons the influence of disorder on the DOS is milder in the AFVH model, where a large peak can still be seen in Fig. 9c even with a broadening $\Gamma = 0.02$ eV.

V. CONCLUSIONS

We have shown that the qualitative behavior of the magnetic susceptibility of the one band Hubbard model strongly depends on the coupling *U/t*. Experimental results are well reproduced using couplings equal to or larger than *U/t*=10 in the one band Hubbard model.

We presented evidence that in the 2D Hubbard model the strong antiferromagnetic fluctuations produce hole pockets at momenta $\mathbf{k}=(\pm\pi/2,\pm\pi/2)$. Monitoring the behavior of $n(\mathbf{k})$ along the line $(\pi,0),(\pi,0)$, where the electronic momentum distribution is constant in the noninteracting system, we have observed a minimum at $(\pi/2,\pi/2)$ compatible with the existence of a hole pocket. Along the $k_x = k_y$ direction no local minimum is numerically observed but by comparing our results to a SDW-MF approximation we have shown that this feature is due to finite temperature effects, as predicted by Dagotto et al.[26] For the values of *U/t* here studied, i.e. 4 and 6, the pockets exist for $\langle n\rangle > 0.75$ and the maximum effect is observed at $\langle n\rangle =0.9$. Larger couplings and lower temperatures would certainly increase the range of existence of these hole pockets. Thus, we have shown that the recent experiments of Aebi et al. [25] are compatible with results obtained for the 2D Hubbard model giving support to theoretical descriptions of the cuprates based on antiferromagnetic correlations.

Finally, simple and *quantitative* ideas for high-T_c superconductivity have been discussed combining for the first time two apparently different proposals for the cuprates i.e. the antiferromagnetic and van Hove scenarios. We have claimed that these approaches are actually deeply related with *both* the pairing mechanism and the vH singularity in the DOS of the high T_c materials caused by antiferromagnetic correlations. This theory explains in an economical way the $d_{x^2-y^2}$ superconducting state apparently observed in several experiments, and goes beyond previous work showing that the critical temperature is maximized at a particular optimal doping where the d.c. resistivity is linear with temperature. An excellent quantitative agreement between theory and experiments is reported for the critical temperature, optimal doping, superconducting gap and specific heat.

ACKNOWLEDGMENTS

This work is supported by the Office of Naval Research under grant ONR N00014-93-0495. We thank SCRI and the Computer Center at FSU for providing us access to the Cray-YMP and ONR for giving us access to the CM5 connection machine.

REFERENCES

1. For a complete list of references see E. Dagotto, Rev. Mod. Phys., **66**, 763 (1994).
2. A. Moreo, *Phys. Rev. B* **45**, 5059, (1992).
3. C-H. Pao and N. E. Bickers, Phys. Rev. Lett. **72**, 1870, 1994; P. Monthoux and D. Pines, Phys. Rev. Lett. **72**, 1874, 1994.
4. J.G. Bednorz and K.A.Müller, Z.Phys., **B64**, 188, (1986); C.W. Chu et al., *Phys. Rev. Lett.* **58**, 405, (1987).
5. J.B. Torrance et al., *Phys. Rev. B* **40**, 8872, (1989).
6. D. Johnston., *Phys. Rev. Lett.* **62**, 957, (1989).
7. W.W. Warren et al., *Phys. Rev. Lett.* **62**, 1193, (1989); R.E. Waldstedt et al., *Phys. Rev. B* **41**, 9574, (1990); R.E. Waldstedt et al., *Phys. Rev.*, 7760, (1991); M. Takigawa et al., *Phys. Rev. B* **43**, 247, (1991); H. Alloul et al., Phys. *Rev. Lett.* **63**, 1700, (1989).
8. A. Moreo, et al., *Phys Rev. B* **41**, 2313 (1990).
9. A. Moreo, E. Dagotto, J. Riera, F. Ortolani and D.J. Scalapino, *Phys. Rev. B* **45**, 10107, (1992).
10. Hybertsen et al., *Phys. Rev. B* **41**, 11068, 1990.
11. E. Dagotto and J. Riera, *Phys. Rev. Lett.* **70**, 682, 1993.
12. A. Moreo, *Phys. Rev. B* **48**, 3380, (1993).
13. N. Bulut, D.W. Honre, D.J. Scalapino and N.E. Bickers, *Phys. Rev. B* **41**, 1797, (1990); J. Lu, Q. Si, J.H. Kim and K. Levin, *Phys. Rev. Lett.* **65**, 2466, (1990).
14. R. Singh, and R. Glenister, *Phys. Rev. B* **46**, 11871, (1992).
15. Studies for χ for the t-J model based on a slave-boson mean-feild approximatio do not agree with the HTE results. See T. Tanamoto, K. Kuboki and H. Fukuyama, J. *Phys. Soc. Jpn.*, **60**, 3072, 1991.
16. R. Blankenbecler, D.J. Scalapino and R. Sugar, *Phys. Rev. D* **24**, 2278, (1981).
17. F.C. Zhang and T.M. Rice, *Phys. Rev. B* **37**, 3759, (1988).
18. A. Moreo, *Phys. Rev. B* **45**, 4907, (1992).
19. G. Gomes-Santos, J.D. Joannopoulos and J.W. Negele, *Phys. Rev. B* **39**, 44435, (1989); S. Bacci, E. Gagliano and E. Dagotto, *Phys. Rev. B* **44**, 285, (1991).
20. T. Usuki, N. Kawakami and A. Okiji, J. *Phys. Soc. Jpn.* **59**, 1357, (1990).
21. A Moreo and D. Duffy, preprint (1994).
22. J.R. Schrieffer, X.G. Wen and S.C. Zhang, *Phys. Rev. B* **39**, 11663, (1989).
23. J.C. Campuzano, et al., *Phys. Rev. Lett.* **64**, 2308, (1990) and *Phys. Rev. B* **43**, 2788, (1991); Liu Rong, et al., *Phys. Rev. B* **45**, 5614, (1992) and *Phys. Rev. B* **46**, 11056, (1992).
24. W. Stephan and P. Horsch, *Phys. Rev. Lett.* **66**, 2258, (1991); R.R.P. Singh and R.L. Glenister, *Phys. Rev. B* **46**, 14313, 91992).
25. P. Aebi et al., *Phys. Rev. Lett.* **72**, 2757, 1994.
26. E. Dagotto, A. Nazarenko and M. Boninsegni, Phys. Rev. Lett. **73**, 728 (1994).
27. The data for U/t=6 were obtained using $\Delta\tau$=0.0625 while for the remaining values of U/t we used $\Delta\tau$ =0.125. The number of Monte Carlo iterations used depended on the filling. For each point we did at least two independent runs with 2000 warm-ups and 8000 measurement sweeps.
28. For values of U/t smaller than 6, i.e. for U/t=4, the mean field result gives a reasonable qualitative fit to the Monte Carlo data. Quantitative agreement is obtained by reducing Δ.[8]
29. In small lattices, i.e. 4×4, 8×8, not even at T=0 are the pockets observed at the mean field level along the $k_x = k_y$ direction.
30. We also studied couplings ranging from U/t=0 to 8 at T=t/3 and we observed that at $\langle n\rangle$=0.9 the depth of the pocket increased with U/t. However, the pocket is deeper for U/t=6 and T=t/4 than for U/t=8 and T=t/3. This can be understood by noting in Fig. 4b that the first case is closer to the pocket region than the second.
31. E. Dagotto, A. Nazarenko and A. Moreo, Phys. Rev. Lett. **74**, 310, (1995).
32. N. E. Bickers, D.J. Scalapino, and S.R. White, *Phys. Rev. Lett.* **62**, 961, 1989.
33. P. Monthoux and D. Pines, *Phys. Rev. Lett.* **69**, 961, 1992.
34. E. Dagotto and J. Riera, *Phys. Rev. Lett.* **70**, 682, 1993; E. Dagotto, et al., *Phys. Rev. B* **49**, 3548, 1994. See also Y. Ohta et al., Nagoya preprint; E. Heeb and T.M. Rice, ETH preprint.
35. C.C. Tsuei et al., Phys. Rev. Lett. 65, 2724, 1990, D.M. Newns, P.C. Pattnaik and C.C. Tsuei, Phys. Rev. **B** 43, 3075, 1991; and references therein.
36. K. Gofron et al., J. Phys. Chem. Solids **54**, 1193, 1993; K.J. Gofron, thesis, 1993; J. Ma et al., UW-Madison preprint, 1994.
37. D.S. Dessau et al., *Phys. Rev. Lett.* **71**, 2781, 1993; D.M. King et al preprint; B.O. Wells et al., Phys. Rev. Lett. **74**, 964 (1995).
38. N. Bulut, D.J. Scalapino and S.R. White, Phys. Rev. B **50**, 7215 (1994).

39. A. Kampf, and J.R. Schrieffer, Phys. Rev. **B42**, 7967, 7967, 1990, and references therein.
40. A. Moreo and D. Duffy, preprint, in preparation.
41. A.A. Abrikosov, J.D. Campuzano, and K. Gofron, Physica **C 214**, 73, 1993.
42. J. Carlson, Phys. Rev. **B40**, 846, 1989.
43. Note that the procedure followed to obtain the potential is not restricted to the large J/t limit. As an example, consider the attractive Hubbard model which is superconducting: the potential can be read from the bound state formation at large U/t while the kinetic energy term can be obtained from the opposite, non-interacting) limit, and then both used in combination for any ratio U/t. A similar approach in the t-J model was followed by J. Inoue and S. Maekawa, Prog. of Theor. Phys. 108, 313, 1992, although these authors studied a dispersion proportional to, $(\cos k_x + \cos k_y)^2$.
44. D.J. Scalapino et al., preprint.
45. J. Riera et al., work in progress.
46. We explicitly checked that corrections at the Hartree-Fock level are small, and do not alter the qualitative features of the dispersion Eq. (6).
47. S.I. Vedeneev et al., Phys. Rev. **B 49**, 9823, 1994.
48. Jian Ma, C.Quitmann, R.J. Kelley, H. Berger, G. Margaritondo, and M. Onellion, Madison preprint, May 1994.
49. B. Batlogg, Springer Series in Solid-State-Sciences, Vol. 106, Eds.: S. Maekawa and M. Sato, page 219, 1992.
50. The theoretical calculation is done at fixed volume, while the experiments are at fixed pressure. However, the difference is small and not qualitatively relevant.
51. N. Phillips et al., Phys. Rev. lett. **65**, 357, 1990; J.W. Loram and K.A. Mirza, Physica **C 153-155**, 1020, 1988.
52. C.C. Tsuei et al., Phys. Rev. Lett. **69**, 2134, 1992.
53. This interaction does not alter the d-wave condensate, and actually it helps in stabilizing it against phase separation.
54. It has been shown in the context of a modified 2D Hubbard model that corrections beyond the one loop diagram, preserve the linear behavior of $1/\tau(\omega)$ at the van Hove singularity [J.W. Serene and D.W. Hess, *Recent Progress in Many Body Theories*, Vol. 3, Eds. T.L. Ainsworth et al., Plenum Press (NY), 469 (1992)]. We expect the same qualitative behavior in our model.

List of Participants
Electronic Structure of Solids Using Cluster Methods
July 16-19, 1994

Paul S. Bagus
National Science Foundation, Room 1055
4201 Wilson Blvd.
Arlington, VA 22230

Aniket Bhattacharya
Department of Physics and Astronomy
Michigan State University
East Lansing, MI 48824-1116

Randy Boehm
Department of Chemistry
Michigan State University
East Lansing, MI 48824-1322

Tina Briere
Department of Physics
State University of New York at Albany
Albany, NY 12222

Hyunju Chang
Department of Physics and Astronomy
Michigan State University
East Lansing, MI 48824-1116

Hwasuck Cho
State University of New York at Albany
1400 Washington Ave.
Box 3620 D2 102
Albany, NY 12222

T.P. Das
Department of Physics
State University of New York at Albany
1400 Washington Avenue
Albany, NY 12222

Debasis Datta
Department of Physics
Michigan Technological University
1400 Townsend Dr.
Houghton, MI 49931-1295

M.A. Garcia-Bach
Departament de Física Fonamental
Universitat de Barcelona
Av. Diagonal, 647
08028 Barcelona
Spain

G. Gowri
Department of Physics
State University of New York at Albany
Albany, NY 12222

Jeff Grossman
Department of Physics
University of Illinois at
Urbana–Champaign
1110 W. Green Street
Urbana, IL 61801

J.F. Harrison
Department of Chemistry
Michigan State University
East Lansing, MI 48824-1322

Kenneth Hass
Ford Motor Co.
SRL MD-3028
Dearborn, MI 48212-2053

F. Illas
Facultat de Quimica
Departament de Quimica Física
Universitat de Barcelona
C/Marti i Franques 1
08028 Barcelona
Spain

Koblar Jackson
504 S. Fancher
Mt. Pleasant, MI 48858

Thomas A. Kaplan
Department of Physics and Astronomy
Michigan State University
East Lansing, MI 48824-1116

Dmitry Lebedenko
Department of Physics
University of Illinois at
Urbana–Champaign
1110 W. Green Street
Urbana, IL 61801

J.E. Lowther
Department of Physics
University of the Witwatersrand
PO Wits 2050
South Africa

S.D. Mahanti
Department of Physics and Astronomy
Michigan State University
East Lansing, MI 48824-1116

Richard Martin
Theoretical Division, MS B268
Los Alamos National Laboratory
Los Alamos, NM 87545

A.K. McMahan
U.C. Lawrence Livermore National
Laboratory–L-299
P.O. Box 808
Livermore, CA 94550

Luboš Mitáš
National Center for Supercomputing
Applications
University of Illinois at
Urbana–Champaign
405 Matthews Avenue
Urbana, IL 61801

Adriana Moreo
Department of Physics and National
High Magnetic Field Lab
Florida State University
Tallahassee, FL 32306

W.C. Nieuwpoort
Laboratory of Chemical Physics
Nijenborgh 4
9747 AG Gronigen
The Netherlands

Stacie Nunes
Department of Physics
State University of New York at Albany
Albany, NY 12222

Pablo Ordejon
Department of Physics
University of Illinois at
Urbana–Champaign
1110 W. Green Street
Urbana, IL 61801

Ranjit Pati
Department of Physics
State University of New York at Albany
Albany, NY 12222

David C. Patton
Physics Department
University of Alabama at Birmingham
Birmingham, AL 35294

M. Pederson
Complex Systems Theory Branch–6692
Naval Research Laboratory
Washington, DC 20375-5342

Bijan K. Rao
Virginia Commonwealth University
Department of Physics
1029 W. Main St., Box 824000
Richmond, VA 23284-2000

Douglas Ritchie
National Research Council
100 Sussex Dr., Room 2155
Ottawa, Ontario K1Y 4S7
Canada

Christine Russell
Department of Physics
State University of New York at Albany
Albany, NY 12222

Bill Schneider
Ford Motor Co.
P.O. Box 2053, MD 3083/SRL
Dearborn, MI 48121

Koun Shirai
Department of Physics and Astronomy
Michigan State University
East Lansing, MI 48824-1116

Sudha Srinivas
Department of Physics
State University of New York at Albany
Albany, NY 12222

E. Stechel
Sandia National Laboratories
Department 1153 MSO 345
Albuquerque, NM 87185-0345

Amin Sutjianto
Physics Department
Michigan Technological University
1400 Townsend Drive
Houghton, MI 49931-1295

M.F. Thorpe
Department of Physics and Astronomy
Michigan State University
East Lansing, MI 48824-1116

John M. Vail
Department of Physics
University of Manitoba
Winnipeg, MB R3T 2N2
Canada

J.L. Whitten
Department of Chemistry
North Carolina State University
Raleigh, NC 27650

R. Zeller
Institut für Festkörperforschung des Forschungszentrum Jülich
Postfach 1013
D-5170 Jülich
Germany

H. Zheng
Department of Physics
Virginia Commonwealth University
Richmond, VA 23284

INDEX